by Richard Crossley

Crossley Books

Princeton University Press

Published by Princeton University Press, 41 William Street, Princeton, New Jersey 08540

In the United Kingdom: Princeton University Press, 6 Oxford Street, Woodstock, Oxfordshire OX20 1TW

nathist.princeton.edu

ISBN 978-0-691-14778-9

Library of Congress Control Number: 2010934303
British Library Cataloging-in-Publication Data is available

This book has been composed in Gill Sans

Printed on acid-free paper. ∞

Printed in Thailand

10 9 8 7 6 5 4 3 2 1

CONTENTS

For Brian and Margaret Crossley

A lifetime of encouragement and caring.

Thank you.

PREFACE

This book is for beginners, experts, and everyone in between; indeed anyone who loves to identify birds, or just look at them, or simply enjoys the beauty of the outdoors. Its main goal, using unique photographs and page layouts, is to show birds as we really see them in the field. Straight-talking text, in conjunction with abundant and painstakingly selected images, provides keys to improve and hone basic and advanced identification skills and enhance all-around enjoyment of birds in their varied habitats.

I'm not exactly sure when this book was first conceived. I suppose, like other serious birders, I always try to weigh the pros and cons of bird books, to determine which is the best and how I might improve on any or all of them. I have always loved photographs but felt their potential was never fully realized in field guides. This lifelong interest was heightened during my work as co-author of *The Shorebird Guide*. From that simple starting point, this project has developed—sometimes painfully—into its final form as this book. I'm proud of what I have achieved, but I'm also sure that this is just the first in a new generation of nature guides. As technology advances, Crossley Books will be looking to push the boundaries and help promote birding and the outdoors. So watch this space!

My goal for the book was to make it visually striking, educational, innovative, entertaining, and comprehensive. I wanted to create a book that replicates the world of birds as I see it. Of course, I'm not sure if I have managed to achieve all this, but my hope is that this book will help to change the way we look at birds, and, more importantly, the way we look at and use books. If this is successful, it will help us become better birders, and also increase our appreciation for our beautiful surroundings.

This book has been an exciting—sometimes overwhelming—challenge. I have taken it from inception, virtually a blank sheet of paper, through many frustations and failures, to its final design. I photographed all of the approximately 10,000 images in this book (I consider over 99% to be all!). Unlike other photographic guides, I wanted this to be entirely mine: photographs, words, design—everything.

I regard this book as one part of a broader aim to both serve and expand the world of birding, make it more fashionable, current, and exciting. You, too, can be part of the journey. See additional text for this book and our other adventures at WWW.CROSSLEYBIRDS.COM.

Whether I succeed or not, I continue to learn so much—and it's quite a ride.

I hope you enjoy and learn from all of this!

Richard Crossley
Cape May, NJ

Swimming Waterbirds pp.38–96

Size Guide: DOVE–8.25in; MUSW–60in

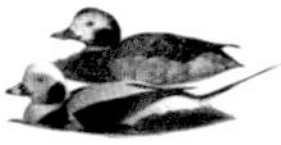
LTDU p.72

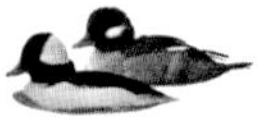
BUFF p.73

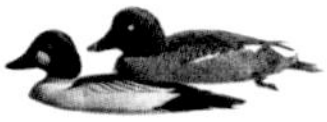
COGO p.74

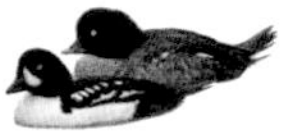
BAGO p.75

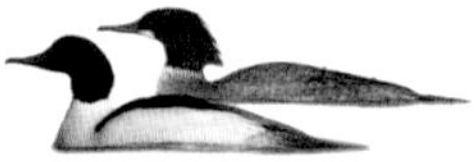
COME p.76

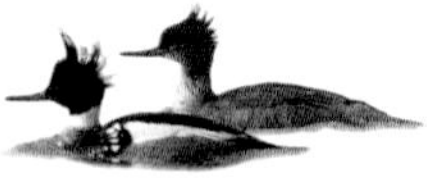
RBME p.77

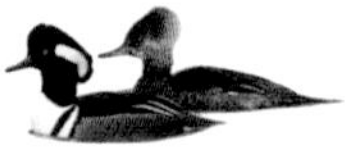
HOME p.78

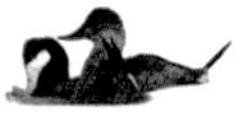
RUDU p.79

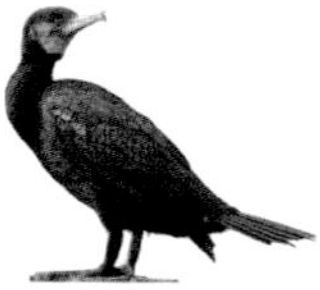
GRCO p.80

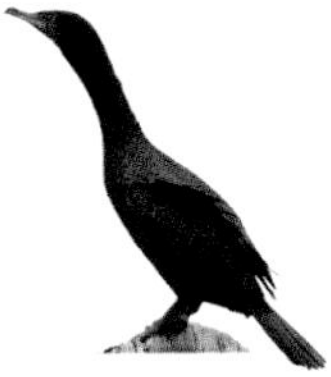
DCCO p.81

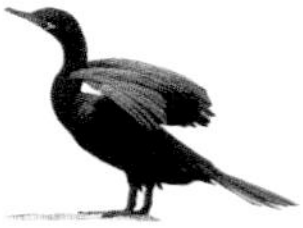
NECO p.82

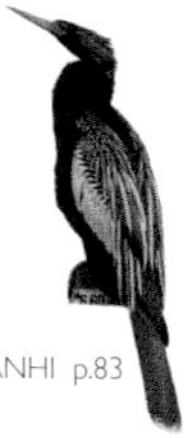
ANHI p.83

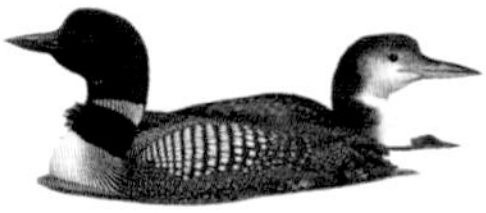
COLO p.84

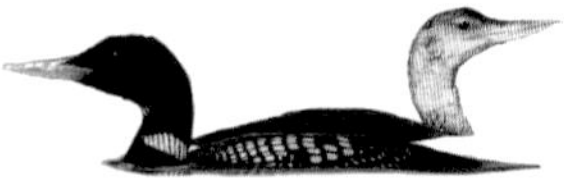
YBLO p.85

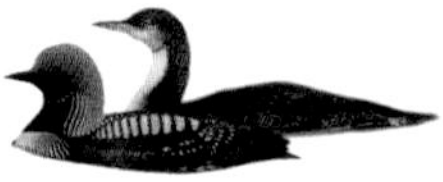
PALO p.85

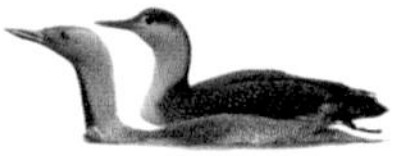
RTLO p.86

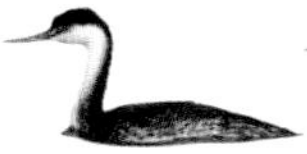
WEGR p.87

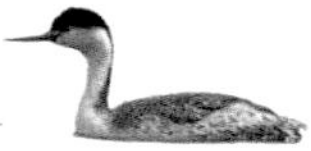
CLGR p.87

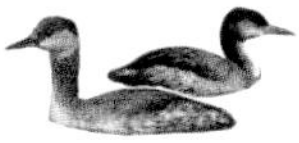
RNGR p.88

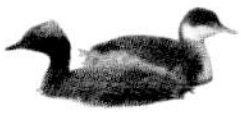
HOGR p.89

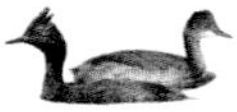
EAGR p.90

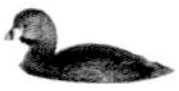
PBGR p.91

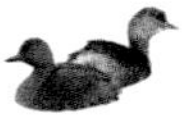
LEGR p.92

BLGU p.92

DOVE p.93

COMU p.94

TBMU p.94

RAZO p.95

ATPU p.96

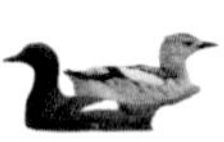

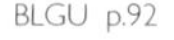

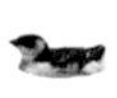

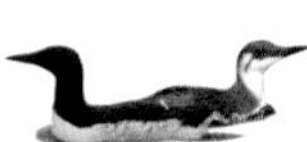

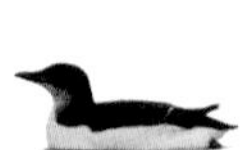

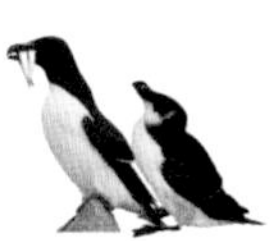

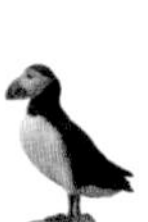

Flying Waterbirds pp.100–143

Size Guide: WISP–7.25in; AWPE–62in

WTTR p.114
RBTR p.114
BRNO p.115
SOTE p.116
BRTE p.116
BLTE p.117
FOTE p.118
COTE p.119
ARTE p.120
ROST p.120
LETE p.121
GBTE p.122
SATE p.123
ROYT p.124
CATE p.125
BLSK p.126
SPSK p.127
GRSK p.127
PAJA p.128
POJA p.129
LTJA p.129
LAGU p.130
FRGU p.131
BOGU p.132
LIGU p.133
BHGU p.133
SAGU p.134
BLKI p.135
RBGU p.136
CAGU p.137
HERG p.138
GBBG p.140
LBBG p.141
GLGU p.142
ICGU p.143

Walking Waterbirds pp.146–217

Size Guide: LESA–6in; WHCR–52in

BBPL p.146 AMGP p.147 KILL p.149 WIPL p.150 SEPL p.151 SNPL p.152 PIPL p.153

AMOY p.154 BNST p.155 AMAV p.156 SPSA p.158 SOSA p.159 LEYE p.160 GRYE p.161 WILL p.162

UPSA p.163 WHIM p.164 LBCU p.165 HUGO p.166 MAGO p.167

RUTU p.168 REKN p.169 SAND p.170 DUNL p.171 PUSA p.172 PESA p.173

WRSA p.174 BASA p.175 WESA p.176 SESA p.177 LESA p.178 BBSA p.180 STSA p.181

LBDO p.182 SBDO p.183 WISN p.184 AMWO p.185 WIPH p.186 REPH p.187 RNPH p.188

AMBI p.189 LEBI p.190 GRHE p.191 YCNH p.192 BCNH p.193 TRHE p.194

LBHE p.195
REEG p.196
SNEG p.197
CAEG p.198
GREG p.199
GBHE p.200
ROSP p.202
WHIB p.203
GLIB p.204
WFIB p.205
LIMP p.206
WOST p.207
SACR p.208
WHCR p.209
CLRA p.210
KIRA p.211
VIRA p.212
SORA p.213
YERA p.214
PUGA p.215
COMO p.216
AMCO p.217

Upland Gamebirds pp.220–229

Size Guide: MONQ–8.75in; WITU–♂ 37in

Raptors pp.232–267

Size Guide: NSWO–8in; BAEA–31in

Miscellaneous Larger Landbirds pp.270–313

Size Guide:
DOWO–6.75in; CORA–24in

BLJA p.304

WESJ p.306

FLSJ p.306

GRAJ p.308

BBMA p.309

AMCR p.310

FICR p.311

CORA p.312

CHRA p.313

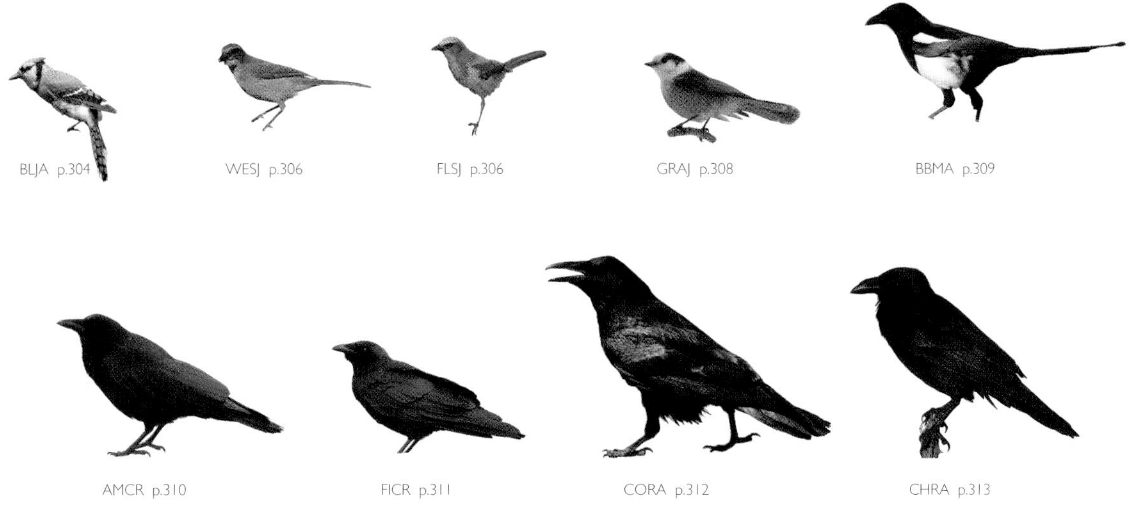

Aerial Landbirds pp.316–330

Size Guide: CAHU–3.25in; PUMA–8in

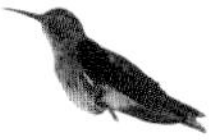

BBEH p.316

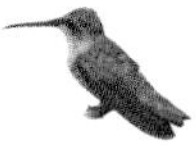

BCHU p.318

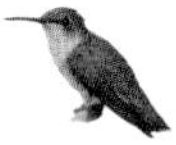

RTHU p.319

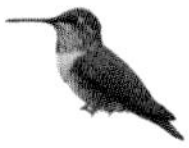

RUHU p.320

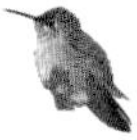

CAHU p.321

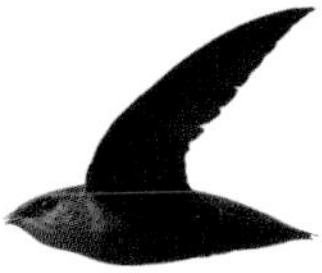

CHSW p.322

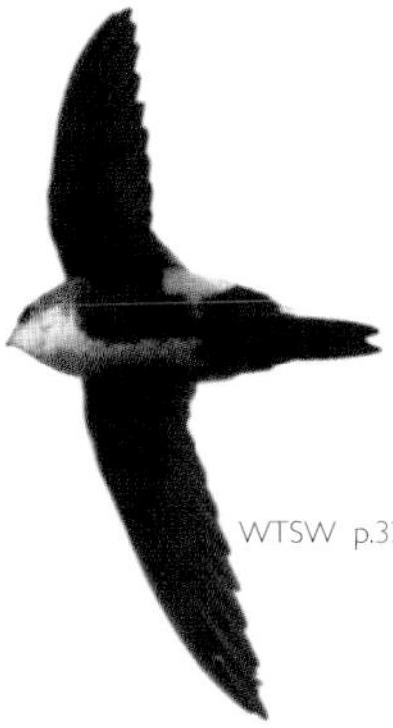

WTSW p.323

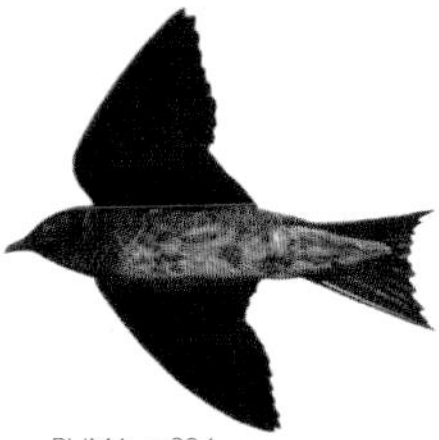

PUMA p.324

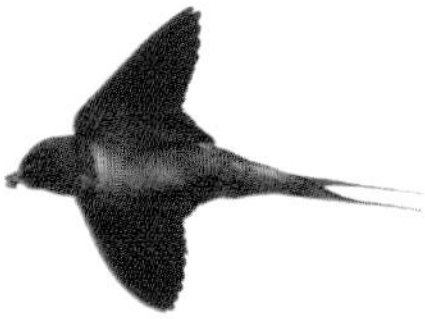

BARS p.325

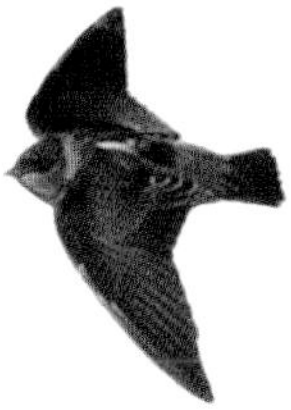

CASW p.326

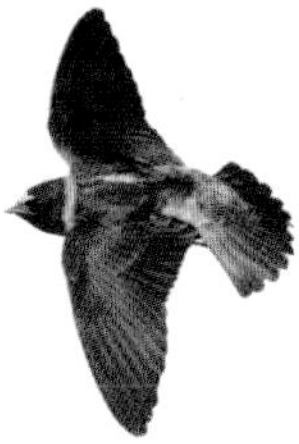

CLSW p.327

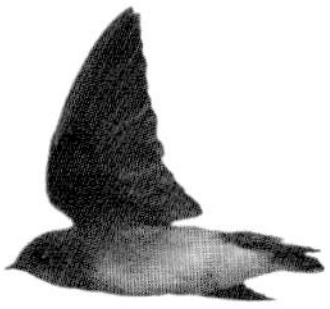

NRWS p.328

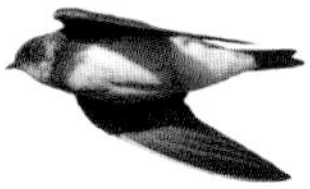

BANS p.329

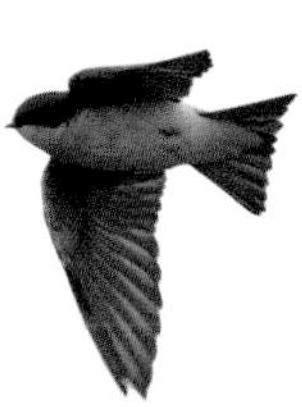

TRES p.330

Songbirds pp.334–515

Size Guide: WIWR–4in; BRTH–11.5in

CEDW p.370
BOWA p.371
BCCH p.372
CACH p.373
BOCH p.374
BCTI p.374
TUTI p.375
BHNU p.376
WBNU p.377
RBNU p.378
BRCR p.379
BUSH p.380
VERD p.380
ROWR p.381
CANW p.381
CARW p.382
BEWR p.383
CACW p.383
HOWR p.384
WIWR p.385
SEWR p.386
MAWR p.387
BGGN p.388
BTGN p.389
NOWH p.389
GCKI p.390
RCKI p.391
WEVI p.392
BEVI p.394
BCVI p.394
YTVI p.395
PHVI p.396
WAVI p.397
BHVI p.398
REVI p.399
BWVI p.400

NOPA p.401
BWWA p.402
GWWA p.403
TEWA p.404
OCWA p.405
NAWA p.406
YWAR p.407
KIWA p.408
CSWA p.409
MAWA p.410
CMWA p.411
BTNW p.412
GCWA p.413
YRWA p.414
YTWA p.415
BLBW p.416
CERW p.417
PAWA p.418
PRAW p.419
PIWA p.420
BBWA p.421
BLPW p.422
BAWW p.423
BTBW p.424
AMRE p.425
PROW p.426
WEWA p.427
SWWA p.428
OVEN p.429
NOWA p.430
LOWA p.431
KEWA p.432
CONW p.433
MOWA p.434
COYE p.436
HOWA p.437
WIWA p.438
CAWA p.439
YBCH p.440

HOLA p.441
AMPI p.442
SPPI p.443
SMLO p.443
MCLO p.444
CCLO p.444
LALO p.445
SNBU p.446
EATO p.448
SPTO p.449
OLSP p.449
DEJU p.450
GTTO p.451
CANT p.451
CASP p.452
BOSP p.452
BACS p.453
RCSP p.453
ATSP p.454
FISP p.455
CHSP p.456
CCSP p.457
VESP p.458
LASP p.459
BTSP p.460
LARB p.460
SAVS p.461
GRSP p.462

Size Guide: LEGO–4.5in; GTGR–15in

SUTA p.495
SCTA p.496
WETA p.497
SCOR p.498
AUOR p.498
HOOR p.499
OROR p.500
BAOR p.501
BUOR p.502
EUST p.503
EAME p.504
WEME p.505
BOBO p.506
RWBL p.507
YHBL p.508
RUBL p.509
BRBL p.510
COGR p.511
BTGR p.512
GTGR p.513
BROC p.514
BHCO p.515

INTRODUCTION

I DON'T LIKE TEXT

Personally, I soon get very bored with the introductory sections of any book. My guess is that many of you feel the same way. But this book is quite different from any that you have used before, so please, please do read on to get a clearer understanding of its overall design and how you might get the most from it.

THE PAST

The past is important because invariably it shapes who we are and how we think.

I grew up in England. I started collecting eggs when I was 7 and was introduced to birding, as so many people are, by an early mentor, my schoolteacher Mr. Sutton, when I was 10. I tried to identify everything I saw from then on. I started chasing ('twitching') rare birds at 16 and became an obsessed and obsessive birder when I went to university. I hitchhiked over 100,000 miles in 3 years looking at birds (perhaps I should have been listening to lecturers!). More importantly, during this time I became part of an incredible British birding culture. Its influence on the way I think about birds and birding—and of course my lifestyle—can't be underestimated.

I first came to America—and Cape May—when I was 21. I loved it so much that I moved here for good in 1991. My experiences in learning a new avifauna have also greatly influenced my thinking. And, thrown into the mix, has been a lot of travel much farther afield, including living in Japan for nearly 2 years. I have tried to draw from this rich and varied experience to produce the book.

THE PRESENT

Birding knowledge has evolved greatly over the last 40 years, but the basics of field identification have remained the same. Photography, however, has changed beyond all recognition. The digital age has revolutionized and popularized bird photography. For one person to make a comprehensive book using over 10,000 images and covering 660 species would have been almost unthinkable until recent times. And as digital cameras continue to improve at a dizzying pace, more birders are becoming confident and proficient photographers, thereby accelerating the trend for more photo-based guides. But, until now, these guides have essentially followed the same 'static' formula as artwork-illustrated books, that is to say they concentrate on individual images isolated from the overall context of habitat—and other birds. *The Crossley ID Guide* changes this approach in a fundamental way.

HOW TO USE THIS BOOK

ORGANIZATION

The first thing about this book that may strike you as different is its organization. Traditional field guides largely use a taxonomic sequence, which, while aiming at scientific accuracy, doesn't always make sense in field birding. It also changes, literally from day to day, making it hard to keep any guide current. This book splits the species into 8 groups, based on habitat and physical similarities. It is no coincidence that a bird's appearance is largely influenced by its environment. It also means that taxonomic order is not broken as often as you might think. Similar-looking species are grouped together so they can be compared easily.

Unfortunately, the largest downfall of all books is the impossibility of portraying size in a lifelike manner. The closest we can come to this is to write down the bird's length, which is included in each species account next to its name. The other is to show its size relative to other species. Because of the importance of this I have done it in 2 places in this guide.

Firstly, the front inside cover has representative species from most families that cover the complete range of birds in this book. For the beginner, this is a good place to start, and it is where you can narrow your search for any bird you are trying to identify.

Immediately preceding this introduction are 16 pages that feature all regularly occurring species, those that don't have a very restricted range. Like the birds on the front inside cover, these have been carefully measured. This is the best place in the book to make direct size comparisons. These pages should also be used to study shape, and, to a smaller degree, patterns of color. In most cases, a species will be identifiable from these plates. Included under the species image is its name in shorthand (alpha code) and its page number. It is on this page that you will find the species plate, the textual account, and the range map.

PLATES

The plates are intended to be the heart and soul of this book. Throughout I have strived to use images that I think represent accurately what each species looks like in the field, or put another way, as I would have painted them. They were chosen because they accurately portray the species' shape, plumage, and behavior. One of the beauties of these plates is that you can control your canvas and include lots more (lifelike) information than is possible with painted plates. These images were 'molded,' often after a considerable period of trial and error, to create an overall scene that is as lifelike as a printed image will allow. Each plate contains a massive amount of identification information within a relatively small area.

Here's the rationale for this approach:

1) *Reality Birding!* One of the most important things to learn in becoming a good field birder is the ability to see the features that remain constant, regardless of distance. By creating depth in the plates, we can see how a bird's appearance changes with distance.

Books typically only show birds close up and in detail. Yet in reality we very rarely see them like this. The plates in this book allow you to study distant birds and compare them to larger images. Seeing the similarities between the different-sized images will help you focus on the features that remain constant.

2) Why didn't I caption or give you all the answers to the identity of the birds in the background? *Because the answers lie in the larger captioned images.*

The book is designed to be interactive. Use these captioned birds in the foreground to try to work out the age and sex of birds in the background. This won't be possible for some birds, but for most it will. At school and in other walks of life, it is not until you practice or get your hands dirty that you really learn. Give it a really good bash! Most people are pleasantly surprised when they make a prolonged effort. If you can't do it here, you will find it very difficult with moving birds. If you can, work out many of the answers yourself. You will be far better prepared for the real world.

This is the first guide that uses lifelike scenes. Take advantage of them to practice so that you are better prepared to identify any bird you see in the field. *Practice makes perfect!*

3) *A picture says 1000 words!* And these plates contain many pictures. The amount of information in these plates is staggering. It is up to you to take advantage of this.

4) It's much easier for the human brain to absorb information from a single image than from many separate images. We are more likely to create a mental picture that we retain. I still remember some images from the first bird book I got as a kid!

The plates are mostly 'in focus' throughout their full depth, unlike other photographic bird guides. This is how most of us perceive the world we see (with the aid of contact lenses or glasses as necessary). We therefore relate naturally to this representational approach because we are most familiar with it—and so we will learn more.

5) All or most plumages are shown. Also included are birds in transition (molting). Many books show birds in breeding and nonbreeding plumage. It typically takes weeks, sometimes months, and occasionally years, for a bird to molt the feathers that change its appearance from one plumage to the next. By showing a more complete picture it is easier to visualize and understand the workings of molt and how it affects appearance. This expanded coverage also gives you a better chance to find a comparable image of the bird you are trying to identify in the field.

6) Behavior is a broad topic yet vital to many aspects of bird identification. Many examples of behavior are included in the plates. All actions and movements shown, such as feeding behavior or unusual poses, are typical for that species. Birds that occur commonly in flocks are usually depicted in this way.

7) There are flight photos for most species; many have multiple images. Please look at these closely as they were extremely challenging to take and almost certainly account for my hair loss over the past few years! You will naturally ask: "Why so many flight shots?" Well, we see birds in flight more than in any other pose. Birds in flight can be relatively easy to identify, but it does require specific focus on size and shape. A bird's structure on terra firma is mirrored in flight.

8) Habitat plays a large role in identification. The plates capture a habitat or environment that is typical for that species. This is sometimes difficult because birds often live in a variety of habitats, and many species breed in very different habitats from those they occupy during the nonbreeding season.

9) Species get proportional representation, i.e., the commoner and/or more widespread species typically get full-page coverage. These species are treated more thoroughly, using additional images and greater variation in image size. Half-page plates are typically used for scarce and more localized species. Species that get only a quarter of a page are rare and are very unlikely to be seen. If you think you have found one of these rarities, look very closely and carefully, then let others know so that you can attain 'hero' status (or not!).

The secret to becoming a better birder (and finding rarities) is knowing how to look at birds, and *to gain an intimate knowledge of common species.* When you know what something isn't (a common bird), it's usually fairly easy to work out what it is (a rare bird)!

10) WWW.CROSSLEYBIRDS.COM will have expanded captions for many of these plates. With limited room for text, and a topic that is constantly changing, this provides the opportunity for me to include additional identification information.

HOW DO I LEARN FROM ALL THIS?

Hopefully, many readers will like the appearance of these plates; others will at first find them overpowering, and perhaps even confusing.

Yes, I could have made them simpler and perhaps more attractive. However, I didn't want to compromise my effort to get people to understand the 'big picture' of bird identification. And, yes, I am asking readers to think of, and use, this book differently from any other guide they may have. There are many different types of guides, but they bear no resemblance to *The Crossley ID Guide* either in their appearance or goals. Although this book will be used for reference, the principal reason for its design is to be interactive with the reader—much like a workbook at school.

When looking at a plate for the first time, try to view it without any preconceived ideas—just an open mind. Simply ask yourself: "What do I see?" Be careful that you do not look at it based on any preconceived notions of "What am I supposed to see." In particular, look at the smaller images in the background, because the chances are that this is what you will actually see in real life. By zooming in and out of the bird images, try to absorb the things that remain constant—shape, patterns of color, and so forth. If you can create a good mental image of these patterns, it will serve you well in the field.

As you look at the plates try to get a feel for this bird, its lifestyle, where it lives, and what it does. In this sense, how you view it should be no different from how people view each other. If you are not interested in getting so involved in honing your skills, see if you can find every bird in the plate. Many are not so easy to locate. Just as in real life, the harder you look, the more you will see.

Keep in mind the points discussed throughout the 'How to Use this Book' section. Moreover, the plates contain a lot more visual information than has been shown in other guides to date. If you can remember details about the plate after you close the book, you almost certainly learned much more than you might think. Please remember, the best field birders keep it simple, so if you can look at a plate and create a picture in your mind of the bird involved, you will begin to think like an expert.

CAPTIONS, TEXT, AND MAPS

Captions, text, and maps are intended to complement the plates, and fill in pieces of information, some of which can't be shown visually.

Below each plate is the species' *common (English) and scientific name.* Nearly all of these follow the names used by the American Ornithologists' Union (AOU). Please remember these can change. Also, different bodies and countries use different common names. Most are similar, but some are significantly different; for example, Common Loon is usually called Great Northern Diver in Europe. If in doubt, you can cross-reference the scientific name. Taxonomy today seems to be in a greater state of flux than ever before, so scientific names, like common names, may change.

Next to the scientific name is an *alpha code*. This is the 'shorthand' used by banders for recording data. I have used this shorthand in the text for 2 reasons. The first is a space issue. Birds' names are often long and take up a lot of space at the expense of itext. The second reason is that alpha codes are already part of many birders' lives. For example, there are several regional 'texted' bird alerts, such as where I live in NJ, that use this shorthand. If you are having problems understanding what the 4-letter abbreviation stands for within a species account, it usually refers to one of the comparative species found on an adjacent page. If not, there is an index at the back of the book (p.518) with a full listing. Typically, single-name species use the first 4 letters Killdeer (KILL), double-name species 2 and 2 (Blue Jay is BLJA), triple-name species 1, 1, and 2 (Great Blue Heron is GBHE).

The *average length* of the species depicted is shown next to the alpha code. Length is measured from tail tip to bill tip on stretched birds. When there is a large difference in size between sexes, as in many raptors, measurements for both are shown.

This book covers eastern North America (USA and Canada). A map showing the area covered by this guide and a key to colors used in the maps is on the inside back cover. Distribution plays a major role in bird identification. Use the maps to check where a species typically occurs. While it is not impossible to find a Black-capped Chickadee well south of its normal range, you will naturally be in the backyard of a Carolina Chickadee. While range may help you to make a positive identification, still study the bird carefully. By looking at where a species breeds you will have a good sense of which migration route it takes. Birds with a wide range can be assumed to migrate on a broad path unless specifically noted otherwise. Relative abundance is discussed in the text.

Maps are sized to maximize information. In many cases the map includes the West (western North America) to give a clearer, bigger, picture of the species' distribution, particularly as a large number of birds that migrate through our region originate in the West. Specific regions, such as the East, the West, the South, are capitalized in the text. When I use the abbreviation 'w.,' I am referring to the western part of our region, i.e., the Great Plains.

Rare species get a written description of distribution because maps are typically of limited value for these.

On most plates, each plumage, age, and sex is labeled once. A label placed between 2 or more images implies it's applicable to both or all. On some plates, where all plumages are similar, for example Barred Owl, there is no label. In these cases, the birds are considered impossible or very difficult to age or sex in the field under usual conditions.

On some plates you will see captions that represent 2 or more plumages, for example 1st-s. male/ad. female. This means this plumage is representative of both plumages. In some cases, it is possible to age and sex the bird from the photograph, but I may not have done so on the plate. This is going to be a bone of contention for some. The reason is that, in many cases, such birds in the field cannot be aged or sexed with certainty all of the time.

One of the benefits of artwork is you can be purposefully ambiguous in such cases and not get flak for it. Where my approach may be particularly controversial will be with labeling of many ducks as 'eclipse/imm male.' Ageing these birds is difficult and still developing. Yes, they are different, but the differences are often subtle and usually too difficult for all but the very best birders to determine. Where they are easier to work out, I have labeled them accordingly. This is a judgment call—not the only one in this book! Trying to cater to a very broad audience (in the case of this book, everyone), and give detailed information on such a huge topic in such a small space is not without its difficulties. This is a major reason for having a website (WWW.CROSSLEYBIRDS.COM), which will have expanded captions for those who really want to get in at the deep end.

Identifying birds is fun and rewarding. Trying to age and sex all the birds you see is just an extension of basic identification and can be equally enjoyable. As mentioned earlier in the introduction, try to use the larger captioned images to age and sex the smaller birds farther away. If you look closely, and carefully, you will find it's often not so difficult.

The clues always lie in the larger images. Using this simple approach, you will be training yourself to analyze and identify the birds you see in the field faster and in enhanced detail. In school we learned by working out the answer on our own. If we are given the answers, as in all the other guides, it is very hard to improve. After all, when we are in the field it is rare, and not particularly rewarding, when we have someone there to spoon-feed us. And, as my kids would tell you, one of my favorite sayings is "Practice makes perfect."

When you read the text, it will hopefully reinforce and enhance all you have learned from the plate. It starts with the bird's status (relative abundance) and preferred habitat. Knowing where a bird lives plays a significant role in successful bird identification.

The text then discusses behavior. Just like humans, birds tend to behave in a certain way. We can invariably identify loved ones by the way they move, by particular mannerisms or quirks they have, or by the places they tend to visit. But just like in humans, it sometimes takes time to learn these. Obviously, the more time you spend around people or birds, the better you get to know them. Try to read the text and get a sense of the bird, what it is likely to be doing and how it will do it. Cross-reference your mental image with the plate. *Get to know the bird inside out*!

Within the '**ID**' section all the important field marks of that species are mentioned, starting with reference to size and shape. From there the text moves to other features that are constant in all plumages. More specific details on the characteristics that enable you to to recognize each plumage are then discussed. These field marks not only help you identify the bird to species but also age and sex it. These can be considered the same as captions pointing to specific field marks on plates in traditional field guides.

Hopefully you will have already noticed many of these field marks if you have looked at the plate carefully! If you have, you will be sure to remember them in the field. If not, again cross-reference the plates. Take your time and go one point at a time. For very similar species, the text will tell you the features to focus on.

Try not to rush to read the text, but when you do, always do it slowly and with reference to the plates.

HOW TO BE A BETTER BIRDER

42 million people consider themselves to be bird-watchers in the USA alone. Most do not subscribe to birding magazines, read birding articles on the internet, or hang out with experts. If you are one of the many who are not familiar with the fundamentals of becoming a better field birder, the following is included in the hope that it will help you.

LOOKING VERSUS SEEING

The biggest mistake most birders make is to rush to get to a name rather than learn the species; how it behaves and what it looks like. *We often look rather than see*!

Do you know what a Blue Jay looks like? Really? Describe its tail. Is it black or blue, or a combination of the two? Is there white in the tail? How do you age the bird? What is its head pattern, and where is the black-and-white barring on the wing?

Okay, you get my point. And don't worry, there are probably only a few artists and bird banders who could answer all those questions correctly so you are in good company.

LEARNING TO LOOK: TAKING FIELD NOTES

Birding, as with most things in life, is about the basics. In this case it's learning to look. We can be told how to learn, but ultimately it's down to us to make it happen. The best field birders in the world, at some point in their lives, were made to look at birds closely, a crucial step in their (self-) training. I believe that all experts have taken detailed field notes, which is simply the best foundation for becoming a good or great field birder. Leading field trips, being a bird artist, or writing books also forces you to look at birds in this concentrated fashion. Only when you are put on the spot do you realize how little information you are really taking in, and this realization compels you to look much more critically at every detail of a bird's plumage and behavior. For those of you who do take field notes, you already understand the value of this as an habititual exercise; you are able to fill in pieces of the identification puzzle more accurately. In this way, you have already become a better field birder.

Funnily enough, I grew up in a British birding culture where you didn't take a guide into the field—only a notebook. You focused on the bird and wrote notes. This was the 'law,' and, if you wanted to be taken seriously as a birder, you simply didn't carry a guide. This fundamental difference in approach certainly affected how I look at birds today. And I'm genuinely thankful for it. Of course, in the end, this is an entirely personal choice for all birders, but I would urge you to give it a shot.

Taking field notes also makes you think for yourself—*to look at a bird for what it is rather than what someone else tells you it is supposed to be*. We are all influenced or biased by the world around us and the things we have read or heard. Remember, the bird in front of you is your immediate reality. Watch it and you will learn it. *Believe your own eyes*! You will understand the bird in a way that books cannot teach you. And remember, a bird's size, shape, and behavior usually do not vary too much from individual to individual, even though the plumage may. The same is true of humans. We change clothes seasonally, but underneath we remain the same.

The following may prove useful:

Take 1: My name is Richard, I am white, pink-cheeked (though my cheeks get a lot rosier after exercise or booze). My complexion is paler in winter and darker after I spend time in the summer sun. My 'derriere' is white because it never sees the sun. I have brown hair and I'm not very tall.

Is this a good description and would you be able to recognize me based on it? The answer is obviously "No."

Take 2: My name is Richard. I am about 5 feet 9 inches tall. I weigh around 185lbs—I could do to lose a couple, though I am fairly broad. I am in my 40s, Caucasian, with short brown hair. You can often find me birding or photographing birds in Cape May.

Is this a better description? The answer is clearly "Yes," simply because it gives more specific information about my size, shape, and other physical attributes, as well as habits. You would probably use a similar approach to provide a decent description of yourself.

Now, a tough question. Do you know what a Northern Cardinal looks like? If so, please get a piece of paper and write down a full description (without recourse to a field guide). Or at least describe it in your head. Let's look at your description of the Cardinal. Was your mental image based on size, shape, and behavior as well as color? Probably not. But if you start to recognize birds as we do people, then you will be using the same set of skills as the best field birders in the world.

I firmly believe field identification of birds can be broken down into these key areas (in my personal order of importance): *size, shape, behavior, probability, color, and sound.*

SIZE

Somewhat surprisingly, perhaps, it turns out that we're remarkably good at judging an individual's height, in fact to within a 2% degree of accuracy on most occasions. The truth is that we spend most of our lives practicing. So, not surprisingly, adults tend to be much better at judging height than children. We all judge relative size in birds to some degree, but often put little emphasis on this in the field. You should always focus on size and try to make as accurate an assessment of this feature as you can. Compare the bird you're trying to identify with other nearby birds that you have already identified. With practice, you can become accomplished at determining size, which is critical since it is the least variable character that birds possess. Sometimes you will know you are very accurate with your assessment of size, but at other times it will only be a rough estimate. Naturally we can also get this wrong, particularly when views are brief or distant. *The secret is to know your limits.*

STRUCTURE AND SHAPE

Along with size, structure and shape are fundamental to the identification of nearly all birds. Shape is remarkably consistent in individual species. Color and lighting have little or no effect in our determination of a bird's shape and structure. Always try to describe a bird's shape in language that makes sense to you. We each interpret or understand words such as fat, rounded, slim, and long differently. While, as an author, I'm compelled to use these terms when describing a bird, you should create your own language and sense of scale to describe the same bird in terms that resonate with you.

BEHAVIOR

Learning the 'personality' of a bird is hugely important. This obviously takes longer to master than assessing a bird's size and shape. Knowing the behavior of birds with which we are familiar is essential in the field. Behavior encompasses many aspects of identification, just as it does with our interrelationship with other humans. For instance, consider the type of habitat a species favors, how it moves, and whether it's a loner or gregarious. For example, a Sanderling is instantly recognizable when it relentlessly chases waves along the beach, a clinching identification feature regardless of color or shape.

PROBABILITY

We use probability in bird identification, sometimes more than we would credit. Does the bird usually or always occur in this location and in this habitat? When you go birding in an unfamiliar area, you always start with this basic question, consciously or subconsciously. On your local patch you would naturally be more confident since you have built up experience of species' occurrence and distribution. For instance, if you come from Massachussetts and find yourself birding on the Delaware River in New Jersey, you need to ask yourself: "Is it Carolina or Black-capped Chickadee I'm likely to see here?"

I estimate that I identify approximately 90% of the birds I see as silhouettes or simple black-and-white images—a flock of European Starlings swirling around, a Cooper's Hawk chasing a Mourning Dove, a Northern Cardinal darting across a road, and a huge, dense flock of hirundines that will almost certainly be Tree Swallows. These are almost subconscious, reflex identifications built on years of careful field observation, and a just reward for learning to look.

COLOR

We love the myriad color of birds, and stunning photographs that capture them in all their astonishing beauty. Often we can't help but be overwhelmed by a blast of color as we chance upon a stunning red-and-black Scarlet Tanager. The problem is that Scarlet Tanager (a bird that is consistent in size, shape, and behavior, and also spends most of its life in uniform habitat) changes its colors. In one season it is usually yellow, green, and black; but it has to change its feathers (molt), and so it has a period when it shows a complex combination of different feathers and therefore a changing pattern of colors. I won't dwell on the challenge of learning plumages of females, juveniles, 1-year-old males, and so forth! And there are other important factors that influence identification such as time of day, whether sunny or cloudy, position of the sun, amount of shade, feather wear and fading, aberrant (abnormal) plumage, and of course just normal variation between individuals within the same species.

Of course, we are naturally attracted by color. Even so, always try to stick to identification basics: "Is the bird in front of me the correct size and shape for the species I believe it to be?" "Does the species I've identified even occur here?" Color can be extremely variable, so it is important to focus less on the tone of the color itself and more on the overall pattern it creates, i.e., the relative colors of different parts of the body. For example, the shades of yellow in a Yellow Warbler are variable from bird to bird, but the lightest and darkest parts on each and every bird are remarkably consistent.

Ultimately, color is undeniably important in bird identification, and for beginners, in particular, it will almost always be the first feature to attract the eye. But the secret is to learn how to use color in combination with all of the other identification factors described above, and to always remember that most misidentifications are made because of a reliance on color as the key to successful field identification.

I have included some basic references to sound in this book but not for all species. The value of written descriptions of bird songs and calls has been debated for a long time. Without being able to listen to these vocalizations, descriptions are of limited use. I initially intended to exclude all descriptions of vocalizations but have found myself including more and more. I have questioned the wisdom of this more than any other aspect in this book. These descriptions can never compete with recordings on an iPod, phone, or the internet, where listening to any sound you want is nothing more than a couple of clicks away. However, even this does not compare to the real thing in the field.

A quick test! Identify every bird in the photo below. At first glance many of you might think this blur of birds is impossible to work out without lots more fine detail. Please look again, and think in terms of size and shape. When considering color, think in terms of patterns—which part of the bird is dark or bright and which parts dull or pale? Look again and see if you can put all the birds into 3 groups.

The birds (about 20) in the foreground with the short red legs, fat body, chisel-like bills, and bold color patterns are Ruddy Turnstones. Center left is a slightly larger bird with a longer but thinner bill, an orange face, and contrastingly gray spangled upperparts—Red Knot. The 2 birds slightly closer to the left have the same gray spangled upperparts, and you can just make out the orange face—therefore also Red Knot. Just behind the clearest Red Knot is another bird with an orange-red head, but the orange is more uniform and extends onto the upperparts—it also looks smaller. This is a Sanderling. This pattern appears to be the same for all the birds in the background and, yes, they are also Sanderling.

At first this may seem tough, but try to remember it is the pattern of color and where it is brightest and darkest on the body that is most important.

VOCALIZATIONS

Songs and calls are a large part of bird identification. Many birds are identified without even being seen. Even if you hear a sound you are not familiar with, it helps you locate a bird that you might not have known was there. For the beginner, or someone starting from scratch, such as myself when I first came to North America, the range of vocalizations can seem overwhelming. I will always remember trying to differentiate all the songs in the spring dawn chorus for a bird race. It is much easier for me now but still a challenge.

BIRD TOPOGRAPHY

Songbird

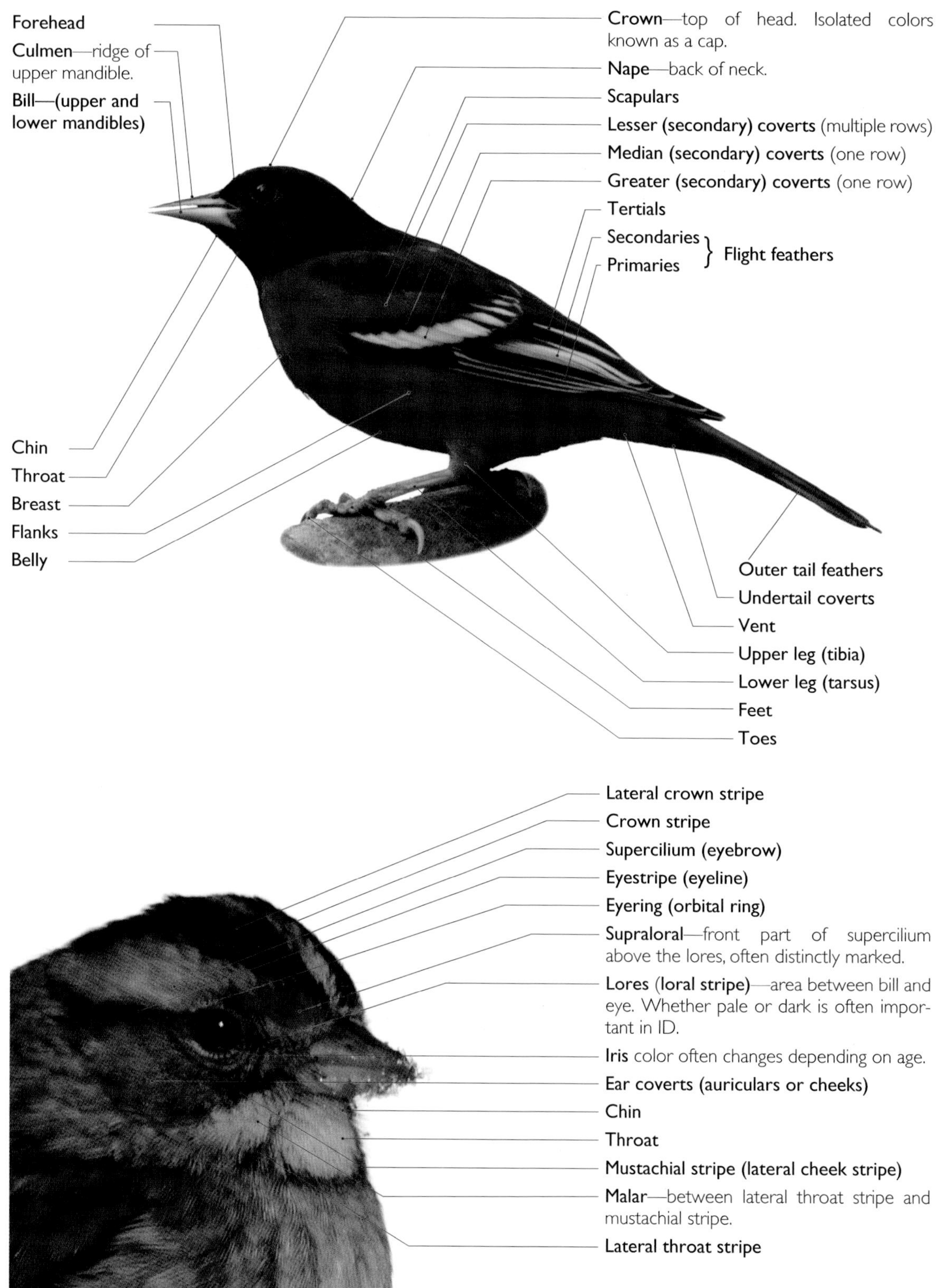

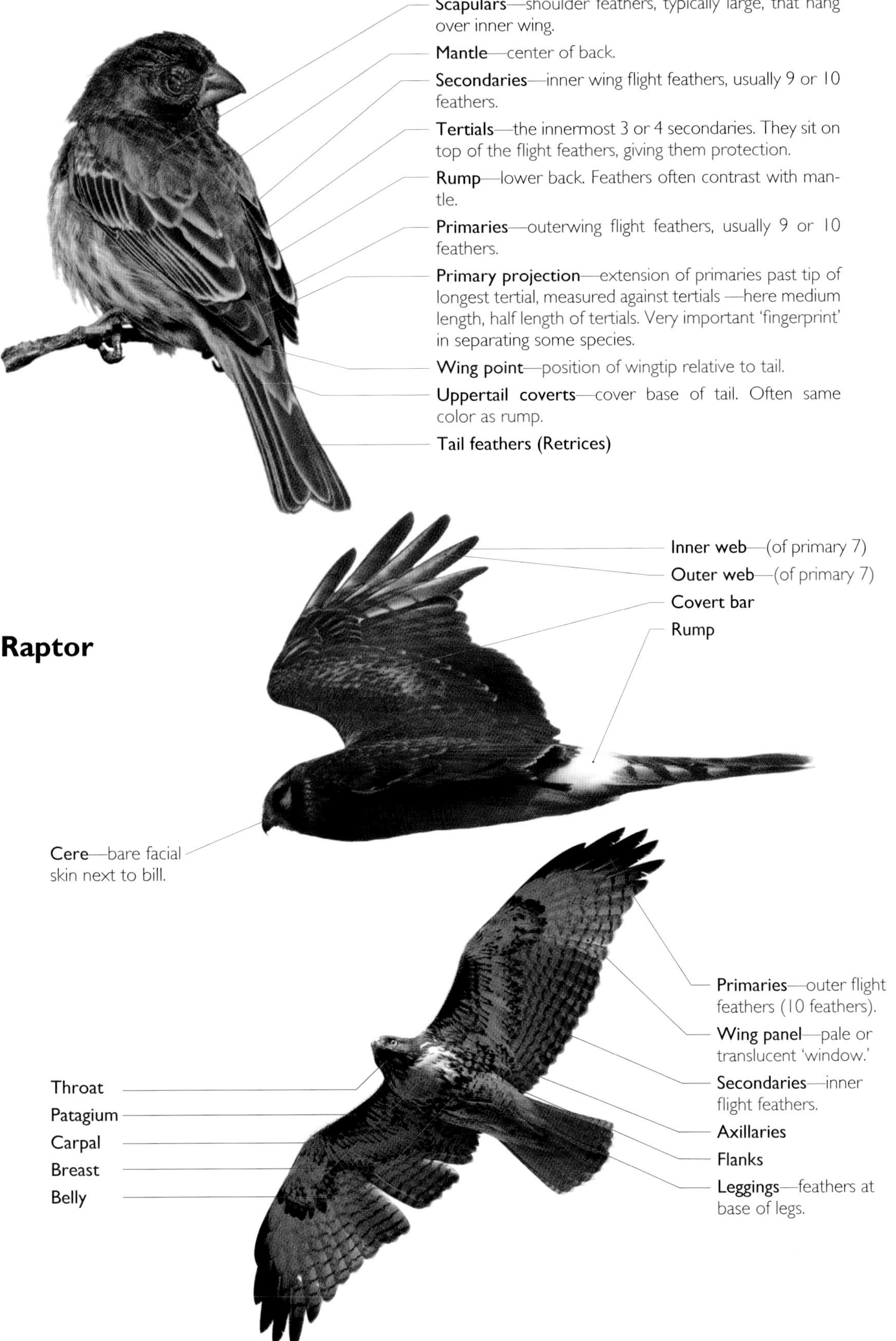
Scapulars—shoulder feathers, typically large, that hang over inner wing.
Mantle—center of back.
Secondaries—inner wing flight feathers, usually 9 or 10 feathers.
Tertials—the innermost 3 or 4 secondaries. They sit on top of the flight feathers, giving them protection.
Rump—lower back. Feathers often contrast with mantle.
Primaries—outerwing flight feathers, usually 9 or 10 feathers.
Primary projection—extension of primaries past tip of longest tertial, measured against tertials —here medium length, half length of tertials. Very important 'fingerprint' in separating some species.
Wing point—position of wingtip relative to tail.
Uppertail coverts—cover base of tail. Often same color as rump.
Tail feathers (Retrices)
Raptor
Inner web—(of primary 7)
Outer web—(of primary 7)
Covert bar
Rump
Cere—bare facial skin next to bill.
Primaries—outer flight feathers (10 feathers).
Wing panel—pale or translucent 'window.'
Secondaries—inner flight feathers.
Axillaries
Flanks
Leggings—feathers at base of legs.
Throat
Patagium
Carpal
Breast
Belly

Duck

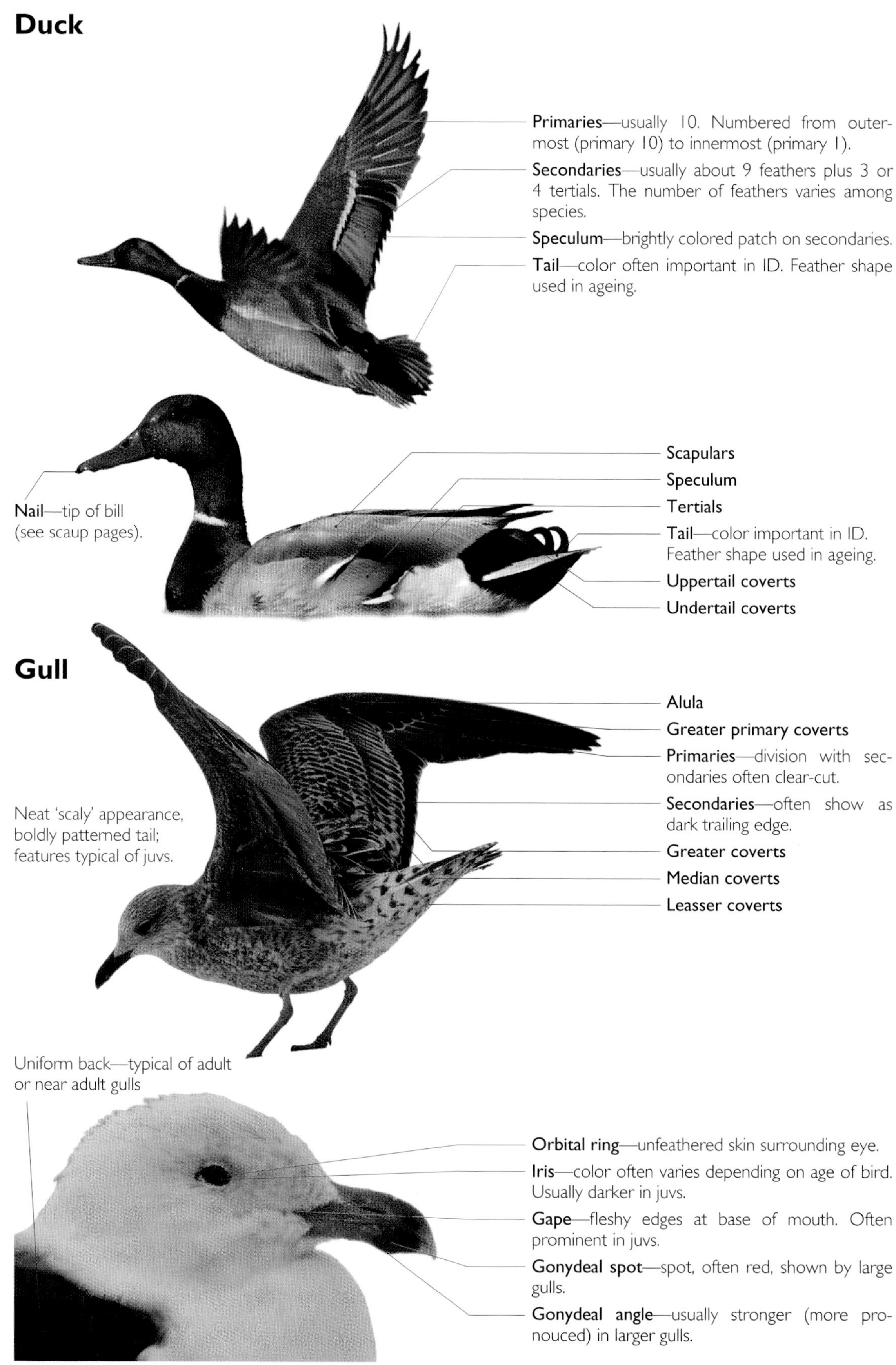

Primaries—usually 10. Numbered from outermost (primary 10) to innermost (primary 1).
Secondaries—usually about 9 feathers plus 3 or 4 tertials. The number of feathers varies among species.
Speculum—brightly colored patch on secondaries.
Tail—color often important in ID. Feather shape used in ageing.
Nail—tip of bill (see scaup pages).
Scapulars
Speculum
Tertials
Tail—color important in ID. Feather shape used in ageing.
Uppertail coverts
Undertail coverts
Gull
Neat 'scaly' appearance, boldly patterned tail; features typical of juvs.
Alula
Greater primary coverts
Primaries—division with secondaries often clear-cut.
Secondaries—often show as dark trailing edge.
Greater coverts
Median coverts
Leasser coverts
Uniform back—typical of adult or near adult gulls
Orbital ring—unfeathered skin surrounding eye.
Iris—color often varies depending on age of bird. Usually darker in juvs.
Gape—fleshy edges at base of mouth. Often prominent in juvs.
Gonydeal spot—spot, often red, shown by large gulls.
Gonydeal angle—usually stronger (more pronouced) in larger gulls.

Shorebird

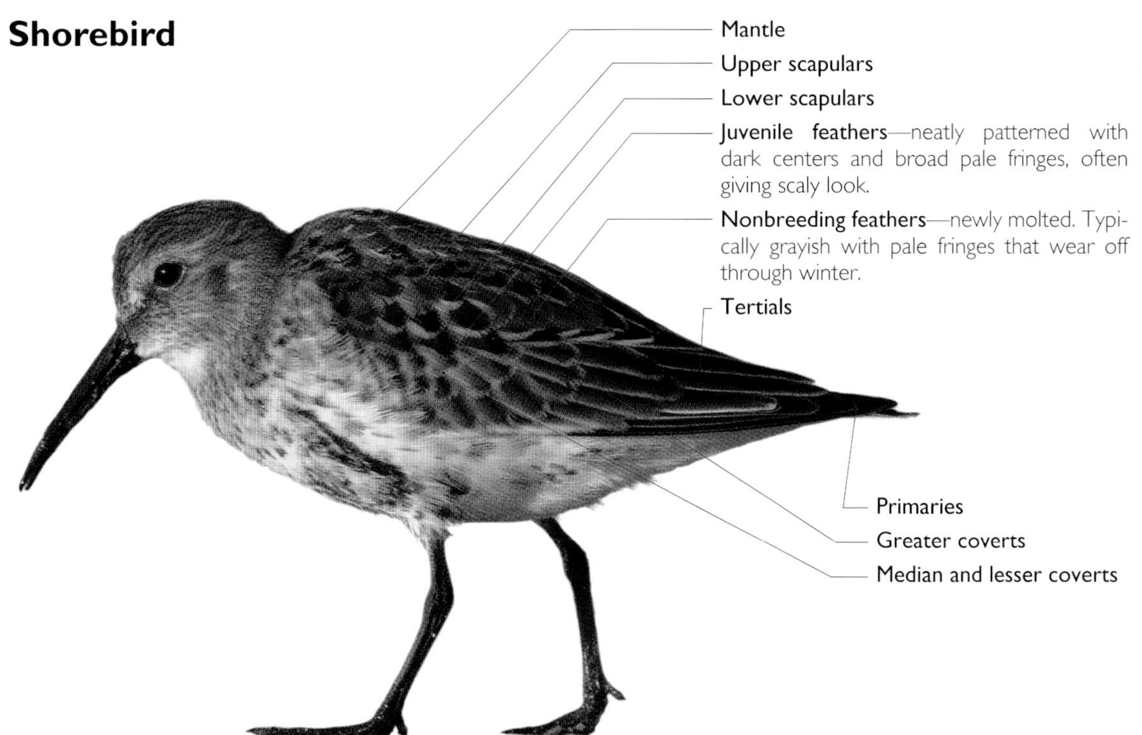

Hummingbird

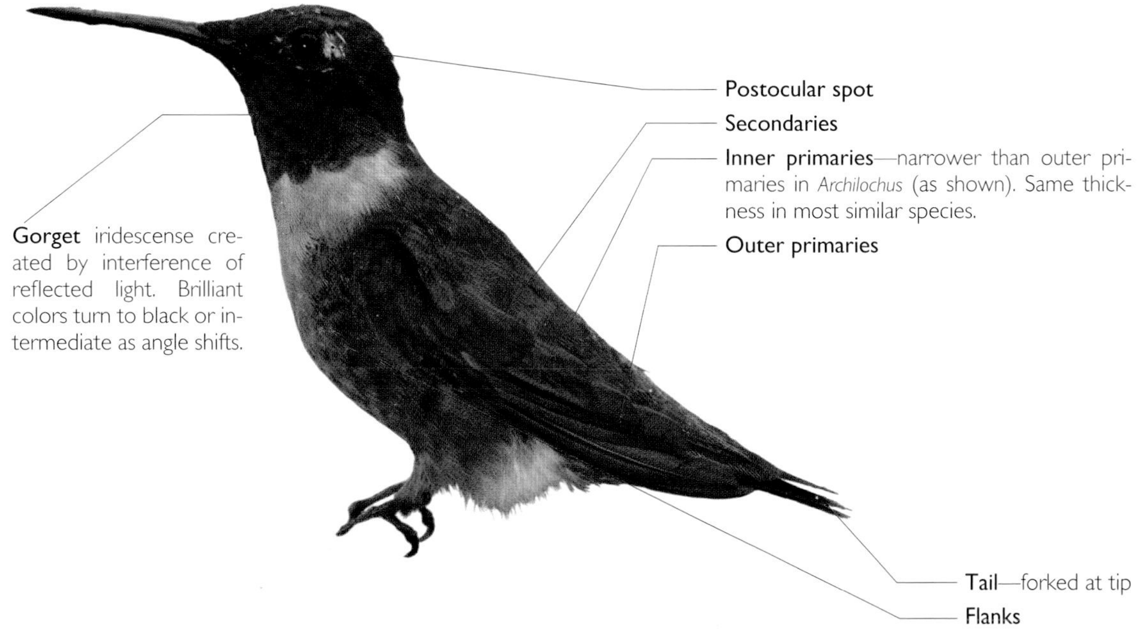

The variation within each species' vocalizations is large—clearly shown in sonograms. The best way that I found to learn songs was to try to describe them to myself. Try to write them down thinking about pitch (high or low, and whether it is going up or down), length of vocalization, and rhythm. A few birds have songs that are phonetically described, for example *witchitty witchitty*—Common Yellowthroat, *sweet sweet I'm so sweet*—Yellow Warbler. I found these useful, but most written descriptions were of little or no use. Listening to tapes helped the most, particularly in confirming identification after listening to a bird. I forced myself to identify every sound I heard. There is no substitute for real-world repetition.

TOPOGRAPHY

Birds are covered in feathers. These are split into batches or groups known as tracts. Knowing these tracts is a great help in bird identification. Color patterns within these feather tracts are usually the same. The major exception is when birds are growing (molting) new feathers.

I remember eavesdropping on a conversation in a pub on the Isles of Scilly when I was 16. The group of top young English birders in the next booth were talking a language I didn't know. They were discussing primary projections, tertial fringes, greater covert bars, alulas. They were using these terms to identify, age, and sex birds. This "tertial talk" as I now call it, at first sounded overwhelming and more like double Dutch. It inspired me to start looking at birds more closely. At first it seemed confusing, but soon the mystery disappeared. As I began to learn the different groups of feathers, how they were used by a bird, and how their appearance changed, it had a big impact on my birding skills. The language of brilliant birders was not so difficult to learn.

One of the keys to learning topography is being able to see birds repeatedly and well. Birdfeeders near your windows are perfect for this. I also find it useful to think of bird topography in human terms. Most of us know words like forehead, crown, legs, feet, breast, belly, and even scapulars. Once you have a good grounding, it is easier to work out the missing parts such as tertials and coverts. Of course, birds change their shape, most notably when they open their wings. Understanding the mechanics of this, and how things fit back into place on the closed wing, is also a great help.

On the confusing side, you will find that tracts of feathers don't appear the same on different groups of birds. For example, scapulars will often cover much of the wing on shorebirds but relatively little on warblers. Learning these differences unfortunately comes down to repetition and experience. However, taking field notes certainly helps you become familiar with them. As a result, when you see an unfamiliar bird, you are able to take in far more information in a shorter time, focus faster on important identification features, and, quite simply, become a much better field birder. *Knowing the name of a bird is not important, but knowing how to look at it is crucial.* The top people in any field are the best because they can analyze things better and faster than others. It is all in the training!

MOLT

Birds essentially do the same things cyclically, this includes molting. They do this for a number of reasons. It is important that feathers be kept in good condition to keep the bird warm, dry, and mobile. Of course, some like to look good to attract a mate! Having an understanding of molt and how it affects a bird's appearance is a large help not only in identifying birds but also in ageing and sexing them.

Molt is complex, but we can try to simplify it. Fundamental to a clear understanding of this process is knowing that birds molt all their feathers at least once a year, replacing them after they have finished breeding. Many also have a molt in spring; these are frequently only partial molts of head and body feathers. These spring molts are often into brighter plumages to attract a mate or help the bird blend into its environment.

A feather starts as a follicle within a sheath (known as a pin)—just as our hairs do. As the follicle grows, it pushes out the old feather. Feathers grow anything from a couple of mm's to a cm per day (larger birds' feathers grow faster, as you would expect). When they are fully grown, they are cut off from the circulatory system so that they can't grow any more. This also means this process doesn't use up any more energy. Molting takes up a lot of a bird's energy and for this reason has a large impact on its lifestyle and behavior. The bird must balance the advantages of having new feathers with the energy required in growing them. Large birds molt much more slowly than small, simply because of this large energy cost, and their molt can therefore be protracted.

Soft downy feathers are quickly replaced by the first strong feathers. These are known as 'contour' feathers, and the first plumage they form is called 'juvenile.' This is the only plumage where all the feathers are the same age and have the same appearance and texture. Adults can also have a uniform appearance, but birds never molt all feathers simultaneously. These adult feathers therefore usually have subtly different colors and patterns of wear.

Small birds usually molt from a briefly worn juvenile plumage to an adult-like plumage. Some of these birds can be aged through the first year of life by looking at retained juvenile feathers. A number of large birds don't start to molt out of their juvenile plumage for over a year. As a broad generalization, the larger the bird, the longer it takes to become an adult. For example, most small gulls take 2 years, medium-sized-gulls 3, and large gulls 4, to reach adulthood. Raptors follow a similar pattern, with Bald Eagle taking 4–5 years to reach adulthood but American Kestrel only 1.

In most birds, the difference in appearance between an often scaly or spotted brown juvenile and a boldly patterned adult is striking. Trying to understand what happens in between these two appearances is very important. For the reason stated above, large birds tend to molt very slowly. In fact, they are in active molt the majority of the time. Their plumage is usually made up of feathers of different generations. The newest feathers have adult-like characters, compared to the oldest. This mishmash of feathers can appear confusing, but understanding what happens during development toward adulthood can help unravel many of the apparent mysteries of a bird's appearance. Also, understanding the sequence in which a bird molts feathers (most follow the same general patterns) is, in most cases, the best way to accurately age it.

In this book, unlike most others, *I have tried to put many of the transitional appearances in the plates*. Birds never change their appearance overnight—it's a prolonged process. If you keep in mind that molt is an ongoing process, it will become easier to understand a bird's appearance.

The timing and patterns of molt vary. Many birds have to make the decision to molt before, after, or during migration. There are no hard-and-fast rules. However, some patterns are clear. If you are a shorebird migrating from the Arctic to Chile for the winter, you can't afford to be using energy for molt, and it would not make sense to have feathers missing in your wings. Hence long-distance migrants either molt before or after migration, this being largely dependent on where food supplies are best. These birds always have long pointed wings—an adaptation to help them fly fast and far. Short-distance migrants, on the other hand, that don't have to fly long distances over water, will often molt while migrating. For example, Western and Least Sandpipers will often show gaps in their flight feathers because they molt while migrating—they are not going far. On the other hand, Semipalmated Sandpiper has to fly to South America so always has a complete set of flight feathers—this can be a great way to differentiate these particular species in flight.

Feathers are molted in different sequences depending on the species. However, it is almost always symmetrical from one wing to the other, and done such that there is balance between the wings.

TERMINOLOGY

One area of confusion in books today is the different terminology used to describe birds. This encompasses 3 different things: age, plumage, and molt patterns. The reason for these differences is in large part due to the lack of a clear and consistent terminology that is a good match for describing each species' appearance.

The following is the most widely used terminology:

The Life Year System

Traditionally this has been the most popular system. I have used it in this book. Fundamental in understanding how it works is knowing that a bird starts its life in the summer (assuming it hatched in the Northern Hemisphere). To calculate a person's age we need to know his or her birthday. Once we know a person's age, this provides a better understanding of behavior and appearance. Birds are the same.

1st-yr: starts as a fledgling in the nest, where its contour feathers grow within the space of a few weeks. This is juvenile plumage. Most birds molt a number of these juvenile feathers in fall, and they are replaced by adult-like or older immature feathers creating the first-winter plumage (1st-w.). In spring, many species, for example Indigo Bunting, will molt in many new feathers, which are often bright, creating the first spring/summer plumage (1st-s.). The term first-year (1st-yr) encompasses all 3 plumages: juvenile, first-winter, and first-summer. This term is used in this book when there is insignificant difference in a bird's appearance through the first year. At the end of the summer, a small bird has gone through its complete annual molt. At this point it becomes an adult. On occasion, a bird will retain older immature feathers after the first year so that, in this case, it can be aged as a 2nd-year bird. For others, mostly larger birds, the cycle is much longer, and they become 2nd-year (2nd-w. and 2nd-s.), 3rd-year, and so on until they reach adulthood.

Many adults have 2 different plumages or appearances in winter and summer (ducks are the major exception). The typically bolder plumage they molt into in spring, and also have during summer, is called 'breeding' (also known as alternate plumage). The complete molt occurs in late-summer/fall and results in a typically drabber appearance called 'nonbreeding' (also known as basic plumage). A few species, such as ptarmigans, have a third plumage called 'supplemental.'

This terminology describes an appearance or plumage. However, these appearances can be variable. Personally, I do not think of these stages as plumages but as a period in time when a bird is a certain age. Knowing the date we are birding (or a photo was taken) is critical. If you are birding on August 2nd and you know the gull you are watching is approximately 1 year old, after a while you will learn the variation in appearance of birds of that age. Personally, I think of it as a 1-year-old gull, just as I would a 1-year-old child.

The Calendar Year System

As the name implies, this system uses language based on calendar dates—January 1st being a new year. An American Robin this year would be in its second calendar year on January 1st (but not in its 2nd year of life). This is at odds with how we age people, pets, and most things. This terminology is not widely used, but often enough to cause confusion with the Life Year System. The Calendar Year System has no advantages over others and has several obvious problems—it is best not used!

The Humphrey–Parkes System

This is based solely on molt patterns. Because of the cyclical nature of annual molt and its relationship to physical appearance, the Humphrey–Parkes system has become more common, especially for dealing with gulls and raptors. It is particularly valuable when using molt patterns rather than physical appearance as the way to age birds. To a large degree, the systems are intertwined and arguably inseparable.

The 1st cycle includes all the different appearances or plumages the bird has until it starts to molt into its second basic plumage (2nd-winter). This includes juvenile, 1st-winter, and 1st-summer. Using the term '1st cycle' is technically correct during this time period. By comparison a juvenile bird, for example Canada Goose on p.44, that has molted a few feathers, is technically no longer a juvenile, though the majority of feathers are still juvenile. On one hand the term 1st-cycle is correct but lacks detail on plumage; but on the other hand you have a more detailed description of the bird's appearance and plumage that is not (arguably) correct. You take your pick!

In this book I use the latter. You will see a number of images, such as Canada Goose, where the bird has mostly juvenile feathers but has molted a few feathers (see if you can tell which they are). This could have been labeled '1st-cycle' if I had chosen to write the book from that perspective, or '1st-winter.' Again, this is a judgment call. Where the bird has primarily juvenile feathers (over 90 per cent), I have usually called it a juvenile because it helps the viewer understand the appearance of the bird at this stage. Now you can look at all the birds labeled 'juv' and see if they have molted out some of their juvenile feathers!

The major drawback of the Humphrey–Parkes system is that it is at complete odds with other terminology. It is also an unfamiliar language for most people and hard to grasp. A bird is in its first cycle until it starts molting into its second basic plumage when it is about 1 year old—usually first seen as newly growing flight feathers (primaries and secondaries). This transition is often impossible to see in the field. Two birds the same age can be in different cycles, and birds that are a year apart in age can still be in the same cycle. For example, yesterday (July 25th) I watched a 2nd-cycle Herring Gull—How old is it and what does it look like? The answer: you don't know. It could be a 1st-summer/1-year-old bird that has just started its second cycle or a 2nd-summer bird that has not started its third cycle.

However, the understanding of these molt cycles and how different generations of feathers have different appearances is often critical in the correct ageing of some birds. At present, this terminology is poorly understood by most and often misused. It is best reserved for those with a solid understanding of molt.

Birds spend varying amounts of time molting from one plumage to another and therefore have feathers of both plumages at the same time. These transitional birds cause confusion. For this reason, I have treated transitional birds extensively in this book. Besides taking note of different colors and patterns, always remember: one feather is new and one old, so different amounts of wear are often easy to see.

FACTORS AFFECTING APPEARANCE

If all this talk of changing plumages, molt, different feathers, and so on, sounds confusing, you are correct, it is—for everyone. But simply knowing and understanding this is a help in itself.

There are other things that change a bird's appearance that we must also consider. When a bird's feathers are old, particularly in the summer months, they have been subjected to many hardships. Sunlight bleaches feathers, making them paler. The wear and tear of time does the same thing. Not only are colors usually faded, sometimes considerably, but the feathers are also, quite simply, beat up and heavily worn. Certain groups, such as gulls, can look remarkably scruffy and disheveled in the summer months. Juveniles have softer feathers than adults so are more prone to wear and fading.

Some birds appear atypical or outside the norm. These are known as 'aberrant.' Although uncommon, they are frequent enough to warrant consideration when things don't seem to add up. Some are particularly pale: leucistic; some completely white: albanistic (see Tree Swallow, p.330); and, very rarely, all-black: melanistic. Often an aberrant bird will still show the basic patterns of 'normal' color, so always look for these. Size, shape, behavior, probability, and vocalization generally stay the same, so, although these birds are initially puzzling, actual ID is usually fairly straightforward. The only feature that has changed is color.

And of course there is the dreaded 'H' word. The word 'hybrid' comes up a lot today in the serious birding world. As we become more knowledgeable, and have technology to record ever more accurately sight and sound, and the ability to dig deep with things such as DNA analysis and blood sampling, the study and understanding of hybridization grows apace. In some groups, such as gulls and ducks, hybridization is fairly common. In others, such as shorebirds, it is very rare. However, the biological imperative to breed is strong. Even in shorebirds, the frequency of known examples of hybridization is increasing, and no doubt will continue to do so as time passses. The appearance of hybrids ranges across the spectrum, from total similarity to 1 parent to displaying intermediate characters of both. The latter occurrence is the most common. Should you find a bird that is puzzling because it seems to show features of 2 different species, you should consider the chance that it could be a hybrid.

Lighting also has a great effect on the appearance of birds. When the sun is high and bright, the contrast is at its greatest: dark colors look blacker and pale colors look whiter. The lower the sun is in the sky, the richer and more saturated the colors become. At sunrise and sunset, whites become creamy or have red tones and contrast is at its lowest. On overcast days ('flat light'), colors tend to have a more 'accurate' representation. Learning to understand the tricks of light is very important.

In this book, most plates are made up of images taken in flat light for a more accurate representation. On a few pages, there are images that have been taken in both lighting situations (sun and shade); for example, White Ibis (p.203). This is lifelike since it happens in the field frequently. When a bird is in the shadow of a tree or building, it is in flat light, while other birds may be in sunlight.

Size illusion can also occur. When looking directly into the sun, birds look slimmer than what they really are. However, while much has been written about size illusion, I personally have rarely thought it significantly impacts field observation, and that the most important thing is still to look longer and more carefully.

Sometimes birds get sick, or their normal way of life is interrupted, perhaps due to unexpected bad weather or shortage of food. As with humans, birds have to adapt to this as best they can. Sometimes they will have to change their normal way of life. For example, if a shorebird that is heading to the Arctic to breed becomes ill, or can't make the journey in a timely manner because there is not enough food to replenish its fat reserves, then it might alter its normal behavior. If it can't breed, why go to the breeding grounds or molt into a bright breeding plumage?—something that takes a lot of energy. If it is sick, it cannot afford to waste energy molting, because this is not vital to its survival. It may therefore stay farther south than normal and not molt into what is effectively an unnecessarily bright plumage.

Some birds show outward signs of illness or abnormality. Conjunctivitis has had a large impact on House Finch and other birds. It is not uncommon in some areas to see the effects of this around the bird's eyes (see p.29). Birds can also have deformed bare parts. This is most obvious in bill shapes; seeing birds with abnormally long bills is not rare.

Cosmetic color changes also occur due to rust staining, particularly common in groups such as cranes and ducks. Diet also has an impact. For example, House Finches in some areas are yellow rather than red because of their specific diet.

There are other factors that can change a bird's appearance; take them into account but don't be sidetracked; *go back to the basics!*

SWIMMING WATERBIRDS

GEESE

There are few sights in birding more inspiring than a huge flock of honking geese 'blasting off.' Because they are large birds, and usually occur in flocks, often in their thousands, they can be easy to find and watch. Geese often return to roost on the same bodies of water every night, and are equally at home in rural areas or in city parks and lakes. They tend to be very vocal and are often heard before they are seen. The sound of migrating geese at night is a great treat. In addition to the spectacle they provide, they are of interest to birders because of the ID challenges they pose.

Recently, Canada Goose was split into 2 species: Canada Goose and Cackling Goose. Moreover, each of these new species has several subspecies, and these may interbreed. Throw into the mix introduced 'European' birds, and you are left with a real mess. Personally, "The more I look the less I know," which is often the case in birding. As you get to know a species better, you find more birds that don't neatly fall into a 'slot.' In many ways, this makes the subject more fascinating, but it means the identification of subspecies is often difficult, and virtually impossible for some individuals outside their normal range.

Large flocks frequently act as 'carriers.' Other species join the flock and start to move with them, ending up many miles away from their usual range. With patience you can often find a Ross's Goose in with Snows; Cackling or Greater White-fronted in with Canadas; a 'Black' Brant with regularly occurring 'light-bellied' Brant; or get really lucky and find a vagrant Pink-footed or Barnacle Goose. It's surprising how hard it is to pick the odd one out in large flocks—so keep looking! However, beware as there are all sorts of oddball escaped captive birds to contend with.

DUCKS

Ducks are a fantastic group of birds. Only when you look at them really closely can you appreciate their intricate markings, iridescent colors, and true beauty. The contrast between gaudy males and, shall we say, 'subtle' females (sexual dimorphism) is no more obvious in the bird world than in ducks. For ID purposes, working out the male can easily be based on patterns of color alone. Ducks are frequently seen in pairs, so a great starting point for identifying the confusing brown duck is seeing if it's the partner of a bright bird close by. However, the basics of size and shape pertain to ducks just as much as to any other group of birds. They are often in mixed flocks or have other groups nearby for comparison for both size and shape. This is particularly important, as female-type plumages are commoner than bright ones, simply because young and eclipse males share this appearance for much of summer, fall, and, sometimes, early winter.

Eclipse is the term used to describe the dull female-like plumage males molt into on their breeding grounds. This molt sometimes takes place near the nesting site, but they often migrate to a different location, which can be hundreds of miles away. During this period they molt all their flight feathers at once and so are flightless for several weeks. The molting site needs to have cover, and plentiful food supplies and allow them to dive to escape predators. One reason for this duller plumage is its value as camouflage while they are flightless. Summer is also a time with plentiful food supplies, so that the disadvantage of being flightless is less serious than when molting later in the year. Males usually molt back into their more easily recognizable bold color patterns (often called 'bright' plumage) during the fall. Females usually molt several weeks earlier than males and are often on the nest during this period.

In recent times there has been discussion whether 'eclipse' is actually a breeding or nonbr plumage. This naming scheme leads to confusion when dealing with ducks. I have described the plumage and refer to eclipse in the accounts when the bird molts out of its bright plumage.

Sexing birds is often difficult at this time and is sometimes only possible by focusing on the upperwing pattern. Adult males retain their boldly patterned coverts, though these are usually not visible on sitting birds. Eye color always stays the same; bill color is often duller.

Watch for distinctively marked bright adult male feathers molting back in. As time passes, they appear as a variable mix of dull brown-gray feathers and bolder bright plumage feathers. Most adults have finished their molt by November.

Telling young males from eclipse males is often particularly perplexing and not well documented in guides, in part because it is genuinely difficult but also because there are still major gaps in our knowledge.

Juveniles look superficially like adult females. Most have warmer brown tones. Juvenile feathers are softer and not so durable as those of an adult. Juvenile feathers also have a different shape from those of adults. Flank feathers and wing coverts are smaller and rounder, and usually have more solidly colored centers (adults often have complex markings). The tail feathers of juveniles are narrower and more pointed than in adults. These become heavily worn through the fall —a feature that can readily be seen with close views. Feathers often have an inverted 'V' chipped out at the tip. Tertials, most noticeable in males because of their larger size, are very broad in adults and more rounded at the tip; juveniles' tertials are shorter, narrow, and pointed. The scapulars are also shorter, smaller, and more pointed in juveniles—again more noticeable in males.

First-winter (1st-w.) males are birds that have a combination of juvenile feathers and newly molted (bright) feathers. This combination, often a mix of brown and brighter feathers, is similar in adults. Working out age and sex is difficult, but with good views and practice, the above criteria will help you identify many birds. This is complex stuff in many cases, some of which is beyond even experts, so do not despair if it seems difficult. Remember, most adults have completed their molt by November (some earlier), so that most birds that are not fully molted into bright plumage after this date are likely to be 1st-w. males. These birds often show signs of immaturity into spring.

Most immature birds winter in the southern portion of the species' range; adults tend to stay to the north.

Ducks are typically placed in 3 groups: Tree/perching, dabbling, and diving. Tree ducks include whistling-ducks, Muscovy, and Wood. They often roost or nest in trees. Dabbling ducks (*Anas*), like tree ducks, rarely dive but do, as the name implies, dabble or tip up, to feed just under the water's surface. Diving ducks include sea (scoter, Eider, Long-tailed, Bufflehead, and goldeneyes), bay (scaup, Redhead, Canvasback, and Ring-necked), sawbills (mergansers), and stiff-tails (Ruddy and Masked). Diving ducks have feet set farther back on the body to help them swim under water. They typically run across the water on takeoff. They sometimes stay under water for over a minute and can dive to considerable depths in search of food such as mollusks and fish. Dabblers 'jump' off the water on takeoff.

Most ducks breed in northern wetlands and through the prairie heartlands, an area often known as the 'Duck Factory.' Some stay farther north in winter on unfrozen wetlands, but most move to the south and coastal areas. Most species are declining, owing mainly to loss of breeding habitat. The conservation of wetlands is fundamental to the well-being of many populations.

Hybridization is relatively common in ducks. Just about any combination is possible, and some such as Mallard X American Black Duck are fairly common. The parentage of many hybrids can often be worked out, as they usually show characters of both parents. It is often easy to work out the ID of one of the parents, but the second has to be left as 'possible' (a chance of) or 'probable' (likely to be). Sometimes it's just not clear if a bird is pure or not.

juv.

1st -w.

1st-w.

ad.

Mute Swan *Cygnus olor* **MUSW** L 60in

Introduced species. Common locally in marshes and on other water bodies. Elegant and beautiful, or big and nasty, take your pick as both can apply. Takes over its patch, immediately attacking any other swan that dares to come nearby. Nest is large and conspicuous. Young are protected zealously; an adult, standing its ground, has hissed fear into many a human. Neck is often held arched rather than straight as in other swans. Wings are sometimes held high and arched, always when bird is in an aggressive posture. Has a running takeoff with wings thrashing water. Wings make a diagnostic whistled hum in flight (silent in other swans). Calls: a variety of hisses, snorts, gurgles, and *keoor*. **ID**: The largest swan with thick neck, square-headed look, and long tail, a very useful ID feature on sleeping birds. Bill pattern usually makes ID straightforward. Adult: white, often with yellowish neck. Striking orange bill with large black knob. Knob averages larger in males. Juv/1st-w: variable, usually medium-brown or white. Molts through fall/winter into whiter plumage. Bill initially dull, soon becomes pink and small knob grows through winter. 2nd-yr: similar to adult but with slightly smaller knob and duller orange bill.

Tundra Swan *Cygnus columbianus* **TUSW** L 52in

Uncommon breeder on tundra ponds. In winter in marshes, ponds, and shallow lakes and on flooded fields. Frequently grazes in agricultural fields. Wintering birds arrive in the S Nov onward, often traveling in high-flying flocks. Mellow hooted *klooh* travels long distances and is a sure sign that winter is on the way. Flock sizes vary from a few birds to dozens, usually in family packs, dingier young birds often equalling the number of adults. Like other swans, staining from water can make them appear darker. Frequently harassed by larger MUSW when present. **ID**: Slightly smaller than MUSW with shorter and straighter neck. Head and bill profile strikingly different. With head tucked in, shorter tail is quite easy to pick out (TRUS is also short-tailed). Adult: bill black usually, but not always, with yellow spot at base. Yellow stripe on cutting edge of lower mandible. Pointed head profile useful in separating from other swans. Juv: dingy gray-white. Bill initially nearly all-pink to base, becomes darker through winter. 2nd-yr: often retains a few gray juv feathers.

Trumpeter Swan *Cygnus buccinator* **TRUS** L. 60in

Introduced into the Great Lakes region on marshes, rivers, and flooded farm fields. Native populations inhabit forested lakes. Perhaps formerly native in MN. Uncommon but numbers increasing and occasionally wanders. Native birds from AK rarely as far s. as TX in winter. Sometimes mixes with other swans. Builds large nest of sticks. Named after distinctive trumpeted nasal *oh oh*. **ID**: Same size as MUSW but much slimmer with different bill pattern. Easy to overlook as very similar and commoner TUSW. TRUS is slightly larger, longer and appears thinner-necked. Always focus on face pattern. Base of bill appears more pointed where it meets the eye with the white forehead coming to a 'V'-shaped point, and white cheek meets the bill at a straighter and shallower line. Bill is black and lacks the small yellow spot of most, but not all, TUSW. Adult: all white. Juv: dingy gray (rarely white). Bill pink, typically with more black at base and along cutting edge, overall impression is not so pale as TUSW. Bill turns darker and plumage fades paler through summer with some adult feathers molted in. 2nd-yr: often retains a few juv feathers.

Greater White-fronted Goose *Anser albifrons* **GWFG** L 28in

Breeds mostly on tundra (*gambeli and frontalis*—'Tundra'). Winters in s. US, where often found with SNGO. Locally common in agricultural areas, prairies, and some coastal regions. Scarce in other areas, usually mixed with other geese. Greenland birds (*flavirostris*) are scarce winter visitors to the E coast. **ID**: Adult: medium-sized, stocky gray-brown goose with orange legs and usually pink bill. Variably mottled black on belly, and white 'face' diagnostic. Domesticated Graylag Goose and interbreeds can look similar. Juv: bill briefly dull yellow becoming pink, with uniform gray head and underparts. Birds develop black belly and white around base of bill and then forehead at variable rates from Sep through first winter. 'Tundra': pink bill sometimes with orange; generally larger, slimmer, paler-headed and-bellied with less black mottling on belly than Greenland birds, though much variation. Mostly orange bill, though some have pink mixed in. Imm: dull orange bill by Oct.. Separation of these two ssp is often very difficult and judging bill color is similarly tricky. Lesser White-fronted Goose (*Anser erythropus*), a common captive, is smaller with yellow eyering and larger frontal shield (see p.47).

Snow Goose *Chen caerulescens* **SNGO** L 29in

Breeds on Arctic tundra. Locally common in winter, in agricultural areas, prairies, or marshland, sometimes in flocks of 10,000s. Large skeins will fill the sky and lead you to where birds are feeding. In large flocks, most birds appear white, with darker birds (juvs and dark morphs) mixed in here and there. Flocks are compact with 'lookouts' at the edges. When flushed, they take off in unison with a loud din—one of the great spectacles for birders. **ID**: Medium-sized with heavy body and sloping head with large bill. Striking dark 'grin patch,' where mandibles meet, enhances 'mean' impression. Noisy. Sexes similar. 2 morphs: White outnumber Blue 100:1 in the E, 40:1 in the Mississippi River Valley. White morph: Adult: white with black wingtips. Juv: variably dingy gray; Often has rusty staining on head and neck (never in ROGO). Blue morph ("Blue Goose"): Adult: gray-blue with white head, white extending onto neck. Breeding of blue and white morphs creates offspring with intermediate features. Juv: variably dingy dark gray-blue; occasionally hybridizes with ROGO, when all features are intermediate. Gradually molts in to adult-like plumage by 2nd fall.

blue morph

1st.-w.

ad. Snow Goose

ad.

ad.

Ross's Goose *Chen rossii* **ROGO** L 23in

Breeds on wet tundra. Locally uncommon in the Great Plains. Rare but regular in the E, mostly in SNGO flocks (1:1000). Like SNGO, population is increasing rapidly. **ID**: Much smaller with proportionally shorter legs and neck than SNGO. Cute look with round head, stubby short triangular bill has straighter border with face. Gray area (wart) at base of pink bill. Adult: white with black wingtips. Juv: as adult except for brown-gray tertials and secondary coverts (often wear white by late winter), duller bare parts and gray wash to crown and rear neck. Often difficult to age by late winter. Blue morph is extremely rare and perhaps a a hybrid with SNGO where the blue-morph gene is dominant. These birds are more contrasting than blue-morph SNGO with white face and belly and almost black darker parts. In flight, the smaller size, short neck and stubby bill combined with shorter wings and faster wingbeats are quite easy to pick out in ROGO in direct comparison with SNGO, though, on the ground, among numerous SNGO, it requires careful scanning to pick out individuals.

juv. Greater

ad. Lesser

ad. Greater

Canada Goose *Branta canadensis* **CANG** L 36–45in

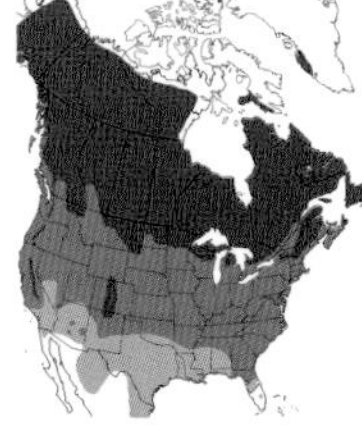

Very common and familiar sight on farm fields, in town parks, on water bodies, and flying overhead. With several races, a large introduced population, interbreeding, and individual variation, flocks can show great variety. Generally split into 2 groups: Greater (*maxima*, *canadensis*, *interior* and *moffitti*) and Lesser (*parvipes*). Lesser is very similar to CACG, some wintering in e.TX. **ID**: Greater forms vast majority of birds in our region. *maxima* is largest and palest but overlap makes subspecific ID impossible other than on breeding grounds. Introduced/feral birds, widespread breeders in the S, are approachable and sometimes considered a nuisance. Migrates only short distances. 'Wild' populations are longer-distance migrants and less approachable. They head s. in late fall in large skeins and back n. as the snow melts. Usually fly in 'V' formations, flocks sometimes in the hundreds—occasionally with other species mixed in. They take over arable fields, often roosting on nearby water bodies, but may all be gone two weeks later. Familiar honking call. Juv: soft-textured brown feathers become frayed and are molted quickly. Sometimes has more extensive white on head recalling BAGO.

with Mallard

ad.

Cackling Goose *Branta hutchinsii* **CACG** L 24–30in

Formerly conspecific with CANG. Breeds on moist coastal tundra. Uncommon winter visitor to wetlands and farm fields in the s. Great Plains. Regular migrant to the N. Rare but regular in the E, mostly in large flocks of CANG. *hutchinsii* only race that occurs in the E, often called 'Richardson's Goose.' Behavior as CANG and frequently seen together, loosely sticking to their own groups. Call: higher-pitched than CANG. **ID**: Can be very similar to small CANG with overlap in all characters, and some individuals are best left unidentified. Most CACG are smaller with a stubby bill, short thick neck, squarer head and longer primary projection. The body tends to be chunkier and the upperparts usually, but not always, have grayer bases to feathers, creating paler appearance overall. Underparts are also typically pale. Most look small and cute compared to CANG, but not always the case. CANG and CACG show great variation in characters such as shape of white throat patch, white ring at base of black neck, and paleness of underparts. These features are too variable to be of significant use in field ID.

ad. Black Brant

juv.

ad.

Brant *Branta bernicla* **BRAN** L 25in

'Pale-bellied' *hrota* ssp breeds in the high Arctic, usually near the coast, and often colonially. Locally common in winter (until late May because the tundra is still frozen where they breed), mostly found swimming in shallow sheltered coastal bays, or barrier islands, or grazing in grassy areas. Typically in flocks of varying sizes. 'Black' Brant, *nigricans*, a rare visitor (3000:1), occurs in 'Pale-bellied' flocks, sometimes returning annually. There are a few claimed records of 'Gray-bellied' Brant that breed on Melville Island and vicinity. These show features intermediate between Pale-bellied and Black Brant. **ID**: Fairly small and compact with short, thick neck and pointed head. Appears dark with contrastingly black neck and white vent. Variable white necklace. Adult: upperparts uniform, lacking distinct pale fringes. Juv/1st-w: broad white fringes to coverts. Initially dark-necked, white necklace grows in through fall. Juv coverts start to be replaced in spring. Black Brant darker-backed and -bellied with contrasting white flanks. Necklace typically large and meets on foreneck. Averages slightly larger, often with more erect posture and longer bill—a more 'muscular' appearance overall.

Barnacle Goose *Branta leucopsis* **BARG** L 27in

Annual winter visitor from Greenland and Europe, mostly in CANG flocks. Fairly common in collections clouding true vagrancy pattern. **ID**: Distinctive but beware of CANG showing similar face pattern (regular in juvs). Juv very similar to adult but flank markings more evenly spread and upperpart markings less boldly contrasting.

Pink-footed Goose *Anser brachyrhynchus* **PFGO** L 26in

Recent annual visitor from Greenland to the NE, usually in CANG flocks. **ID**: Most closely resembles GWFG but slightly smaller, shorter, and slimmer-necked with dark-headed appearance. Legs pink, but judging color can be difficult. Bill smallish and dark with some pink towards tip. In flight dark underwing and extensive pale-gray on upperwing.

Domestic Waterfowl—often found on local ponds. Tame and highly variable in appearance.

Domestic Mallard

Domestic Swan (Chinese) Goose

Exotic Waterfowl—commonly kept in waterfowl collections. Escapes can be found anywhere! Some tame, others less so! Look for bands on legs, clipped wings, or a missing hindclaw.

Mandarin Duck

Eastern Spot-billed Duck

Red-crested Pochard

Common Shelduck

Lesser White-fronted Goose

Egyptian Goose

Ruddy Shelduck

Black Swan

Bar-headed Goose

Emperor Goose

juv.

ad.

Black-bellied Whistling-Duck *Dendrocygna autumnalis* **BBWD** L 21in

Rapidly increasing resident of the Gulf states, prone to wander long distances, particularly in late spring and summer. Found in wet areas such as agricultural ponds and wetlands, particularly where birds can perch in trees. Increasingly common in suburban areas in FL with access to nearby water bodies. Foot paddles and searches for food with its head under water. Nests in tree holes or nest boxes. Sometimes solitary but often in noisy flocks. Noisy 'whistling' carries long distances. Most often heard before being seen. Most active at dawn and dusk but also nocturnal. **ID**: Typical whistling-duck shape: fairly large and elongated with long legs, neck and bill but fairly small head. Erect posture. Striking color pattern in adults allied to shape gives an 'exotic' feel. In flight, obvious white in upperwing easy to see. Striking black underwing. Sexes alike. Adult: striking. Juv: (Jul–Dec): superficially similar to FUWD, but is darker and duller with uniform brown back and less contrasting underparts. The cheeks are grayer and birds lack prominent flank stripes. In mid-winter juv becomes more adult-like but duller, with brown mixed in black underparts and heavily worn tail feathers.

Fulvous Whistling-Duck *Dendrocygna bicolor* FUWD L 19in

Mostly summer visitor. Scarce but sometimes occurs in large flocks in wetter areas such as rice- and flooded fields. Occasionally wanders out of range. Noisy whistling call, birds often heard before being seen, particularly early and late in the day. **ID:** Typical whistling-duck shape: fairly large and elongated, with long legs, neck, and bill, but is fairly small-headed, recalling NOPI. Overall impression is of a uniform rich orange-buff, bird, dark-backed with a bold white vent. In flight, black underwing stands out on an otherwise fairly uniform bird. Upperwing lacks white of BBWD. Ages and sexes similar.

Muscovy Duck *Cairina moschata* MUDU L 25–31in

Rare on the Lower Rio Grande river. Feral populations in s. TX and FL. Wild birds very wary and usually seen flying 'down river.' Crepuscular. **ID**: Very large and heavy with bulging sternum and sloping forehead. Wild birds are blackish with variable iridescent purple or green. ♀: duller. Striking white patches on underwing and forewing (lacking in very young birds). Feral birds variable but almost always with white in body plumage and more extensive red bare skin in face. Common on urban waterways—and near handouts. It's the big ugly one!

ad. ♀

juv. ♂

eclipse ♂

juv. ♀

ad. ♂

ad. ♀

Wood Duck *Aix sponsa* **WODU** L 18.5in

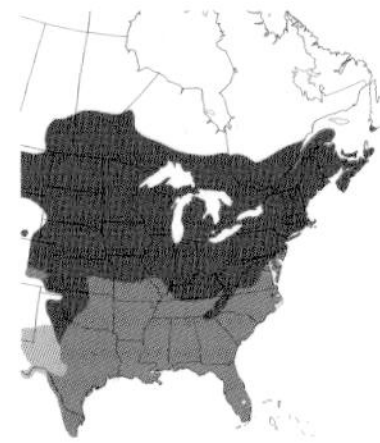

Fairly common but often hard to see. Lives in swampy woodland or any combination of water and trees. Picks food from surface, swimming with head jerking backward and forward. Often perches on branches. Tends to flush easily, trying to gain height as fast as it can. Gives panicked squealed calls—a good way to track it down at any time. Most often seen in flight (particularly dawn and dusk) in compact oval flocks. Usually in small groups out of breeding season, rarely with other species. Nests in cavities or nest boxes, often twice a year. Kept in collections. **ID**: Long-tailed, a particularly useful feature in flight and makes bird identifiable in silhouette. Shape and patterns of color diagnostic. Fairly small. Ad ♂: stunning! Uniquely marked with bushy crest creating bulbous back of head. Head patterns always distinctive. Eclipse ♂ (Jul–Sep): retains distinctive bold red iris, orbital ring, and bill colors. Adult ♀: neatly patterned brown and white with distinctive white tear around and behind eye. Juv: similar to adult ♀ with stronger breast spotting extending onto belly and different face pattern. Juv ♂: has mostly yellow bill and dull red iris, and lacks bold red orbital ring of adult. Mostly adult-like by late Oct.

eclipse ♂

mating

juv

ad. ♂

ad. ♀

Mallard *Anas platyrhynchos* **MALL** L 23in

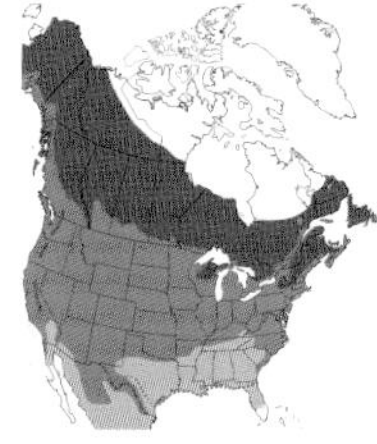

The familiar and common duck, found in city parks, farm fields, including water bodies of whatever size, to the wildest places. The one to learn well, as this species is the basis for learning all dabbling ducks! Often tame and approachable. Ancestral stock of many domestic ducks. Pinpointing truly wild birds often tough. Commonly hybridizes, particularly with ABDU. Call: the stereotypical *quack quack* duck call given by ♀ only. ♂'s call note is quieter. **ID**: Fairly large, well-proportioned, and muscular. Blue speculum (can appear purple) bordered by black and white. Adult ♂: striking green head with white collar; funky upturned black central uppertail coverts. Bill is bright yellow to yellow-olive with black nail. Eclipse ♂ (Jun–Sep): retains yellower bill. 1st-w ♂: identical to adult ♂. ♀: dullish gray-brown, paler on head and neck with dark crown and eyestripe. White outer tail feathers. Variably yellow-orange bill with mottled dark center (lacking on AMBD and MODU). Juv: briefly held plumage similar to adult ♀ but warmer brown with darker centers to feathers. Learning to id ♀s, and knowing the variation they show, makes identifying other duck species much easier.

Mexican Duck *Anas platyrhynchos diazi* **MEDU** L 21.5in

A ssp of MALL (10% smaller) found close to the Mexican border, where MALL is not usually seen. Sexes are alike, resembling ♀ nominate MALL. Brown tail. Overall color intermediate between nominate MALL and ABDU. Very similar to MODU but with grayer leading edge to wing, broader white borders to speculum, subtly bolder face pattern with streaked throat and smaller gape spot. Usually identified by range alone.

ad. ♂

ad. ♀

Mottled Duck *Anas fulvigula* **MODU** L 22in

Replaces ABDU in Gulf states. Similar fondness for coastal marshes; but also common in freshwater wetlands. Pairs stick together most of the year, so rarely seen in larger groups. Intermediate in color between ABDU and MALL with a well-demarcated pale head/neck and a black spot at base of bill (gape). ♂: slightly darker version of ♀ with brighter yellow bill—olive to drab orange in ♀. Shows complex feather-center pattern as ♀ MALL, solid in ABDU. Pale crown. Speculum blue but can appear purple with narrow white border. Throat buff and unstreaked. Shorter wing point than MALL and ABDU.

♂ Mallard hybrid

ad. ♂

ad. ♀

American Black Duck *Anas rubripes* **ABDU** L 23in

A darker cousin of MALL. Scarce in most areas, though locally common in saltwater marshes where they can form sizable flocks in winter. Declining, largely due to encroachment from MALL with which it frequently hybridizes. Working out which are 'pure' birds is difficult. Very wary and easily flushed. Coastal birds often in single-species flocks or with MALL. In Interior, regularly mixes with MALL (mostly) and other puddle ducks. Call: similar to MALL. **ID**: Size and shape as MALL. In flight, bold whitish underwing contrasts strongly with rest of bird. Purple/dark blue speculum bordered by black with narrow or no trailing white edge to secondaries (compare MALL). ♂: darkest with strong head and neck contrast. Dark cap often shows green iridescence when seen well but extent of MALL influence is unknown. Bill yellow with black nail. ♀: paler with duller green bill. Hybrids: true status unknown because of similarity of backcrosses to purebreds. Many hybrids have intermediate characters, notably half a green head and obvious white borders to speculum; others show far more subtle differences, so detection and ID can be tricky.

1st-w. ♂

juv.

ad. ♂

ad. ♀ with chicks

Gadwall *Anas strepera* **GADW** L 20in

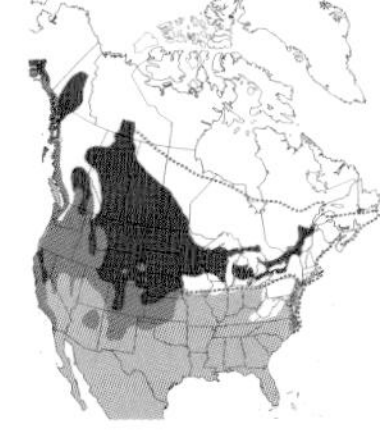

A common duck of shallow wetlands, mostly seen in pairs but sometimes larger groups. Superficially nondescript, so easy to overlook, particularly when mixed in with other more striking dabblers, as often occurs. ♀ gives a dull *quack*, similar to MALL but harder-toned. **ID**: Slightly smaller than MALL. Distinctive nipped-in neck and steep forehead often give a square-headed look. Head is larger in ♂ than ♀. Sits high in the water. Striking in flight: upperwing with white inner secondaries bordered by black, boldest in ♂ which also has chestnut coverts. White patch is often visible on sitting birds. Underparts also striking with distinct white underwing and central belly bordered by dark. Ad. ♂: intricately patterned with striking black butt. Eclipse ♂ (Jun–Aug): similar to ♀ but retains bold wing pattern Ad.♀: often confused with MALL and other ♀ ducks; look at shape and orange-sided bill, pale head/neck contrasting with darker body. Strong eyestripe. Bill often darker in summer. Juv: more neatly patterned and buffier than ad ♀. ♂ grows bold adult-type feathers through first winter; by Dec some still similar to ♀, others similar to adult ♂, but with shorter and more pointed tertials and scapulars.

American Wigeon *Anas americana* **AMWI** L 20in

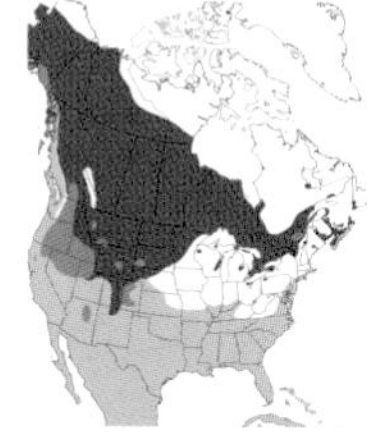

N. breeding dabbler. Widespread and fairly common, often forms large noisy flocks in winter in wet areas or grazing in fields. Its high-pitched and distinctive whistles belong on a soccer field. Can be tame, hanging around local ponds with AMCO and gulls. Will also graze on large lawns and golf courses. Also known as 'Baldpate' for its white/yellow crown. **ID**: Smaller than MALL with peaked forecrown, small bill, chunky body, long tail, and short legs. Flocks appear as a variety of pinks and burnt oranges with variable head patterns. In flight, ♂ has the most white on the upperwing of any duck; ♀ much duller, creating flocks of 2 color patterns. White axillaries. Wingbeats are fast, wings narrow and pointed. Ad ♂: has distinctive head pattern and black rear end bordered by white. Pinker than other plumages. Ad ♀: gray-brown flecked head with smudgy patch around eye—contrasts with warmer colored body. Duller with pale covert panel in flight (absent on most juv ♀s). Eclipse ♂ (Jul–Nov): averages brighter orange than ♀, retains large white covert patches. Juv/1st-w: similar to eclipse ♂ but covert pattern resembles ♀. Most appear adult-like by Feb.

Eurasian Wigeon *Anas penelope* **EUWI** L 20in

Scarce but annual, usually in flocks of AMWI. ♂ is grayer than cousin with red head and yellow forehead. ♀: from similar AMWI by browner head contrasting little with the body. Lacks black gape spot of AMWI. The clincher is the gray armpits (striking white in AMWI). Head shape is subtly different: flatter- and deeper-headed. Eclipse ♂ (Jul–Nov): retains bold white coverts. 1st-yr ♂: similar to adult ♂ by mid-winter. ♂ has higher-pitched whistle than AMWI.

Tufted Duck *Aythya fuligula* **TUDU** L 17in

Very rare. Mainly in the NE among flocks of scaup or sometimes RNDU. Returning individuals occur. ♂, because of its diagnostic dark back, white flanks and head tuft, is usually the one identified. ♀: tougher, similar to ♀ scaup with small but obvious tuft, flatter head, and uniform darker brown plumage. Most show some scaup-like white around base of bill.

White Cheeked Pintail *Anas bahamensis* **WCHP** L 17in

West Indian species. Very rare to s. FL in winter. Common captive bird, so true status unknown. Good numbers of presumed escapes have turned up in many widespread locations. Sexes similar. Distinctive head pattern. If in doubt, check bill pattern.

Masked Duck *Nomonyx dominicus* **MADU** L 13.5in

Rare stiff-tail from farther s. Most regular in TX and FL but also other Gulf states, sometimes in groups (invasive), and has bred. A real skulker in weedy, overgrown ponds. ♀: similar to nonbr ♀ RUDU but slightly smaller with double cheek stripe. Nonbr ♂: similar but with bolder face pattern and stronger colors. Breeding ♂: distinctive. In flight, diagnostic white wing patches.

eclipse ♂

juv.

ad. ♂

ad. ♀

Northern Pintail *Anas acuta* **NOPI** L 21in

Breeds in US heartland n. to tundra. Fairly common in freshwater and coastal marshes though just as likely to be seen grazing in agricultural fields. Sometimes forms very large tight flocks, particularly in the W. Small groups often fly in loose lines or 'V's. **ID**: Usually best to id by structure alone: slim, long-necked, long-tailed with erect posture creating very distinctive elegant appearance—a really beautiful duck! This slim appearance is just as obvious in flight. Relatively plain upper wing. Ad ♂: striking with brown head and white neck stripe—overall impression is of a bird with a very 'thoughtful' and somewhat dapper look. Long tail with black butt distinctive at long range. ♀: plain and pale brown, particularly on face and neck; simple but elegant. Bill blue-gray, darkest in center. Eclipse ♂ (Jul–Oct): appears as ♀. Retains bolder covert pattern, gray and black striped bill and has longer gray scapulars until it starts molting new adult-type feathers in fall. Juv: like ad ♀ with flanks more diffusely spotted and barred. In flight, ♂ is white-bellied with gray flanks, ♀ browner with coarse flank markings. Speculum green (browner with narrower trailing edge in imm ♂) bordered by chestnut at front.

1st-w. ♂

ad. ♂

ad. ♂

ad. ♀

Northern Shoveler *Anas clypeata* NSHO L 19in

Fairly common in a variety of shallow wetlands. Often found at water treatment plants. Usually in small loose groups but can be in larger numbers in suitable habitat. Sometimes feeds in tight packs that move in circles. Bill often swished from side to side as it sieves out food. **ID**: A large heavy duck that sits low in the water, and has a stonking spatulate bill. The bill, combined with short neck and broad head, adds to its distinctive bulkiness and should be enough to id most birds. ♂: unmistakable. Bright white accentuates green head and chestnut flanks. Depending on light, head can look purple or black. ♀: often looks pale due to broad fringes to feathers. Bill variably orange with bright edges. Eclipse (Jul–Nov) and young ♂ (Jul–Mar): variable with multiple spots and arrows on breast/flanks with a range of chestnut, white, and green until after Dec. Ad ♂: retains pale iris year-round, darker in imm ♂. Partial face crescent and upperwing pattern similar to BWTE—but always concentrate on size and shape! In flight, ♂ has a striking blue, green, and white innerwing; duller and grayer in ♀, but still relatively bold.

Cinnamon Teal *Anas cyanoptera* **CITE** L 16in

Scarce visitor to w. part of region. Mega rare in the E. Barely enters region as a breeder in mostly alkaline wetlands. Often seen with slightly smaller but similar BWTE. **ID**: like a BWTE on hormones—slightly larger, longer-bodied, and thicker-necked with warmer colors. The larger bill is slightly wider, the head deeper, and the crown flatter, though beware of variation. ♂: distinctive deep red color. Bold red eye. ♀: similar to BWTE but most are noticeably warmer buff. The face pattern is plainer with indistinct eyeline and less pronounced pale patch at base of bill. Juv: similar to ♀. Wears pale through fall looking BWTE-like. Eclipse ♂ (Jul–Sep)/1st-yr ♂: gets red iris about Oct and molts in brighter plumage through winter. Wing pattern as BWTE. Beware of noticeably larger NSHO and stained BWTE that look reddish on underparts like CITE.

Garganey *Anas querquedula* **GARG** L 15.5in

Extremely rare Eurasian species. Similar to BWTE in size and shape with plumage in ♀ types also similar. Not surprisingly, most records are of ♂s, typically in spring. ♀: strongly contrasting pale-fringed dark brown plumage with a bold horizontal cheek stripe (GWTE can show this, particularly in fall). Upperwing pattern as BWTE but paler gray (♀) or gray-blue (♂) coverts extend onto primaries.

♂

♀ ♂

juv.

1st-w. ♂

ad. ♂

ad. ♀

Blue-winged Teal *Anas discors* **BWTE** L 15.5in

A welcome summer visitor (some winter in the S) usually in pairs or small parties, often announcing itself with a high-pitched *kip*. Can be tame with a friendly demeanor. Fairly common on a variety of shallow marshy wetlands, usually near vegetation. **ID**: Small dabbler; slightly larger than GWTE. Sits low and long in the water and with a wide, flattish head, but looks long-necked when head held high. Fairly large dark broad bill. Ad ♂ br: striking half-moon face. Pale blue coverts (can appear light gray) bordered by a broad white bar and green speculum striking in flight (juv ♂ similar). Eclipse ♂ (Jul–Nov): similar to ♀ but retains bold wing pattern. Ad ♀: neatly patterned brown-gray with well-defined pale fringes to feathers. Distinctive pale area at base of bill. Juv: similar with smaller rounder feathers; ♂ becomes adult-like by spring. Similar ♀ GWTE is darker brown, flank feathers have complex pattern, buff patch on tail sides, different face pattern, is smaller and shaped differently. Often confused with NSHO—check comparative size! Striking upperwing pattern with pale blue forewing, similar to both NSHO and CITE. Dark leading edge to underwing creates bolder underwing pattern than in GWTE.

eclipse/1st-w. ♂

ad. ♂ Common Teal

ad. ♀

ad. ♂

Green-winged Teal *Anas crecca* **GWTE** L 14in

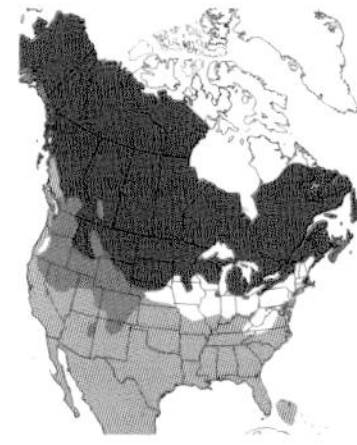

The commonest small duck in many areas. Found in wetlands, agricultural fields, and tidal mudflats. Forms compact groups, easily flushed with fast, vertical takeoff. Fly close together in sync, twisting with very fast wingbeats. ♂: often heard giving football-whistle *krik*. **ID**: Usually the smallest and darkest brown duck in the marsh. Compact and deep-bodied, sits high in the water. Large square head on a short neck with small bill—the rubber duck of ponds. In flight, plain-winged except for green-and-black speculum bordered by white and rufous. Ad ♂: rich chestnut head with green mask, finely patterned body. At a distance, a white vertical flank line and black-bordered yellow butt stand out. Eclipse (Jul–Nov) and 1st-w ♂ (through winter): very similar, molt bright plumage at different times. ♀: distinct dark eyeline, often with cheek stripe on uniform brown speckled face. Buff line on tail side. Eurasian ssp 'Common' Teal is scarce but regular with its cousin (1:1000). ♂ differs by white horizontal breast line (not vertical), coarser flank vermiculations, and bolder face lines. Birds with vertical and horizontal lines and other intermediate characters are hybrids. ♀ has broader white trailing edge to secondaries.

ad. ♀ summer

♂

♀

Canvasback *Aythya valisineria* **CANV** L 21in

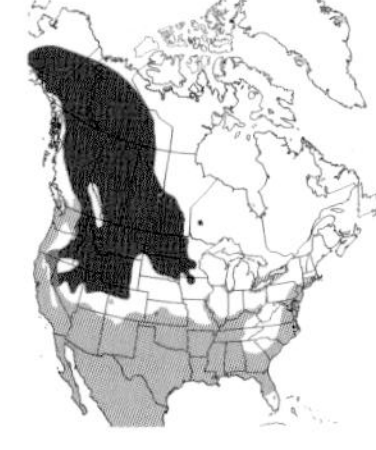

Locally fairly common, often in flocks in deeper water inland and brackish water in coastal areas. Frequently with other bay ducks, particularly scaup and REDH. A favorite quarry of hunters. Nests on small marshy ponds. Wide ranging with favored wintering areas, where flocks in thousands from the Great Lakes to Mexico, but there are also large stretches of land with very few individuals. **ID**: Largest bay duck with a diagnostic slim-necked and angular-headed profile accentuated by black bill. Ad ♂: pale back obvious at long distances, when often looks 2-toned. Striking red eye on chestnut head. Often compared to REDH, but CANV is a totally different shape and much paler. Eclipse birds change little having rufous mixed-in with black on breast. 1st-w ♂: ♀-like until Oct, and then appears similar to ad ♂ with subtly duller wing coverts and tertials. ♀: appears pallid and 2-toned: gray body and brown neck/head, though upperparts take on browner tones in summer. Uniform face with pale around and behind eye. 1st-w ♀: averages more uniform brown than adult. Shape is equally distinctive in flight with large wing bar often not obvious because of overall paleness.

ad. ♀ summer with chicks

ad. ♂.

♀

Redhead *Aythya americana* **REDH** L 19in

Locally fairly common, attractive diving duck, sometimes in big winter flocks on rivers, deeper water inland, and coastal lagoons. Over 75% of N American Redheads winter at Laguna Madre in s. TX. Large flocks also in the Great Lakes area, but there are large areas of the US with only a few individuals. Often with other diving ducks. Breeds in a variety of wetlands. **ID**: Fairly large with big-headed look, steep forehead, flat crown, and thick neck on a body that sits high in the water. Shape is strikingly different from CANV, closer to some other diving ducks, but still the key ID feature when not sure. Ad ♂: distinctive rufous, black, and finely barred gray and white. Eclipse (Jul-Oct)/1st-w ♂ (Jul-Dec): dull brown with reddish-brown head and red iris. Molts into bright plumage by Nov. ♀: brown, grayer above. Often confused with LESC and RNDU—focus on head shape and color pattern. Most similar to RNDU, but REDH is much paler and more uniform brown with a less distinct eyering. Scaup lack pale line behind eye and are darker. Adult and imm ♂ (similar): often compared with CANV because of their chestnut-and-black head and neck. There the similarity ends. If it's not pale, it's not a CANV.

1st-w. ♀

1st-w. ♂

ad. ♂

ad. ♂

ad. ♀

Greater Scaup *Aythya marila* **GRSC** L 18in

Fairly common n. breeder. Forms large flocks in winter in deepwater coastal bays, estuaries, and on the ocean. Frequently mixes with very similar LESC. In flight, scaup form tight flocks with fast wingbeats. **ID**: Sometimes obvious and at other times separating from LESC a real head-scratcher. (See LESC account for comparison.) A large diving duck with broad head and evenly domed rear crown, thick neck, and bulging cheeks. Broad bill flared at tip with black typically extending off the nail onto leading edge of bill (usually restricted to nail in LESC). Sometimes with other diving ducks; comparing size is a good ID starter. Ad ♂ br: iridescent green head often appears black, but rarely purple. Finer vermiculations on back and whiter, looking flanks than LESC. Eclipse ♂ (Jul–Oct)/1st-w (Jul-Mar): molts through winter. ♀: appears fairly uniform brown with variable amounts of gray mixed in. Well-defined white around base of bill. Worn birds (late winter/summer) often show a white cheek patch (more so than LESC). Juv: as adult ♀ but initially darker brown iris and more uniform brown. Less white on face at base of the bill.

ad. ♂ summer

ad. ♀ summer

ad. ♂

1st-w. ♂

ad. ♀

ad. ♂

Lesser Scaup *Aythya affinis* **LESC** L 16.5in

Common. Only scaup breeding in the Great Plains. Widespread winter distribution with preference for fresh water, though more common on salt water in the S. Rarely forms large dense flocks. Often seen in salt water after frozen out in fresh. **ID**: Compared with similar GRSC, roughly 10% smaller and looks slighter-bodied. Different head shape critical: peaked at rear crown (long crown feathers), giving a more balanced square-headed feel than GRSC (very difficult to judge on diving birds). Thin-necked. Shorter and narrower parallel-sided bill with black usually confined to nail. Lacks the striking bulging cheeks of GRSC. Flanks with brown most of winter—cleaner in GRSC. In flight, upperwing has white wingstripe on inner wing, contrastingly browner in primaries, but not always easy to see. GRSC's larger head is sometimes noticeable in flight. LESC has darker gray greater underwing coverts. Ad ♂ br: purple-headed ♂ scaup is usually LESC though latter commonly shows green iridescence, particularly in winter. ♂ LESC averages coarser barring on upperparts. Ad ♀: coloring essentially as GRSC. Imm: darker eyes (paler in adults), less white at base of bill, and fewer gray feathers.

1st-w. ♂

ad. ♂

ad. ♀

Ring-necked Duck *Aythya collaris* **RNDU** L 17in

Fairly common. Usually in small groups, sometimes with other diving ducks. Found in a variety of freshwater wetland areas but often in ponds surrounded by trees. Maroon 'ring neck' on ♂ hard to see and not a good field mark. **ID**: A variable-looking duck that often gets people scratching their heads. Focus on the distinctive sloping forehead and peaked rear crown but beware of 'flattened-out' wet diving birds. Perhaps 'Ring-billed Duck' would be a better name, as all birds have white ring emphasized by black bill tip. Adult ♂ br: black and gray with a diagnostic white patch at the front of the flanks that's easy to see at distance. Iridescent blue-purple head usually appears black. White around base of bill, including eclipse (Jul–Nov). 1st-w. ♂: a ghost pattern of breeding ♂ but looks 'grungy' with brown mixed in flanks, duller bill lacking white around base. ♀: darker back and cap than similar species and a more contrasting appearance overall. Distinctive 'spectacles' around eye. Color is variable; 'gray-and-brown' can often be the best term to describe it. In flight, has broad uniform gray wing bar, but also look for the distinctly patterned ♂ in the flock.

Harlequin Duck *Histrionicus histrionicus* **HADU** L 16.5in

Uncommon and tied to fast-flowing streams and rivers on breeding grounds; rocky shores and jetties in winter. Seems to embrace bad weather. Often seen standing on rocks or swimming in small self-contained groups. Appears 'gentle,' and ♂ is a real stunner. **ID**: A uniquely shaped small duck. Stubby bill on a small head with thick muscular neck and long tail. Ad ♀: appears uniform dark brown with bold pale areas on cheeks and in front of eye. Ad ♂: beautifully patterned in stunning colors. 1st-year ♂: subdued version of adult. Upper- and underwing all dark, distinctive in flight.

King Eider *Somateria spectabilis* **KIEI** L 22in

Scarce in winter on rocky coasts or jetties. Usually with COEI; easily overlooked. **ID**: Shorter and more compact than COEI. Focus on head and bill: shorter, thicker, neck, rounded crown, and noticeably shorter bill that looks deep-based. Ad ♂ br: stunning. Has prominent 'sails' (scapulars that protrude from back). ♀: warmer brown than COEI with prominent pale areas above eye, at base of bill and a line across the ear coverts. Bill dark. 1st-year ♂: dark with pale breast and orange bill. Eclipse ♂ (Jul–Oct): all dark brown (keeps bright bill). In flight, white underwing, dark above, like COEI.

partial eclipse ♂

♀s with chicks

1st-yr. ♂

2nd-w. ♂

juv.

ad. ♂ Atlantic

ad. ♀

Common Eider *Somateria mollissima* **COEI** L 24in

Very common n. sea duck of rocky coastlines and jetties. Occasionally seen in other areas. Usually in tight flocks. Nests colonially on ground on rocky islands s. to MA. Large groups, ♀s with chicks, ♂s separate, a common sight close to shore. Formerly hunted for 'eider down.' Powerful flight, birds typically in lines with heads at 45 degrees. Long takeoff with water splashing everywhere. **ID**: Large and heavy with unique profile created by long sloping bill, long neck, and peaked forehead. 3 races separable by ♂ lobe (bill base) shape: *dresseri* (Atlantic Eider) the common bird from NL s. (shown in plate); *borealis* (Northern Eider) of the high Arctic has much shorter and narrower lobe; *sedentaria* (Hudson Bay Eider) is intermediate and grayer, Hudson Bay area. Large variation in color tones, though Atlantic birds usually the richest. Ad ♂ br. boldly pied. Eclipse ♂ (Jul–Oct)/♀: intricately marked brown with 2 narrow wing bars. Juv: rufous-brown with darker feather centers and narrower wing bars than ♀: 2nd-yr: brown-tipped tertials and brown flight feathers. In flight, ♀ appears uniform brown with a striking white underwing.

1st-yr. ♂

ad. ♂

♀

Black Scoter *Melanitta americana* **BLSC** L 19in

Common sea duck, often in large flocks mixed with other scoter and ducks. Some migrate inland and are sometimes seen on larger bodies of inland water. Regular in winter on the Great Lakes, rare elsewhere in interior. On migration, or moving between feeding areas, birds form large, variably patterned lines, often with SUSC. In winter/spring ♂ gives incessant wailing call; often in small groups wooing a single ♀ (lucky thing!). **ID**: Smallest and chunkiest of the scoters with diagnostic bulky round head allied to distinctive bill shape and color. Ad ♂: glossy black with yellow-orange knob on bill. 2nd-w ♂: may show browner throat and slightly reduced knob but most are as adult. ♀/1st-yr: conspicuous pale throat and face similar to smaller RUDU. This area becomes darker through winter in 1st-yr ♂, and bird develops yellow-orange knob on bill. 1st-yr has a pale belly that is easy to see in flight. In flight, relative to SUSC, has an evenly domed undercarriage, broader, less swept-back wings, shorter, thicker, neck, big head and less pointed 'face' that's discernible at long range with practice—a chunkier bird! Lands with a belly flop with water spraying everywhere!

Surf Scoter *Melanitta perspicillata* **SUSC** L 20in

Common sea duck, rare inland. Behavior similar to BLSC —tends to wander into bays and harbors. Like other scoter, actively dives for mollusks. Often seen bobbing around offshore. **ID:** Slightly larger than BLSC but noticeably smaller than WWSC. Appears long-bodied and long-necked with long pointed face. Ad ♂: incredible bill pattern suggests colors of a clown's costume. Striking white nape and forehead on black can be seen at long distance. 2nd-w ♂: has a little less white on bill and head and is subtly duller than adult. Ad ♀: dark brown with a capped appearance. Usually 2 pale patches on face. Some ♀s have pale napes. 1st-yr: averages slightly paler; most easily aged by pale belly—easy to see in flight. 1st-yr ♂ starts to grow in darker feathers through winter and develops bolder bill pattern. ♀: similar to ♀ WWSC, but latter has a less pronounced cap, subtly different head shape, slimmer neck, and larger bill with a square edge at the face; they are easy to overlook! In flight, weight is at the back of the undercarriage. Neck is long and slim, head pointed, wings longer, slimmer, angled-back, and pointed. Lands with wings held pointing skyward!

White-winged Scoter *Melanitta fusca* **WWSC** L 21in

Fairly common on the ocean in winter; prefers to be farther offshore, but also seen in bays and harbors. When close to shore, outnumbered by other scoter. Common in the Great Lakes, often close to shore. **ID**: White secondaries hard to miss in flight and usually visible on sitting birds with patient observation. Usually picked out from other scoter by larger size. Big, with a neck on steroids, particularly broad at base and usually at an angle to body. Overall 'mean' demeanor. Neck held straight and tilted forward in alert posture. Long pointed face, concave from forehead to bill (straight in SUSC). Adult ♂: surprisingly small-billed on close inspection. White eye patch usually visible at long range (smaller in 2nd-yrs with duller bill). Oft-cited contrasting brown flanks easy to overlook. Ad ♀: usually uniform dark brown. Small bill with large feathered area between bill and eye (different shape from SUSC). 1st-yr: similar to ad ♀. Pale loral and cheek patches variable, sometimes absent (1st-yr ♂) to bold (1st-yr ♀); pale belly. ♂ gets pink in bill through winter and darker feathers (beware, 1st-yr ♀ can also show dark head feathers). Lands with wings at 45 degrees.

ad. ♀ summer

1st-s. ♂

ad. ♂ summer

1st-w.

ad. ♀

ad. ♂

ad. ♀ transitional

Long-tailed Duck *Clangula hyemalis* **LTDU** L 16.5in (ad. ♂ 21in)

Common on the ocean, and in bays and river mouths. Very rare inland. Formerly called 'Oldsquaw.' Usually in small scattered flocks. Groups often chase each other around in agile pursuit flight, twisting from side to side. Makes a belly flop and big splash in water when landing. Very vocal with beautiful yodeling call. Size, shape, behavior, and color all contribute to its very dainty but elegant appearance. **ID**: Complex molt with several color patterns. Smallish with thick neck but small head and bill. General impression is usually of a flock of very tastefully pale-colored dainty ducks, with a lot of variation, some with long tails. Adult ♂: has long tail and subtle browns, grays, and whites that any decorator would be proud of. Much darker in breeding plumage. Ad ♀: similar to 1st-yr with warmer brown upperparts and breast (grayish in young birds). 1st-yr: variable with ♂ tending to have grayer scapulars. 1st-yr ♂: becomes more obvious later in winter and summer as pink develops on the bill and new longer and whiter scapulars are molted. In flight, forms tight groups; look for all-dark wings with fast wingbeats on pale birds. Often rocks from side to side, enhancing similarity to alcids.

ad. ♂

ad. ♀

Bufflehead *Bucephala albeola* **BUFF** L 13.5in

Widely distributed throughout on water bodies from larger lakes to flooded farm fields. N. breeder near small lakes and ponds, nesting in tree cavities. Most concentrated in winter on coastal bays. Forms small, loose-knit flocks. Reminiscent of a little black-and-white rubber duck, bobbing up and down but often out of view. Brown ♀ is less conspicuous. **ID**: Very small and squat with oversized head often punked-up at rear crown. Ad ♂: striking with beautiful iridescent green-and-purple head though usually appears black. Ad ♀/1st-yr: same appearance. Head darkest brown with obvious white cheek patch, distinctive at distance. They also look the same in flight though imm ♀ often lacks white on greater coverts. In spring, some 1st-yr ♂s get whiter on breast and cheek. Adult-like after summer molt. In flight, BUFF often appears as a blurry ball rocking from side to side with wings moving 100mph. Alcid-like in some ways, they also make conspicuous crash landings on the water.

1st -w. ♂

ad. ♂

ad. ♀

Common Goldeneye *Bucephala clangula* **COGO** L 18.5in

Common and widespread. Breeds on deep lakes and rivers in forested areas, nesting in nearby tree holes. In winter often forms flocks with many dispersing to the coast, but quite at home in cold landscapes. Found in similar habitat to COME. ♂'s wings in flight produce loud whirring whistle noise. Often seen flying overhead as they move between feeding areas. Very wary and easily flushed. Puts head down before jumping up to dive. Animated displays in late winter: throws head back and also points bill skyward. **ID**: Medium-sized duck with an oversized head that juts out from the back of the neck, but not so extreme as in the scarce BAGO. Large black triangular bill with variable amounts of yellow adds to distinctive head shape. Always alert and ready to dive, but still keeps a beady eye on you. Ad ♂: at a distance appears white-bodied with dark back and upperparts. Oval cheek patch on green head diagnostic. Ad ♀: brown head contrasts with gray body. Juv/1st-yr: as ad ♀ with darker iris. Eclipse and 1st-w ♂: similar but has extensive white in wing and develops a partial white face patch. Adult-like in summer. See BAGO for comparison.

hybrid ♂ x COGO

ad. ♀

1st-yr. ♂

ad. ♀ summer

ad. ♂

Barrow's Goldeneye *Bucephala islandica* BAGO L 18in

Scarcer cousin of COGO. Similar in all ways and usually found while searching through flocks of COGO in winter. **ID**: Similar to COGO. Shape is vital in ID of all female goldeneyes. BAGO is shorter-billed, often giving a stubby appearance. Head shape often striking with steeper forehead and much more puffed-out nape area when neck is held erect. Ad ♂ br: shape, typically purple iridescence to head (not green), black vertical breast-side spur, smaller white scapular spots and white face crescent are all easy to see. ♀: sometimes easy but shape is not always easy to judge. Generally, subtly more compact than COGO. Bill averages more extensive yellow than COGO, often noticeable in a flock, so a good starting point, but variable, and always confirm bill and head shape. Darker-billed in summer. Less white in coverts, noticeable in flight, but varies by age and sex. Young and eclipse ♂: follow similar molt pattern as COGO. But beware of hybrids which occur regularly enough to be a concern. The latter show intermediate features.

1st-s. ♂

1st-w.

ad. ♂

ad. ♀

Common Merganser *Mergus merganser* **COME** L 25in

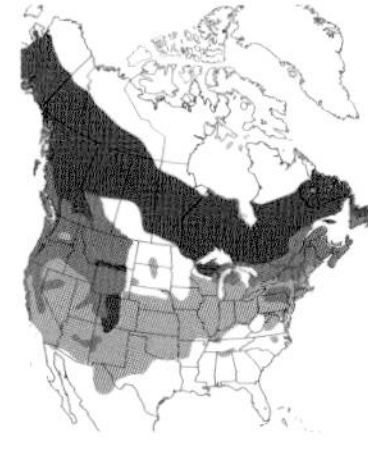

Widespread and fairly common on larger lakes and rivers in hilly areas, and other areas of open water. Nests in tree holes. Very hardy. In winter can gather in large numbers where food is plentiful, such as power stations. Rarely near salt water. Like other mergansers (sawbills), has serrated bill for holding fish. **ID**: Largest merganser. Slim but muscular with large head and thick neck giving front-heavy look. Long red bill broadening onto large head. Typically sits low in the water, head held down and with long tail giving a stretched-out look. Crest usually 'flattened down'. Ad ♂ br: appears mostly white. Dark back and green head that usually appears black at distance. Eclipse ♂: as ♀ but keeps white coverts. ♀ and 1st-yr: rich chestnut head with sharp border and well-defined white chin. Breast pale. 1st-w: usually paler loral area. RBME lacks this clean-cut appearance and has darker neck, is smaller, has different shaped head with spiked hairdo. 1st-yr ♂: does not develop grayer scapulars until spring. In flight, different wing pattern from RBME, but best to look at size, shape, and color pattern with cleaner appearance. Like other mergansers, slim with shallow fast wingbeats, and often flies in lines.

Red-breasted Merganser *Mergus serrator* **RBME** L 23in

Common n. breeder in wet wooded areas. One of the commonest ducks in winter on the coast, both in sheltered and exposed areas. In the Interior, mostly on larger lakes. Usually in small groups. Often migrates later than other ducks in spring, and ♂s can be seen displaying. **ID**: Noticeably smaller and slighter than COME. Punky crest, thin bill, steep forehead, smaller head, thin neck and overall shape give distinctive profile. Complex body patterns never match 'cleanness' of COME. Ad ♂: green, spiked head bordered by a white collar. A mosaic of black, gray and white lacking the clean creamy white appearance of ♂ COME. Eclipse ♂ (Jun–Oct): similar to ♀ with retained white coverts. ♀/1st-w: all similarly plumaged lacking clean-cut appearance of same-age COME. 1st-w ♂: develops dark around eye and bright red iris in late winter. Molts into adult-like plumage through summer.

Hooded Merganser *Lophodytes cucullatus* **HOME** L 18in

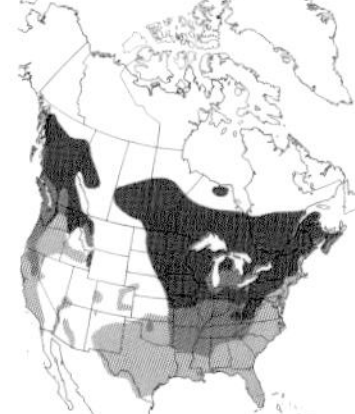

Fairly common and widespread, particularly as you go e. In summer, found on marshy ponds and streams with trees available for nesting. In winter, more often in coastal harbors, and on marshes, ponds, and creeks. Usually in small groups but often gathers to roost in large flocks at favored location. The striking white in the adult ♂ is highlighted by the brown of the rest of the bird, and is the most visible feature at distance. Mobile and often seen in distinctive flight: small and slim but compact with very fast shallow wingbeats. **ID**: The odd merganser out! Clearly the smallest of the 3 merganser species and distinctly different from the others. Ad. ♂: strikingly patterned with crest that opens like a Victorian fan. Long tail sometimes held cocked. ♀/eclipse ♂ (retain gray coverts,)/1st-w: brown overall appearance, warmest on crest, darkest on forehead. Pale base to bill stands out next to such a dark head. 1st-w ♂: darkest around eye. White in crest and black bill develop through winter. Molts into adult-like plumage through the summer.

Ruddy Duck *Oxyura jamaicensis* **RUDU** L 15in

Common stiff-tail with more s. distribution than most other ducks. Breeds on ponds, lakes, and in well-vegetated marshy wetlands. In winter, forms compact flocks, some moving to rivers and coastal bays. Often sits idle and usually swims away from trouble. In display, bangs the water and blows bubbles. In flight, has long, running takeoff with tail and big feet dragged behind. Uniform upperparts, no wing bars. Rarely mixes with other species. **ID**: Small, chunky, and punky little duck with distinctive shape—big broad head and bill. Long tail often held cocked. ♀ and young birds: dark brown with horizontal dark cheek stripe. Winter ♂: similar with white face lacking cheek stripe. ♂ molts in spring to a whole new set of colors but with the same basic ground-color pattern, and retains distinctive white cheek patch.

ad. br.

nonbr.

juv./1st-yr.

juv/1st-yr.

ad. nonbr.

Great Cormorant *Phalacrocorax carbo* **GRCO** L 36in

Uncommon in rocky coastal areas. Very rare on fresh water. Often seen sitting on rocks, sand banks, buoys, and favored exposed perches. Hardy. Like other cormorants, a great fisherman, and can often be seen trying to swallow large fish. **ID**: Largest cormorant, 10% larger than DCCO with which it often mixes. Focus on head; appears thicker-necked and block-headed, with fierce expression enhanced by pale gray bill, which is thicker and longer than in DCCO. Chin patch (gular/bare skin) yellow, small, and pointed contrasting with white throat. Ad br: appears shiny black (feathers have black border, paler centers), white head plumes and diagnostic white flank patch in late winter/spring. 1st-yr: brown neck and whiter belly (DCCO has the reverse pattern). 2–3-yr-old and nonbr adults (4 years): similar, with varying amounts of brown mixed with black feathers.

juv./1st-yr.

ad. nonbr.

ad. br.

Double-crested Cormorant *Phalacrocorax auritus* **DCCO** L 33in

Widespread and very common. Often approachable, and regularly found near humans. Occasionally forms massive feeding flocks on lakes where food is plentiful. Also happy in salt water. Creates long, haphazard lines in flight, sometimes in 100s, a familiar sight on migration. Often glides and occasionally soars ('Fool's Anhinga'). **ID**: Very variable in color. Easily mistaken for both GRCO and NECO, though it is only species found over much of its range. Always check size, tail length, head pattern, and underpart color (1st-yrs). Ad br. shiny black. 'double crest' hard to see and varies from white to black. Gular orange, and bird has beautiful blue eye. Ad nonbr. duller with neck browner than rest of body. Juv/1st-yr: shows great variation from white to dark brown underparts and everything in between, but importantly it is always paler on neck than belly. Gular patch and loral skin varies from deep orange to yellow-orange and is large with rounded rear edge. 2–3-year-old imm: as nonbr adult with varying amounts of brown mixed with black feathers.

2nd-yr.

ad. nonbr./2nd-yr.

1st-yr.

ad. br.

ad. br.

Neotropic Cormorant *Phalacrocorax brasilianus* **NECO** L 25in

Locally fairly common on lakes, rivers, and in coastal bays. Rare wanderer n. Behavior much as DCCO and often found together. **ID**: Small (15–20% smaller than DCCO) with thinner neck, smaller head, and longer tail, all features noticeable with practice. In flight, the neck and tail length are about equal—the tail appears shorter in other cormorants. With good views, pointed small gular patch with white border is diagnostic. Ad br: black with white flecks on side of head. Duller in winter. 1st-yr: fairly uniform brown; darker and lacks the contrast of most similar-aged DCCO where ranges overlap. Older imm: similar to nonbr adult. Always confirm plumage differences by checking gular pattern, size, and shape.

Anhinga *Anhinga anhinga* **ANHI** L 35in

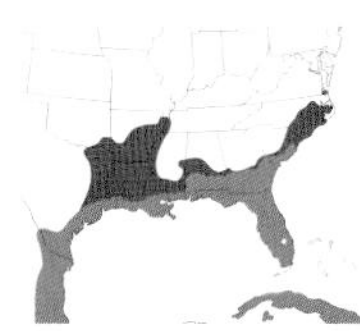

Fairly common locally in freshwater areas with trees such as swamps, mangroves, and canals. Like other cormorants, often colonial with other species (usually cormorants and herons). Wanders out of range, though high-soaring DCCO is frequently mistaken for this species in the field. Sits in the open, usually in trees, sunning itself to get dry. **ID**: Aptly known as 'Snakebird,' given its neck shape. Swims with only head and neck above water. Has wicked pointed bill for spearing fish. Often seen in flight and frequently soars high in the sky, when is often confused with other species. Appears as a flying cross; long, thick tail appears to propel a bird that can look almost neckless at distance. Striking; if in doubt it is probably a cormorant. Adult: shows striking off-white scapulars and coverts. ♂: black-necked, tan in ♀. Breeding birds get brighter bare parts and head plumes. 1st-yr: as adult ♀ with brown neck, however, the body is browner with smaller dull coverts and scapulars. 2nd-yr: more adult-like but retains some signs of immaturity.

Common Loon *Gavia immer* **COLO** L 32in

Widespread and common. Perhaps its European name, Great Northern Diver, is more fitting. Breeds on large tree-lined lakes. Its famed, almost mystical, yodeling call travels long distances. Calls night or day, but also away from breeding grounds, often in flight. Winters on large bodies of water, many moving to the ocean and back bays. 1st-s birds often stay in the S for summer. **ID**: The large commonly occurring loon. Big, variably colored bill held straight out, and has protruding forehead (head can be rounded) and thick neck. Ad br: beautifully marked. Many wintering birds and spring migrants have partially or completely molted before reaching the breeding grounds. Ad nonbr: brown above, pale below with distinctive bulging dark neck sides. Large pale bill always with dark tip and culmen. Juv/1st-yr: distinct pale fringes to upperparts, become worn by spring. 1st-s: worn and faded juvenile plumage. Pale fringes are usually worn-off so looks a lot like adult nonbr. Often misidentified as PALO: look carefully at bill shape and neck pattern. In flight, large and powerful with slower wingbeat than RTLO. Focus on size, structure, head, and neck pattern, and, in particular, the large protruding feet.

Yellow-billed Loon *Gavia adamsii* **YBLO** L 35in

Breeds in Arctic Canada. Very rare to the s. on open lakes and rivers. **ID**: Slightly larger than similar COLO. Has more upright posture with neck held back and appears slimmer, bill tipped up, giving regal appearance. Nonbr: paler, particularly around eye. Ad br: larger white checkers on upperparts and differently shaped white collar. Bill shape and color is always critical. Straight culmen with strong gonydeal angle, ivory in nonbr (named White-billed Diver in Europe) and yellow in breeding (hence American name). Culmen always pale; COLO always shows some dark.

Pacific Loon *Gavia pacifica* **PALO** L 25in

Uncommon on breeding grounds. Most winter on the W Coast. Rare to the S, often with COLO. **ID**: Between COLO and RTLO in size and shape. Bill intermediate. Often confused with round-headed and small-billed nonbr COLO. A dark bird with clean-cut border down neck sides and across face, lacking pale area around eye of COLO. Nape looks 'smooth,' paler than neck sides. Many have chinstraps—a thin dark line across throat. Juv: as nonbr with bolder fringes to upperparts. 1st-s (nonbr): pale and washed out. Breeding birds striking. Often in breeding plumage on spring migration.

ad. br.

juv.

ad. nonbr.

Red-throated Loon *Gavia stellata* **RTLO** L 25in

Common tundra breeder. Winters on the ocean, occasionally coming into bays. Scarce inland. Has traditional spring gathering locations. Rarely flies overland away from breeding grounds, unlike COLO. **ID**: Smallest and slimmest loon. Easiest to confuse with cormorants. Small pointed head with thin bill characteristically tilted upward, obvious at long range. Ad br: more uniform than other loons. RTLO's throat often appears black. PALO always shows white patches on upperparts when dark-throated. Ad nonbr. a strongly contrasting 2-toned bird even at long range. Often appears capped, more white than dark and with a white area in front of the eye. Thin and straight-necked, can look grebe-like. Upperpart pattern never shown by other loons. Juv/1st-w: similar to adult but grayer around throat and neck with contrasting white chin. Upperparts have narrower fringes. Rarely seen in breeding plumage on wintering grounds. In flight, small size, slim build, slim angled-back wings, face pattern, faster (and deeper) wingbeats, and lack of obvious protruding feet, are key features. Migrates in loose flocks, often circa 100ft above ocean, frequently moving its head to look around as it flies.

ad. nonbr.

ad. br.

Western Grebe *Aechmophorus occidentalis* WEGR L 25in

Locally common breeder in the W, most moving to warmer areas in winter. Rare but annual to the E, in coastal bays or on the ocean. Large, erect, and extremely elegant. Bill looks deadly. Famous courtship behavior: pairs contort and entwine, then hit the water before running away in tandem. Formerly considered conspecific with CLGR. **ID**: Year-round 2-toned color always striking. Breeding birds average brighter bills, with cap and lores blacker than nonbr but differences marginal. Double-syllabled call, single in CLGR. For other specific differences from CLGR see that species.

Clark's Grebe *Aechmophorus clarkii* CLGR L 25in

Same as WEGR for most features, but about 10 times scarcer. Often found together. Intermediates occur, and hybrids or individual variation arguably make some birds impossible to id. However, most can be separated. **ID**: CLGR has clean, deep yellow to orange-yellow bill (WEGR usually dull yellow with green element), and pale area around eye, differently shaped border to cap and narrower nape line. Also paler upperparts and flanks with smaller, more defined dark centers at rear. CLGR has a strongly contrasting appearance even at distance. In flight, white extends to outer primaries.

Red-necked Grebe *Podiceps grisegena* **RNGR** L 18in

Uncommon breeder in marshes, on water bodies that are often small but usually have extensive reeds. Nests are not much more than floating debris but can be surprisingly large. Hardy, winters offshore on the Great Lakes, closer to land on the ocean. Extreme winters freeze birds out, resulting in southward movement. **ID**: Size and shape distinctive—large, thick-necked with a stout, straight, yellow-based bill tilted slightly down. Capped appearance in all plumages. Ad br: both sexes have lovely head and neck pattern from Apr onward. Ad nonbr: a dulled-down version of breeding with grayer cheeks, and a pale patch at rear. Neck and bill are dull with limited yellow at base of bill. 1st-w: very similar to nonbr. Size, shape, and bill always diagnostic. In flight, bold white secondaries and white leading edge to wing.

transitional

feeding chick

ad. nonbr.

ad. br.

Horned Grebe *Podiceps auritus* **HOGR** L 14in

Some breed in nw. marshes in the E. In winter, scarce on inland lakes but fairly common on the ocean and in coastal bays. Sometimes in loose roosting flocks. Can mix with other birds such as RUDU. Dives "with a jump," disappearing for long periods of time, and it can be frustratingly difficult to get prolonged views. **ID:** Fairly small, even proportions. Vertical posture. Bill is straight, broad-based, and helps to give head a pointed shape. Bill often pale-tipped. Ad br: colors striking but rufous can look black at distance. Stunning red eye. Molting birds from late Feb through Apr are very variable and often confused with EAGR. Nonbr: appears black-capped with clean horizontal divide through eye. This is emphasized by bright white cheeks—much 'dirtier' in EAGR. Peculiar and distinctive shape in flight with head held higher than body.

ad. nonbr.

ad. br.

Eared Grebe *Podiceps nigricollis* **EAGR** L 13in

Common breeder and migrant in the W, increasingly rarer in the E. Found on lakes, ponds, and at water treatment facilities. In winter most move to milder areas and coastal bays. Early fall migrant often in Aug–Sep (HOGR rare before Oct). Breeds in colonies. Nest is simple, consisting of floating vegetation. In winter often found in groups. **ID**: A fluffy-looking grebe with habit of sitting very high in the water, showing large exposed gray rear end. Strikingly fine bill is slightly upturned but is often held pointed slightly downward. It has a very steep forehead with a 'bump' on top. Thin-necked compared to large head. Shape is always the most important factor in differentiating from similar HOGR. Ad br: bold yellow plumes on black neck and head (beware, HOGR often appears black-necked). Black back, grayer in HOGR. Nonbr: lacks clean-cut appearance of HOGR, shows darker neck and cheek patch with white throat wrapping around ear coverts.

ad. with young

ad. nonbr.

ad. br.

Pied-billed Grebe *Podilymbus podiceps* **PBGR** L 13in

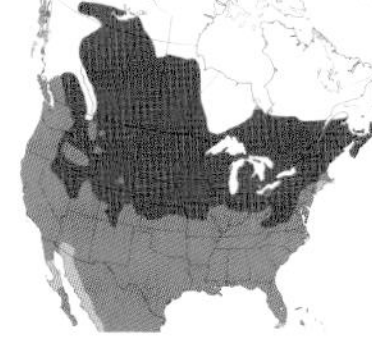

Widespread and fairly common but usually in singles or pairs. Found wherever there is water and adequate food, the generalist among N American grebes. Prefers areas with vegetation but can be found anywhere in winter. Although the commonest grebe in many areas, can be elusive and easily overlooked due to size and behavior. When approached tends to submerge quietly, like a submarine, rather than fly. ♂ gives bizarre loud hooting vocalization: rapidly given *gawup* notes becoming longer and more urgent at end. Also a descending chatter. **ID**: A small compact grebe with broad, short, pointed bill. Brown appearance overall, slightly darker on back as year progresses. Ad br: develops black ring on bill (bill and ring larger in ♂), dark on crown, and has striking black bib. Juv: striped head; a few keep pattern through winter, but most molt and appear as nonbr after several weeks. Rarely flies. More often seen running across water with wings flapping. No wing bars. Vocal.

juv.

ad. nonbr.

ad. br.

Least Grebe *Tachybaptus dominicus* **LEGR** L 9.5in

Fairly common in heavily vegetated marshes and ponds in s. TX. Very small with slimmish neck and thin bill. Sometimes confused for the larger and heavier-billed PBGR. Appears as a ball of fluff when hunkered down. Note brilliant yellow eye in breeding birds. Tends to be active looking for food. Aggressive. Variably dark in all plumages. Pale flight feathers obvious in flight and when wing is stretched.

Black Guillemot *Cepphus grylle* **BLGU** L 13in

Occurs in singles and pairs, sometimes small groups, evenly spread out on rocky coasts and islets. Prefers exposed areas coming to more sheltered bays in bad weather. Little seasonal movement for some. Rare to the S. **ID**: Medium-sized, dainty with slim proportions and fine bill. White underwing and ovals on upperwing. Striking red legs. Ad br: solid black with white wing patches obvious at any distance. 1st-s ♂: as adult but retains old brown coverts and flight feathers. Ad nonbr: dappled gray and black but still with white oval patch. 1st-w: similar but dull-legged, darker with dark spots in coverts.

Dovekie *Alle alle* **DOVE** L 8.25in

Winter visitor from Greenland. Sometimes seen from shore, generally farther n. Easiest to see from boats offshore, often hundreds in a day. Occasionally sick birds wash up and are then approachable. Often drags wings between dives. **ID**: Tiny, particularly in a big ocean. Short, plump, and neckless with stubby bill. In flight, seen as a chunky blur of wings with white trailing edge. Nonbr: striking partial collar and white underparts. Ad br: dark head and neck. ATPU can look small but not so tiny as DOVE, and ATPU never has white neck sides.

Long-billed Murrelet *Brachyramphus perdix* **LBMU** L 10in

Mega-rarity from Siberia, mostly Nov/Dec on large lakes. Formerly ssp of Marbled Murrelet. **ID**: Small slim alcid with a broad-based pointed bill. 2-toned head with clean divide extending down neck. Pale ovals on nape sometimes obvious. Striking white scapular line extends across back—obvious at distance. Ad nonbr: similar to juv, more solidly black above, cleaner white below.

Ancient Murrelet *Synthliboramphus antiquus* **ANMU** L 10in

Mega-rarity from the Pacific NW that turns up mostly in late fall on large lakes. **ID**: Small chunky alcid with tiny pale bill. Gray back contrasts with black head and white underparts. Dark ear coverts contrast with distinctive white rear wraparound; compare DOVE.

Common Murre *Uria aalge* **COMU** L 17.5in

Nests on rocky outcrops, with birds crammed on ledges in their thousands. Disperses in winter to open ocean. Often found in 'packs' hunting small fish. Rarely seen from land. **ID**: Large alcid. Similar to TBMU but subtly slimmer-necked and flatter-headed with thin pointed bill. Differences more striking compared to RAZO. Dark flank markings are hard to see. Brown-backed but often appears black. Nonbr: distinctive dark line on white face (lacking in RAZO). Young bird typically has a slightly shorter bill and duller face pattern. Breeding birds are dark-chested; 'bridled' form has white spectacles.

Thick-billed Murre *Uria lomvia* **TBMU** L 18in

Habits as COMU. More likely to come to shore in winter. **ID**: Tricky, and made worse by individual variation in bill size and head pattern. Many characters intermediate between RAZO and COMU. Head pattern with relatively dark, dingy throat (nonbr) is the most striking feature. Also from COMU by thicker neck, different bill shape, white line at gape (adults), darker upperparts and clean white flanks. RAZO has larger bill, different head shape, white behind ear coverts (nonbr) and longer tail. Trailing feet often visible on swimming birds.

ad. nonbr.
ad. br.
1st-w.
ad. nonbr.

Razorbill *Alca torda* **RAZO** L 17in

Breeds in large spectacular colonies on rocky islands, often with other alcids. Winters closer to shore and is the alcid most likely to be seen from land. When swimming sits high in the water, strikingly black and white. Dives, reappearing minutes later a long way away but sometimes never to be seen again! Forms groups, flying fast in direct lines. **ID**: Always look for the striking large, ridged bill with white lines. Bigger-headed and thicker-necked with longer tail than other similar species. Black-backed. Adult at 4 years of age, when has larger bill with more ridges. ♂ larger than ♀. Dark throated in breeding. In winter is pale throated with a diffuse face pattern, the white extending behind ear coverts, lacking bold pattern of COMU. Younger birds have smaller all-dark bills, broader and blunter than in TBMU. Alcids are often seen in flight, and the 3 larger species are very similar. Use upperpart color, face pattern, tail and bill shape for comparison. Subtly flatter-backed, RAZO has bolder white underwing and whiter armpits, with more extensive white on tail sides. Feet don't extend past tail. Many birds best left unidentified.

Atlantic Puffin *Fratercula arctica* **ATPU** L 12.5in

Uncommon colonial nester on rocky islands, e.g., Machias Seal Is. Winters far offshore. Rarely seen from shore. **ID**: Mindblowing bill. Medium-sized and compact with big neck and head, between DOVE and RAZO. Black collar and cap set off face (white in summer, gray in winter). Adult at 4 years, bill becoming larger with more ridges with age. Smaller and duller in winter. 1st-yr: smallest and darkest bill. In flight, look for darkish underwing, no white trailing edge to upperwing, bill size and pattern, and compact shape. Often flies higher above water, other alcids at surface level. Orange feet striking.

FLYING WATERBIRDS

SEABIRDS

Though birding logistics may sometimes be difficult when trying to see these birds, this group in general is always intriguing. Some species can be seen from land, but typically only in the worst weather conditions. Usually a telescope is necessary for seawatching. Most species breed on rocky islands, some of which are accessible by boat, particularly in the NE. Often seabirds only come ashore at night. In part because of this, there are big gaps in our knowledge of some of these species, and in some cases we just don't know where they breed. Birds such as Black-capped Petrel may in reality be more than one species. Recent studies of Fea's Petrel in Europe have shown that it is actually 3 different species.

In most cases, the easiest way to see seabirds is from a boat, particularly for those species that travel far offshore in search of areas with plentiful food. Water temperature and food supply are connected and therefore play a significant part in distribution. The ocean is continually moving, and so too are the birds. There are a number of organized boat trips for birders and these are highly recommended for a good look at pelagic birds. Boats that 'chum' (throw out food, such as fish oil) often get the best results.

Views of seabirds are often very brief and distant as birds disappear below the waves. Flight style is a key element in the ID of this group of birds. Size and color patterns are also important, though often difficult to determine. Given the difficult viewing conditions, never be in a rush to put a name to a bird; many are best left unidentified.

Large storms, especially hurricanes, transport birds to the coast and sometimes far inland, particularly those trapped in the 'eye.' Occasionally, strong onshore winds will bring birds close to shore—watching the weather forecast is key for the serious seawatcher. Very hot summer weather creates a pressure gradient between land and sea, generating quite strong onshore winds in the afternoon. This, and concentrations of fish, will also bring seabirds closer to shore than usual. A large feeding group of terns and gulls will attract jaegers and shearwaters.

Never knowing what will turn up is one of the great fascinations of seabirding!

TERNS

A group of slim angular-winged birds with long, forked tails. Because of their shape and buoyant flight style, they tend to look very graceful. Some species, such as Black Tern and Gull-billed Tern, only swoop and pick food off the surface; however, most hover and plunge-dive. Some do both. Most species are white with gray backs. They travel long distances and thus can turn up anywhere, particularly as they are quite easily impacted by wind and rain. Hurricanes frequently displace birds. Terns are colonial, often breeding on rocky islands or beaches, and always close to fishing grounds. They often compete for beaches with humans and are often protected behind 'roped-off' areas. Get too close to nests and they will dive-bomb you. They mean business and do occasionally draw blood. They form mixed flocks, particularly on beaches where they gather to rest after fishing. Sorting through birds can initially be confusing, but with practice and understanding this becomes great fun and always offers the chance of finding 'the goody.' ID is made trickier by the wide variety of plumages. It's easiest to break them down into 3: juvenile, nonbr/immature, and breeding.

Juv plumage is only usually held for several weeks in most species. These birds tend to be on or near their breeding grounds—a great clue! Juvs can often be seen begging, since adults frequently continue to feed young after they leave the nest. Juvs usually have scaly brown or black on upperparts, bills are usually dark and often paler at base. During fall many of the darker feathers are molted or fade so they become similar to adult nonbr.

Adult nonbr and imms generally have partial caps, develop dark on bend of wing (carpal bar), and have darker bills. Most terns don't become fully adult until they are about 3 years old. 1st-s birds are as nonbr, but have old, unmolted feathers that are worn and faded, giving them a scruffy look. 2nd-s birds usually as adult with white flecking on crown and darker bill, but variable and probably similar to some sick or failed breeding birds. An understanding of molt sequence is fundamental in correctly ageing these birds.

Breeding birds have solid black caps, brighter bills and legs, longer tail streamers, and clean gray upperparts. This is the most frequently viewed plumage, particularly in the North, where very few terns are seen in winter.

The extent of dark in the primaries is often used for identifying terns. There is a large amount of variation and the exact patterns are often difficult to determine in the field. To add to the confusion, breeding birds get a 'fluffy' filamentous bloom that makes the flight feathers pale (see Forster's Tern p.118). This wears off through the summer, and as the primaries become more worn they are noticeably darker, often causing confusion. Most plumages show secondary bars in flight; but these are variable and of little use in ageing birds.

Learning to id terns is, as always, a question of getting back to basics. Learn the familiar species well, based on size, shape, and behavior, and study the patterns of color that remain consistent. Beware: if you get caught up in the minutiae of color, you will become very confused!

JAEGERS AND SKUAS

Arguably, jaegers and skuas have created more heated debates about identification than any other group of birds—and not just in North America.

Jaegers are tundra breeders that spend the rest of the year at sea. Some jaegers migrate overland and can show up on reservoirs, beaches, ploughed fields, even parking lots, loafing around with gulls. Any species is possible. Perhaps it is lack of context and practice that adds to the difficulty of identifying them. And, to add to this, is the tricky variety of plumages (it takes these birds approximately 3 years to mature).

GULLS

Love them or hate them, gulls are always fascinating and one of the great challenges in birding. A Larophile as a student, I find, among other things, their major appeal is that they can be found anywhere. They love densely populated urban areas, and are highly mobile (some will travel 30–40 miles to feed), so there is always the possibility of something new and possibly rare. Add to this their intrinsic ID challenges: multiple plumages, reasonably frequent hybridization, huge individual variation, taxonomic uncertainties (several have been or will be split into new species), and you have the perfect recipe—depending on your personal birding likes and dislikes! My wife puts it more bluntly: "I hate them. I have to stand out in the cold for hours, and they all look the same." Maybe having our first date at a sewage outfall didn't help! We agree to disagree: perhaps you can relate? Although gulls frequently inhabit unattractive, often dirty, places, I believe that it's the overwhelming plumage variation that is key to many birders' unease with, and apparent lack of interest in, this group.

With practice, many species can be relatively easy, though a fair number are difficult to identify, and several are essentially impossible. This last category produces an assessment qualified by the more realistic terms 'probable' or 'possible.' To demystify gulls, there are some things that you need to know. First, different species take different lengths of time to develop adult plumages. As a generalization, small gulls take 2 years, medium-sized gulls 3 years, and large gulls 4 years. As a result, birds often take on different appearances that can be very misleading in the field. However, all is not lost.

I usually break gulls into 2 groups: (1) younger imms, and (2) adults or near adults (subads). This works for most species. They also have breeding plumage, but this is essentially the same as nonbr, the major difference being head pattern: hoods develop in some of the smaller species, and others become white and look cleaner for breeding.

The most confusing species are generally thought to be the large gulls because they show the widest spectrum of variation. They can be overwhelming in the field at first. As you look at the plates, try to break the color patterns into the 2 groups: younger imms and older adult-type birds. Also go back to the basics of size and shape with a lot of emphasis on the bill. Herring Gull is the 'default' gull in most areas and the one species you have to get familiar with to significantly improve your field skills. For basic ID purposes, Herring Gull can be split into: (1) those that are variably smudgy brown or gray with an overall grungy look (the first two years of their life), and (2) adults or near adults; cleaner-looking birds with a uniform gray back (adult feathers molted in third summer (2nd-s)).

For the beginner, I would recommend trying to simplify things as much as possible, just as experts do. In our region it's best to learn a few species that make up the vast majority of all the gulls you might encounter: Bonaparte's, Laughing (coastal), Franklin's (Great Plains), Ring-billed, Herring, and Great Black-backed Gulls (coastal). If you can get a good grasp of identification of the very common Herring and Ring-billed Gulls, a large number of the problems of overall gull identification will disappear.

Northern Fulmar *Fulmarus glacialis* **NOFU** L 18in

Common but local. Breeds on cliff ledges—visits these in daytime. Winters well offshore. Solitary but found in numbers if food is plentiful. Often follows fishing boats. **ID**: Heavy-set bird. Broad tail and neck with wings and body held stiff and straight. Shallow wingbeats followed by glides and high arcs in bad weather. Sits high in the water. Stout yellowish bill. Colors vary from light to dark, light birds much the commoner. All have dark smudge around eye (giving mean look), pale area at base of primaries, and gray tail.

Herald (Trinidade) Petrel *Pterodroma arminjoniana* **HEPE** L 15in

Very rare in the Gulf Steam. Most birds either all dark or brown with white belly and breast; some intermediate. **ID**: Dark birds easy to dismiss as SOSH, but slightly smaller and more compact with straighter wings and shorter neck. All have pale bases to flight feathers and greater coverts forming pale panel in otherwise dark underwing. Light birds often appear hooded, though many have a pale throat. Uniform upperparts. Sits high in water with relatively dove-like head.

Black-capped Petrel *Pterodroma hasitata* **BCPE** L 16in

Fairly common in the Gulf Stream in summer, scarce in winter. **ID**: Probably at least 2 species. One with extensive white around forehead, other limited white and averages darker-naped. Variable in amount of white on rump, darkness of upperparts, wing-covert contrast, and size. Thick-necked and muscular. Bright white underparts striking at distance with bold contrast on underwing. Upperparts dark with white collar creating 'black cap;' striking white rump. Stiff snappy wingbeats. Arcs high in bad weather. Sits in small groups, often with storm-petrels. Also sits over food with wings raised.

Fea's Petrel *Pterodroma feae/madeira* **FEPE** L 15in

Very rare summer visitor to the Gulf Stream. Palest *Pterodroma*. **ID**: Gray upperparts and neck sides. Darkest around eye. Tail noticeably paler than back. Dark carpal 'M' sometimes obvious. Look for dark underwing contrasting with pale body. Usually solitary. Flies in high steep arcs in strong wind. Slighter than BCPE. In Europe, split into 3 species: Desertas, Fea's, and Zino's Petrels.

Bermuda Petrel *Pterodroma cahow* **BEPE** L 15in

A dream bird rediscovered in 1951 in Bermuda (now 250 individuals). Annual in the Gulf Stream off NC. **ID**: Similar to BCPE but first impression is of a smaller, duller bird. It is slighter, smaller-headed, and proportionally longer-tailed. Dark upperparts with restricted white in tail; dark head usually lacks white (hooded look). Small bill. Like other gadfly petrels shows significant variation.

Cory's Shearwater *Calonectris diomedea* **COSH** L 18in

Common summer visitor, rare in winter in warm ocean waters. Usually sits on water in flocks, and congregations occur where food is plentiful. Occasionally seen from shore, usually when windy. **ID**: Largest shearwater. Heavy build with broad wings held hunched over and with emphasis on downbeat, giving pterodactyl-like feel. All have pale upperparts with little head contrast. Underparts mostly clean white. Many are in active molt in summer, a mixture of browns and grays with patches of white at the base of the coverts. 2 races: Cory's (*borealis*) breeds in the Atlantic, and Scopoli's (*diomedea*) breeds in the Mediterranean. Many treat these as separate species. Very difficult, often impossible, to separate and complicated by the reality that most birds being seen are imms. Scopoli's is scarcer, averages smaller. The bill averages smaller, narrower-based, and duller. Most importantly, has extensive white on inner web of primaries, and one (often 2 on COSH) dark spot on outer greater coverts. Adult COSH usually shows solid dark primaries, but generally paler and more Scopoli-like in imm birds. Much variation and overlap. Takes 5 or so years to reach adulthood.

Great Shearwater *Puffinus gravis* **GRSH** L 18in

Common summer and fall visitor to colder ocean waters, particularly in the NE. Breeds in the S Hemisphere. Sometimes sits behind boats, eats fishing bait, and gets hooked. Occasionally seen from shore, usually when windy. **ID**: Slimmer than COSH with wings held straighter and with stiffer snappy wingbeats. Darker and more contrasting above with dark cap accentuated by white neck sides—striking at distance whether sitting or flying. This is always the easiest field mark to see. White uppertail coverts usually contrast more strongly with tail than in COSH. Underparts usually appear white at distance; closer views show a smudgy belly and vent and intricately marked underwing. Slender bill is always black. BCPE is stockier, usually with white uppertail and nape, and different underwing pattern. Most birds seen are juvs with immaculate plumage and darker collar. Older birds have a more extensive pale collar.

Sooty Shearwater *Puffinus griseus* **SOSH** L 17in

Fairly common summer visitor, particularly in late spring. Breeds in the S Hemisphere. Sometimes seen from land, particularly in windy conditions. **ID**: Fairly large—between GRSW and MASW in size. Fat-bodied with a thick neck but slimmer head and slender black bill. The tail, like the head, gets narrower toward the tip. It is the wing shape that is most striking: narrow at the base, slim and pointed, giving distinctive profile. Arches high in windy weather, wings angled back and held stiff, but flaps and glides in calmer weather on bowed wings. Flight is direct, purposeful, and fast. At a distance, brown plumage usually appears black with (but not always) pale or silvery central panel in underwing (HEPE and jaegers can have similar color pattern). The pale underwing is often visible at long range.

Audubon's Shearwater *Puffinus lherminieri* AUSH L 12in

Fairly common in the Gulf Stream and warm tropical water. Unusual in cold water, though some wander n. in late summer. A small black- or brown-and-white shearwater that is often seen sitting and diving with wings open. Loves lines of sargassum (gulfweed). Distinctive, fast snappy wingbeats with fairly long glides. Doesn't arc so high or so steeply as MASH. **ID**: Slim build with proportionally long tail. Dark undertail coverts noticeable with care on flying and sitting birds. MASH has shorter tail with white undertail coverts. Capped appearance, usually with straight line through eye, often with white smudge above it. Upperparts vary from appearing shiny black (very dark brown with good views) in fresh plumage to paler brown when worn. These plumages can be seen on accompanying late summer birds—you could easily be forgiven for thinking they were different species. Often mistaken for *Pterodroma* petrel on pelagic trips because of its size, shape, fast wingbeats, and direct flight.

Manx Shearwater *Puffinus puffinus* **MASH** L 13.5in

Uncommon in cold ne. waters (breeds) with a few wintering farther s. Sometimes seen from shore. Summering flocks occur, and is presumably breeding in the NE (Boston, MA). **ID**: Black-and-white shearwater. Similar to AUSH but slightly larger, with dusky face and with white wrapping around ear coverts. White undertail coverts can be seen in flight and on sitting birds. More compact than AUSH with shorter tail that often looks broad and square-ended. Wings often slightly angled back with bulging secondaries. Tends to arc steeper and higher than AUSH with faster, 'hurried,' wingbeats.

White-faced Storm-Petrel *Pelagodroma marina* **WFSP** L 7.5in

Very rare far offshore late summer/fall in cold waters in the NE. A wacky-looking storm-petrel with long legs with which it 'pogo-sticks' off the water. Glides for long periods without flapping. Distinctive plumage. Often found near WISP.

European Storm-Petrel *Hydrobates pelagicus* **EUSP** L 6in

Mega-rare but probably annual far offshore. **ID**: Smaller and slimmer than WISP with narrower white rump. A very dark bird lacking obvious pale carpal bar. White line on underwing is essential to clinch ID. Very light butterfly-like flight. Easy to overlook.

Wilson's Storm-Petrel *Oceanites oceanicus* **WISP** L 7.25in

Very common summer visitor offshore; hard to miss from a boat—a fisherman's friend! By far the most likely storm-petrel to be seen. "Small dark things with white" often swarm around food and 'patter' on or 'pogo-stick' off the water. Seen close to shore in strong winds or following food (often in very calm, hot weather). **ID**: Smaller and more compact than other storm-petrels with shorter but broader wings and tail. Tail appears square or slightly rounded (tail feathers often held in 'U' shape, which can appear as notch). Striking broad white rump and white wraparound. Fluttery swallow-like flight followed by flat-winged long glides, often back and forth across water. It will suddenly stop with wings open, and, like LTJA, drop to the water to pick something up. Wing profile changes based on behavior, but usually straight trailing edge and evenly curved leading edge. Legs and feet (yellow webs) extend past tail, sometimes easy to see. In wing molt in summer and in fresh plumage by early fall. Learning the size, shape, and flight style of this bird makes picking out other storm-petrel species straightforward.

Leach's Storm-Petrel *Oceanodroma leucorhoa* **LHSP** L 8in

Uncommon, more regular farther n. in colder waters. Breeds in island burrows, which they only enter at night. Rare close to shore. **ID**: The largest and palest brown of the dark storm-petrel with slim proportions and long, strongly angled wings. Strong pale carpal bar reaches leading edge of wing. Broad white in rump forms a shallow 'V'—narrower and straight with larger wraparound in BSTP. Tail forked, but can be difficult to see. Sometimes hint of central line, also on some BSTP. Bounding nighthawk-like flight with deep, jerky wingbeats with strong downstroke. Rocks from side to side and zigzags.

Band-rumped Storm-Petrel *Oceanodroma castro* **BSTP** L 9in

Uncommon in Gulf Stream waters. **ID**: Size and features fall between LHSP and WISP, but appearance much closer to LHSP. In comparison with LHSP, less stretched out and angular. Darker with less distinct carpal bar, and narrower parallel-sided rump wraps onto side of tail. Square or very slightly notched tail. May fly like LHSP, but usually with shallower and steadier wingbeats, rocking from side to side with bowed wings (like COSH). Recently split into 4 species in Europe; 'Grant's,' presumed most frequent in N America, is usually in wing molt. An interesting challenge for the future!

2nd/3rd-yr.

ad. ♀

ad. ♂

1st-yr.

Magnficent Frigatebird *Fregata magnificens* **MAFR** L 40in

Uncommon and local s. breeder that often travels as far afield as Canada. Usually nests and roosts communally. A huge and prehistoric-looking aerial specialist that hangs in the sky with the ability to drift over large areas, apparently without effort. Catches fish and is piratic, stealing from others. **ID**: Long and slim with deeply forked tail and long, hooked bill. Deep wingbeats with wings that look too long relative to slight body. Adult ♂ (5+ years): appears all black, including upperwing. Bright red throat sac (incredible in display), often difficult to see. Adult ♀: dark head with white on breast—note the shape. 1st-yr: white head and breast with brown patches on breast sides. Sexes alike until 3 years old. Older imms have intermediate amounts of dark on head and breast (♂) as new feathers grow in en route to adult-like plumage. Other frigatebirds have been recorded—study size, shape, and extent of white on underparts and presence of rufous. In adult ♂, double check that it is all black.

Northern Gannet *Morus bassanus* **NOGA** L 37in

Nests in massive colonies on rocky cliffs. Common visitor to most coastal areas, near or offshore, scarce only in summer. Spectacular angled plunge-dives to catch fish. A powerful flyer with stiff shallow wingbeats followed by short glides. Typically travels in groups of 3 to 20 that create undulating evenly spaced lines. Mostly seen in flight, it invariably confuses when seen sitting on water. Sick birds occasionally sit on beach. Large flocks appear where food is abundant, creating impressive feeding frenzies. **ID**: Very large with long slim pointed wings, tail, and head—but all evenly proportioned. Size and shape distinctive from all but smaller localized boobies. Large, pointed bill gives a snouted look. Adult (4+ years): strikingly white with black wingtips and gold-toned head. Juv: all dark with pale spots and streaks. Starts to molt out of juv plumage and molt in progressively more white adult-type feathers over next 2–3 years. Boobies are similarly shaped but are much rarer and always noticeably smaller with differences in bill and plumage patterns.

Brown Booby *Sula leucogaster* BRBO L 30in

Regular in small numbers on the Dry Tortugas; rare elsewhere. Sometimes seen from shore but often misidentified as NOGA. Sits on buoys, channel markers, etc. **ID**: As with other boobies, smaller, slimmer, and longer-tailed than NOGA. Stiff wingbeats. All plumages have uniform dark brown upperparts, clean-cut hooded appearance, and paler belly extending onto underwing coverts. Adult: clean white lower breast and underwing. ♂: duller-billed than ♀. Juv (1st-yr): darker than adult but hooded effect and paler underwing still visible. 2nd-yr: intermediate between juv and adult.

Red-footed Booby *Sula sula* RFBO L 28in

Very rare, mostly on the Dry Tortugas. Polymorphic. 2 morphs in the E: white-tailed brown (most) and brown-tailed brown. **ID**: Small, slim, with buoyant flight. Records either juv (1st-yr) or 2nd-yr birds. 1st-yr: palest on belly, creating hooded look. Blue face. Morphs similar. 2nd-yr: pink-billed. Central tail feathers whitish in white-tailed. Adult: blue bill, pink face, red feet. Tail white or brown.

Masked Booby *Sula dactylatra* MABO L 32in

Regular in small numbers on the Dry Tortugas. Rare far offshore in warm waters. **ID**: NOGA look-alike but notably smaller and slighter. Ad/3rd-yr: yellow bill with extensive black at base. Always look for distinct black tail and trailing edge across wing (beware 3rd-yr NOGA). Juv/1st-yr: dark, hooded look with pale underparts and collar. 2nd-yr: intermediate between 1st-yr and adult.

juv.

ad. postbr.

ad. br.

ad. nonbr.

American White Pelican *Pelecanus erythrorhynchos* **AWPE** L 62in

Fairly common but local. Regular out of range, often in small groups. Breeds on large inland water bodies with many coastal in winter. Like a Hercules aircraft, big and cumbersome. Often seen in organized groups flying between feeding areas, soaring in unison or working together to herd fish, which they shovel into pouch-like bill. Never plunge-dives. Sits around in groups on sand bars or islands. Uses thermals for soaring. Beware high-flying WHCR and WOST—they are large and white with black flight feathers, but always have a different shape. **ID**: Black-and-white in all plumages. Ad br (spring): horn-like plate on top of bill. Post-breeding birds develop gray on head. Ad nonbr: white-headed with slightly duller bill. Juv: similar to adult but cleaner looking with pale brown-gray coverts that are hard to see at distance. Difficult to separate from adult by late fall. Occasional escape white-type pelicans have all-dark trailing edge to wing and different bare-part coloration.

Brown Pelican *Pelecanus occidentalis* **BRPE** L 51in

Increasing in range. A familiar bird of s. coastal areas and nearby lakes. Wanders n. in summer, numbers varying from year to year. Frequently found sitting around marinas and on pilings. Tame and a fisherman's favorite (and vice versa when fed). Often soars over beach in groups of 5–25, just high enough to not affect sunbathers, seemingly unfazed by wind. Flies in formation, either in a line or 'V.' Spectacular plunge-dives when fishing but also feeds from sitting position. Large and cumbersome, but always attracts attention. Rests on isolated sand bars, often with larger AWPE. **ID**: Ad br: dark rear neck and variable bill colors. Ad nonbr: white neck and duller bill. 3rd-yr: as adult with a few imm feathers on underparts. 2nd-yr: appears grayish with streaked underparts (dark in adult, white in 1st-yr), gray on crown and bill. Juv/1st-yr: neat brown upperparts and neck, pale below.

White-tailed Tropicbird *Phaethon lepturus* **WTTR** L 15in

Rare summer visitor offshore in warmest waters. A loner, it is the white high-flyer of the ocean. Usually seen circling briefly above fishing boats, before flying off high into the sky. Frequently sits on water. **ID**: Tern-like but fatter-bodied and narrow-winged. The fast, shallow wingbeats are most distinctive—much like ROTE. Adults seen most often. Yellow bill and diagnostic sharp, black diagonal wing bar (also seen in 2nd-yr); white-backed. Striking, long tail streamers. 2nd-yr: as adult with a few retained imm feathers. Juv/1st-yr: coarsely barred upperparts, white primary coverts, short tail.

Red-billed Tropicbird *Phaethon aethereus* **RBTR** L 18in

Found offshore, and a very rare summer visitor. Adults have summered on ne. islands. Gulf Stream birds usually 1st-yrs. **ID**: In all plumages from similar WTTR by larger size, heavier build, and broader-based bill. Also focus on extensive black around eye that goes back to nape. Black on outer primaries extends onto primary coverts—can be surprisingly easy to see. Ad (and similar 2nd-yrs): red bill and finely barred back. Juv/1st-yr: pale yellow bill and short tail.

Brown Noddy *Anous stolidus* **BRNO** L 15.5in

Common summer breeder on the Dry Tortugas; rare elsewhere in warm waters. Very rare onshore after hurricanes. **ID**: May be confused with juv SOTE, but BRNO is uniform brown, including the underwing, and has long graduated tail. Pale-capped and dark in front of eye with white eyelids. Juv: dullest head. Adult: brighter white forecrown. 1st-s: a mixture of imm and adult-type feathers. Dark underwing (see SOTE). Black Noddy (*Anous minutus*), nearly annual on the Dry Tortugas. Like BRNO, but smaller, slimmer, and darker with more extensive white crown and thinner bill. Most are sub-adults.

White-winged Tern *Chlidonias leucopterus* **WWTE** L 9.5in

Very rare where BLTE occur. Actually very similar to BLTE but slightly chunkier with shorter thicker bill. White tail in all plumages. Adult br: noticeably 2-toned: black underwing coverts, whitish tail, and white upperwing coverts contrast with dark body. Red legs. 1st-s/nonbr: paler upperparts than BLTE with whitish rump, white underparts lacking breast patch or dusky flanks. Isolated cheek spot reminiscent of BOGU.

Sooty Tern *Onychoprion fuscatus* **SOTE** L 16in

Breeds on the Dry Tortugas, uncommon summer/fall visitor to warm waters. Hurricane-blown birds come onshore or inland in numbers (also similar BRTE). **ID**: Larger, more powerful than BRTE, easily confused with a small jaeger. Flies high in small groups (2–5) but forms bigger concentrations near food. Adult: appears strikingly 2-toned black-and-white. Flight feathers black on underwing, paler inner webs on BRTE. White forehead eye to eye. Narrow loral stripe. Worn birds browner, like BRTE. Juv: often with adults, brown with white spots, appears all dark at distance. Pale vent and underwing.

Bridled Tern *Onychoprion anaethetus* **BRTE** L 15in

Summer/fall visitor to warm waters offshore. Some to the NE late summer. Loves weed lines. Frequently perches on floating wood/debris; SOTE almost never perches this way. Often found near similar SOTE, but smaller and shape more reminiscent of COTE, with narrower and more angled wings. **ID**: White forehead 'V' extends behind eye (diagnostic). Adult: paler upperparts than SOTE though often difficult to judge and some SOTE similar. 1st-s: commonest age group with narrow pale fringes to back and pale head. Juv : similar to 1st-s with broad pale fringes to all upperpart feathers.

Black Tern *Chlidonias niger* **BLTE** L 9.75in

Locally fairly common in suitable freshwater marshes in summer. Colonial. Migrants found anywhere, often with other tern species. Commonest offshore, often in flocks. Only regular 'marsh' tern. Buoyant and bouncy flight, tilting from side to side. Never dives but gracefully swoops down to pick food off surface. **ID**: Small, compact, with short square tail and broad-based wings. Gray upperparts in all plumages noticeably darker than other regularly occurring terns. Ad br: striking black head and body with contrasting white vent. Some white-faced in summer (these can be 2nd-s or birds beginning to molt). By late summer, molting birds are blotchy black-and-white. Nonbr: dark ear spot and partial cap. Pale underparts with dark patch on breast sides. Juv: similar to nonbr but has pale-fringed upperparts. Color varies from gray to brown. 1st-s: as nonbr with mixture of old and new feathers (old outer primaries). At sea, distant tropical terns (SOTE/BRTE) are much larger with different shape and relatively slow wingbeats.

Forster's Tern *Sterna forsteri* **FOTE** L 13in

The most numerous tern in many areas. Breeds in freshwater marshes, widespread on migration. The only wintering smaller tern in many areas (mostly coastal). Hovers and dives for fish, often in concentrated or mixed flocks. Fond of perching on posts. **ID**: Variable patterns of color make ID tricky. Subtle but noticeable differences in shape. Compared with COTE is slightly larger, bigger-billed, longer-legged, flatter-headed, thicker- and longer-necked with a less rounded undercarriage. Difference in shape shows better in flight, particularly flat undercarriage. Wings broader-based and less angled. A more powerful look overall! Diagnostic dark mask in most plumages. Ad br: compared with COTE, black cap extends farther down nape, bill/legs more orange and upperparts paler gray with white outer primaries (wear dark through the summer). In flight, coverts form striking pale area, even at distance. Long tail streamers. White breast—COTE usually gray in summer. Nonbr: cap becomes diagnostic black eye patch by late summer. Dark bill. 1st-year: as nonbr with dark tertials and outer primaries. Juv: ginger-brown dark 'saddle' and blackish eye patch. Quickly molts to 1st-w.

juv.
ad. br.
juv.
1st-s.
ad. nonbr.
ad. br.

Common Tern *Sterna hirundo* **COTE** L 12in

The common tern in the NE and offshore. Widespread on migration. Colonial. Nests on sandy or rocky beaches or islands. Hovers and dives for fish, often in concentrated flocks. Variable. **ID**: Compact and well proportioned. Rounded undercarriage (flat in FOTE). Narrow-based wings strongly angled back—very different profile from FOTE. Very light, buoyant flight. Darker upperparts than FOTE. Ad br: blood-red bill with black tip. White cheek accentuated by neat black cap and gray breast/belly (wears paler through summer). Dark wedge to outer primaries gets darker through summer. Nonbr: as adult with white on forecrown, large carpal bar (can appear small), dark bill, duller legs and darker secondary bar. 1st-s: similar and stands out among breeding and fresh juvs. Older worn median coverts create a white panel. 2nd-s: typically as breeding but with white flecking in forecrown, dull bill, and hint of carpal bar. Beware!: sick and failed breeders may look similar. Juv: upperparts dark with subterminal marks, and initially warm brown. Brown fades or feathers molted so fall birds are much grayer. Bill darkish with orangey bases becoming all dark through fall. Different head pattern from FOTE.

Arctic Tern *Sterna paradisaea* **ARTE** L 12in

Only tern in the far N. Coastal islands to ME. Migrates offshore. Rare migrant elsewhere. Migrates late in spring. **ID**: Very similar to COTE. ID based largely on size and shape—ARTE slightly smaller, 'neckless,' with more rounded head and very short legs. This gives front-loaded feel in flight. Ad br: from COTE by shorter all-red bill, paler flight feathers, longer tail projection, and narrower black trailing edge to upper- and underwing in flight. Primaries translucent. Nonbr/1st-s: best differentiated by shape, paler appearance. Juv: white secondaries striking in flight (compare to dark bar in COTE).

Roseate Tern *Sterna dougallii* **ROST** L 12in

Scarce, except on breeding grounds. Nests on rocky islands and sandy beaches. Plunge-dives at varying angles. **ID**: The palest tern in the flock with fast, shallow but stiff, tropicbird-like wingbeats that can be seen at distance. Long tail, exaggerated by short slender wings. Narrow dark wedge to outer-wing. Broad white inner web to primaries. Strikingly fine black bill, sometimes red at base. Ad br: pale upperparts, white underparts often with pink flush. Legs orange-red. Nonbr: speckled crown, white forehead. Juv: Unique bold black subterminal markings to upperparts (lacking brown of COTE). Darker-headed than other juv terns. Call: a distinctive *chivick*.

Least Tern *Sternula antillarum* **LETE** L 9in

Locally fairly common in summer. Breeds on sandy beaches just above tide line in competition for space with beachgoers. Interior birds nest on sandbars. Often the species protected in the 'roped-off' area. Has adapted to dredge spoils and other sandy sites near food (small fish) through necessity. Noisy, a distinctive 'squeaky-toy' voice. **ID**: Strikingly small compared with other terns. Big-headed and thick-necked with narrow-based long wings strongly angled back; very fast deep wingbeats ("at 100mph"). All these features create distinctive appearance even at long distance. Everything is done quickly, even repeated dives for fish. Yellow legs. Ad br: obvious triangular white forehead and long yellow bill. Distinct black wedge created by black outer 3 primaries. Nonbr (Aug–Mar) and 1st-s (retain old outer primaries): different head pattern with streaks extending onto crown. Darker bill, and dark carpal bar on upperwing. Juv: pale-fringed dark coverts contrast with pale inner primaries and secondaries—like a mini SAGU. Size, shape, call, and behavior always distinctive.

Gull-billed Tern *Gelochelidon nilotica* **GBTE** L 14in

Scarce but widespread over fields, coastal marshes, and flats. Sometimes on beaches with other terns. Usually seen alone or in pairs, occasionally small groups. The only tern in some habitats. Gracefully meanders with bouncy, buoyant flight, and dips to pick up insects, crabs, and other prey. Wingbeats are deep. Doesn't hover or dive. Its *kiwik* call is often a giveaway—it sounds more like LAGU than a tern. **ID**: Medium-sized but sturdy and can appear larger. Obvious short stout black bill (hence name). Long black legs. Adult: strikingly uniform pale gray upperparts, including rump and tail. Black trailing edge to outer primaries (wear more extensively dark through summer). FOTE shows contrast in upperwing—noticeable at long range. Black cap in breeding, becomes mottled in fall and appears as postocular spot in winter. 1st-s: similar to molting/nonbr with some retained juv wing and tail feathers. Juv: pale-fringed and browner but usually best picked out by lack of cap and stubby bill.

Sandwich Tern *Thalasseus sandvicensis* **SATE** L 15in

Fairly common in more s. coastal areas. Often with larger ROYT. Noisy, sounding like a weak, high-pitched ROYT: *ka-reek*. **ID**: A very pale 2-toned tern, only ROST is paler. Distinctive by combination of long and narrow yellow-tipped black bill, shaggy crest, fairly large size, and shape. Very slim, particularly in flight, with long angled wings. White rump, black legs, and dark primary wedge (5 pps). Ad br: crested black cap. Often has pink wash to white breast. Nonbr: reduced cap by mid-summer. Black eye bold on white face. 1st-yr: as nonbr with dark-centered tertials. Juv: black subterminal markings to upperparts (like ROTE). Bill usually with yellow or orange, quickly becomes black. Yellow tip by spring.

Elegant Tern *Thalasseus elegans* **ELTE** L 17in

Very rare W Coast species. **ID**: Smaller and slimmer-winged than ROTE. Bill long, narrow, and slightly downcurved. Lollipop-orange but often darker (reddish in summer) toward base, and paler (yellowish in winter) toward tip. Crest longer than ROYT. Ad br: solid black cap. Breast often has strong pink flush. Nonbr: usually solid black around eye; similar species have white postocular. More extensive black on rear crown than ROYT. Juv: bill is shorter, straighter, and yellower than older birds. Lesser Crested Tern (*Thalasseus bengalensis*) is possible.

Royal Tern *Thalasseus maximus* **ROYT** L 20in

Common in coastal areas on ocean beaches, tidal flats, and roost areas close to fishing grounds. Colonial nester on sand bars. Noisy: loud piercing *keerak* can be heard at great distances. Tends to be the highest flyer in feeding flocks. **ID**: Large size, shape, and big orange bill usually make ID straightforward. Apart from CATE, always the largest in the flock. Big and sturdy, yet remains graceful and attenuated. Lollipop-orange bill always noticeable. Pale gray upperparts with narrow black trailing edge to outer primaries in flight. Compared with CATE, lacks the bulk, is slimmer with smaller orange bill lacking a black tip, and has paler outer primaries on underwing. White forehead is diagnostic for most of the year. Adult: solid black cap and bill often deeper red-orange in spring. It can be very CATE-like—underwing pattern remains bold! Nonbr: bald-headed, white with black line/crest behind isolated big black eye. Juv: variable. Legs initially orange turning to black with yellow bill turning orange. Bill sometimes noticeably shorter than in adults Strong contrast in upperparts. 1st-s: similar to ad nonbr, usually with some retained juv flight feathers.

juv.

1st-w.

ad. nonbr.

ad. br.

Caspian Tern *Hydroprogne caspia* **CATE** L 21in

Widespread in small numbers. The only large inland tern. Breeds on large lakes and in marshes (usually on islands). Often migrates over land. Winters in s. coastal areas. Less sociable than most terns and usually hunts alone. Patrols over water, often covering large areas before doubling back to cover same ground. A real bruiser—its mean appearance matches its screaming call: *aargh*. Juv has wheezy whistled *sweeyer*. Largest tern in the world with a thick, killer-looking black-tipped blood-red bill. Heavy-set, short-tailed, thick-necked, with broad head. Wings broad and less angular (more gull-like) than in other terns. Darker outer primaries create large wedge on underwing, obvious at distance. Capped appearance year-round. Ad br: solid black cap. Nonbr: unlike other terns, cap remains mostly dark. Nonbr/1st-s: as breeding with more crown flecking, but overall impression is of a dark cap. Juv: striking black subterminal 'V's on upperparts become less distinct through fall. Legs yellow turning to black. Dark cap extends onto lores and cheeks. 1st-s: as ad nonbr but heavily worn.

juv.

1st-s.

ad. br. with chick

Black Skimmer *Rynchops niger* **BLSK** L 18in

Locally common on open beaches and back bays. Very rare n. of normal range and inland, sometimes after hurricanes. Colonial and generally roosts in large, dense packs. Weird and unmistakable, it always draws comment. Lower mandible much longer than upper. Feeds by keeping lower mandible under water for several yards as it sieves small fish. Also picks from surface. Often in tight flocks that make sharp turns in unison with dramatic effect. Bouncy flight with high exaggerated raising of wings. Sometimes rests with head flat on ground as if sulking. Most active at night and twilight. Yelping *yip*, usually repeated, is a common sound night and day. Often mixes with standing terns. ♂s bigger with larger bills, sometimes noticeable in the field. **ID**: Adult: black upperparts fading to brown. Nonbr: back of neck white. Juv: pale-fringed brown upperparts give scaly appearance. Bill small and dark initially, grows and becomes 2-toned. Orange legs. 1st-w: black adult-like mantle and scapular feathers grown in through winter. 1st s: old juv coverts are replaced through summer so appears as adult by fall.

South Polar Skua *Stercorarius maccormicki* **SPSK** L 21in

Scarce summer pelagic species (GRSK winter). Very rare close to shore. The shout of "Skua" causes a commotion on pelagic trips. Often seen sitting on water. Tends to fly purposefully in a straight line, often above skyline. Usually alone, is attracted to concentrations of birds. **ID**: Large and bulky with striking wing flashes obvious at great distance. Colors vary. Uniformly colored upperparts lack streaking, slightly paler underparts with contrastingly dark underwing coverts. Often paler nape. Can be confused with dark POJA, but SPSK is larger, heavier-bodied, thicker-billed, and has short broad tail.

Great Skua *Stercorarius skua* **GRSK** L 23in

Winter visitor from Europe. Rarely comes close to boats. Most identified by season/probability. **ID**: Very similar to SPSK but larger, slimmer-bodied, longer-tailed, bigger-billed, and slimmer-necked. Longer wings and rangier, a more powerful looking bird. Usually warmer cinnamon-brown than SPSK but difficult to see at distance. Some juvs duller and similar to SPSK. Underwing coverts have warmer brown feathers. Back has paler flecking and most lack uniform cold brown-gray tones of SPSK. Breeds in the N Hemisphere (SPSK in the S Hemisphere) so has different molt schedule.

Parasitic Jaeger *Stercorarius parasiticus* **PAJA** L 16.5in

The common coastal jaeger, rarest far offshore. Rare inland. Harasses gulls and terns, catching dropped or regurgitated food. **ID**: Between LTJA and POJA in size. Larger than LAGU, smaller than HERG. Heaviest in the belly. Wings often seem pointed and swept back. Fast direct flight with snappy wingbeats. Pointed central tail feathers in all plumages diagnostic, but also found in adult breeding LTJE. Long slim bill. Ad light br: dark cap; lacks really 'mean' look of POJA. Pale area at base of bill. Paler upperparts than POJA, browner than LTJA. Pointed tail, shorter than LTJA. Dark birds brown, often looking black at distance, palest around neck. Adult nonbr: similar to imm but underwing solid brown. Usually has flank and vent barring with some pale fringes to upperparts, particularly mantle and rump. 2nd-yr: as adult nonbr with some retained juv feathers, notably on underwing coverts. Juv/1st-yr: incredibly variable. Colors range from tan to ginger to dark brown. All have paler bill base and narrow pale fringes to feathers, including primary tips. Nape streaked, but difficult to see. Diagnostic pointed central tail feather. Jaeger ID is very difficult, largely because of variations in plumage and size.

Pomarine Jaeger *Stercorarius pomarinus* **POJA** L 18.5in

Fairly common offshore, rare elsewhere. The winter jaeger, 'Poms' often migrate in packs. **ID**: Largest, most powerful jaeger, a bruiser. Long-winged and -tailed with flat undercarriage and bulging sternum. Powerful flight. Long thick bill. Large white wing flashes. 2nd pale flash formed by pale bases to greater coverts on underwing. Colors vary from light (90%) to dark (10%) Adult: long central tail 'spoons.' Dark malar region gives 'mean' look. Broad breastband. Nonbr: central tail feathers shorter and blunt-tipped. Barred rump and underparts. Juv/1st-yr: barred underwing. POJA has same molt strategy as PAJA so ageing birds is the same. Only fall jaeger in wing molt—others on wintering grounds to the S.

Long-tailed Jaeger *Stercorarius longicaudus* **LTJA** L 15in

Migrates far offshore, rare inland. Rarely chases other birds. Characteristically, suddenly stops with wings raised and drops to water. Eats insects, unlike other jaegers. **ID**: Smallest jaeger. Some small and buoyant with tern-like flight, others larger, heavier, and easily confused. Long tail, narrow-based wings with flattish undercarriage, deepest at sternum (like POJA). Pigeon-like head shape with small bill. Typically only outer 2 primary shafts white. Ad br: white-breasted, neat cap, and paler gray-brown back. Long central tail feathers. Juv: variable. Colder tones than other jaegers. Blunt central tail feathers.

1st-w.

juv.

1st-s./2nd-w.

ad. nonbr.

ad. br.

Laughing Gull *Leucophaeus atricilla* LAGU L 16.5in

Common on coast, regularly wanders inland. Found on beaches, in public areas, and anywhere people are eating! This is the bird that is always screaming (hence the name), checking you out as you work on your suntan, and always trying to pinch your food. Very approachable. Colonial, nesting in back-bay marshes. **ID**: A 3-yr, dark-mantled, fairly small gull. Rangy with long neck, wings, and legs. In flight, slim with lazy buoyant flight. Bill long and drooping. Ad br: dark-mantled. Dark brown hood (often appears black). White eyelids. Bill and legs usually bright red but can be dark. Extensive dark in outer primaries merges with gray (very little white). Ad nonbr: duller bare parts. Smudges on head strongest on ear coverts and rear edge of hood. 2nd-w: similar with more black in wing and some black in tail. Juv: appears warm brown with broad buff fringes to upperparts. Broad black tail band. Gray rear neck and breast sides. Gray mantle and scapular molted in early winter. Some juv feathers remain by spring. Neck and breast whiter. Most develop partial hood.

Franklin's Gull *Leucophaeus pipixcan* **FRGU** L 14.5in

Abundant in the Great Plains and Canadian prairies, often in dense 'packs' on farmland and lakes. Commonly found with similar LAGU in TX, less regularly elsewhere. Covers large areas looking for food, often soaring fairly high. Can be found following the plough. Colonial, nesting in marshes. **ID**: 3-yr gull with dark mantle like similar LAGU. Smaller, more compact with shorter bill. Short, thick neck. Faster and shallower wingbeats. Larger white eyelids. Smaller hood and whiter neck in summer. Bolder partial hood in nonbr/imm. Less black in wingtips, usually bordered by white—particularly striking overhead. Ad br (Apr–Sep): often has pink flush on breast. Large white wingtips, but highly variable with white tips quickly worn away in summer. 2nd-s: subtly different from adult, averages less white in wing, has darker bill and duller hood: the most similar plumage to LAGU. Nonbr: boldly patterned partial hood. Juv (Jul–Aug): plumage held briefly. 1st-w (Sep–Apr): similar to LAGU but white neck and underparts with narrower dark tail band and bolder head pattern. 1st-s: variable head pattern, gray upperparts, much as ad nonbr. It has 2 complete molts per year.

Bonaparte's Gull *Chroicocephalus philadelphia* **BOGU** L 13.5in

Common locally where plentiful food supply. Often near fast-moving water on rivers, ocean, and lakes but can be found anywhere. Birds follow circuits in flocks while feeding before sitting down in groups. Picks food off surface. Has favored sites, e.g., Niagara river, where may gather in thousands when river is not frozen over. Breeds in boreal marshes, usually nests in conifer trees. Tends to not mix with other species. Tight flocks often fly in unison. Prefers sitting on water to standing on land. **ID**: 2-yr gull. Small and dainty, almost tern-like, with light, buoyant flight, and very clean appearance. Pointed wings. Striking wing pattern with translucent outer primaries, neat black trailing edge obvious even at long distances. Pale pink legs. Thin, pointed black bill. Gray extends from hind neck to breast sides. Ad br: slaty blackish hood. Legs bright pink to red. Ad nonbr: dark spot behind ear, pale pink legs. Juv: often orange base to bill; soon darkens. Mantle and head have warm brown tones. 1st-w: mantle becomes pale gray and head whiter. Dark carpal bar, tail band, trailing edge to wing and primary coverts distinguish from adult. Some birds in 2nd-w still show signs of immaturity and so can be aged.

Little Gull *Hydrocoloeus minutus* **LIGU** L 11in

Rare breeder in the Great Lakes and James Bay. Scarce in winter. Usually found with BOGU. Habits similar. **ID**: 3-yr gull. Buoyant and erratic flight. Watch for size, compact shape, head and wingtip pattern. In flight, wing pattern should be obvious. Ad br: dark hood. Pale upperwing lacks black (can make wing look rounded). Stunning dark underwing. Nonbr: smudgy crown and dark ear spot. 2nd w: as adult but paler underwing and some black in wingtips. Juv: extensive dark markings on upperparts with 'W' across wings. Black tail band. 1st-w: gray mantle.

Black-headed Gull *Chroicocephalus ridibundus* **BHGU** L 16in

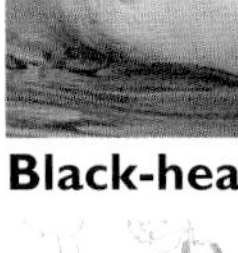

Rare, except in Canadian coastal areas. Often found with BOGU. Habits similar. **ID**: 2-yr gull. Larger European counterpart of BOGU. Similar but slightly larger, paler back, white neck, darker legs, broader head, and fatter body. Bicolored bill can be a giveaway, but some very dull and dark and not obvious. Like LIGU, can disappear in BOGU flocks. In flight, focus on extensive dark wedge on underwing. Ad br: brown hood, smaller than BOGU. Pale eyelids. Bill often bright red. Ad nonbr: less isolated dark ear spot. 1st-w: orange-and-black bill, orange legs, more dark on primaries, brown carpal bar.

ad. br.

juv.

2nd-w./ad. nonbr.

Sabine's Gull *Xema sabini* SAGU L 13.5in

Scarce high-Arctic breeder. Typically migrates far offshore, but some close to shore or on inland lakes, particularly in Sep. Few winter records. A beautiful gull with tern-like qualities and striking upperwing made of triangle patterns. Forked tail. Often mistaken for larger and broader-winged juv BLKI, which has narrower 'M' pattern across back. Ad br: black-bordered gray hood kept into fall. Pale-tipped black bill. Ad nonbr: dark nape on pale head. 2nd-s: as nonbr with heavily worn flight feathers. Juv: dark neck and browner overall. Color can be difficult to judge at distance. Black tail band.

Ross's Gull *Rhodostethia rosea* ROGU L 13.5in

High-Arctic breeder, previously at Churchill, MB. Mystical mega-rarity in the US. **ID**: Small but heavy-chested. Thick-necked, pigeon-headed with long wedge-shaped tail (appears pointed) and stubby bill. Ad nonbr: pale face contrasts with gray neck. Gray underwing. Ad br: black collar on white neck. Underparts often pink. Juv: similar 'M' pattern to juv LIGU.

Ivory Gull *Pagophila eburnea* IVGU L 25in

Uncommon and declining breeder in far n. pack ice. Winters s. to NL. Several recent records to Mid-Atlantic and as far as s. Interior. **ID**: Only all-white gull. Adult: "as white as snow" with unique bill color. Juv: equally beautiful with smudgy black spots on upperparts and face. Scavenger, feeds on dead carcasses. Often remarkably tame. Always a crowd-pleaser!

1st-w.

ad. br.

1st-w.

Black-legged Kittiwake *Rissa tridactyla* **BLKI** L 17in

Forms large, spectacular colonies on rocky cliff faces. Pelagic in winter, sometimes within sight of land, particularly in bad weather or where there is plentiful food. Plunge-dives into water for fish. Named after call. **ID**: An attractive, dainty, clean, and well-proportioned gull. Medium-gray upperparts. Short black legs, yellow bill, and dark eye. Buoyant but powerful flyer with long, slim body and wings. Wings usually held straight and stiff. Sometimes elbow is pushed forward creating 'W' along leading edge of wing, making wings look slimmer. Stiff shallow wingbeats. Easy to overlook as RBGU, but slimmer, slightly darker, with distinctive squared-off black wingtip and different flight style. Ad br: white head. Nonbr: gray spot to rear of ear coverts. 2nd-year: as adult but some have more extensive black in wing. Juv/1st-yr: bold 'W' pattern on upperparts. Some have dark ear spot and collar, others all white as adult. Black tail band. Black bill becomes yellow in spring/summer.

1st-w.

2nd-w.

ad. nonbr.

2nd-w.

juv.

1st-w.

ad. br.

ad. nonbr.

Ring-billed Gull *Larus delawarensis* **RBGU** L 17.5in

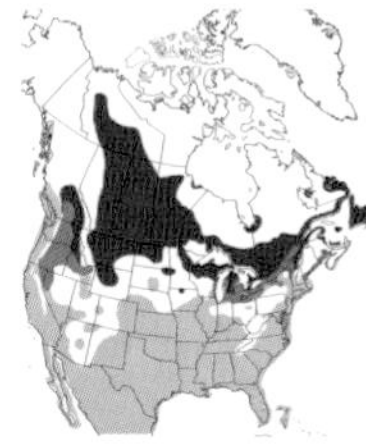

Very common. Breeds colonially in marshes. In many areas, particularly inland, commonest gull and is good starting point for becoming comfortable with gull ID in general. The infamous parking-lot 'beggar,' will eat just about anything so occurs anywhere: farmland (follows the plough), lakes, beaches, dumps. **ID**: 3-yr gull. Medium-sized with small head, lacks bulk and aggressive appearance of larger gulls. Slim and attenuated, has well-defined markings and a 'cleaner' appearance than larger but similar HERG. Bill always smaller than larger gulls with well-defined colors. In flight, long and slim-winged with extensive, well-defined black wingtips (older birds) with 2 small mirrors (ad). Ad br: pale gray upperparts. Yellow legs. Bright yellow bill with distinct black ring. Red orbital ring and gape. Pale iris. Nonbr: duller bare parts, well-defined blotches on crown and nape. 2nd-yr: as adult with more black in wing, typically 1 mirror, greenish legs. Partial tail band. Compare 3rd-yr HERG. Juv: all-dark bill, scaly brown plumage. 1st-w: pale gray mantle. Pink legs. Pink bill with black tip. Black tail band. Compare 2nd-yr HERG. 1st-s: grayer upperparts, changing bare parts suggest older birds. Molting birds can appear pale-winged.

California Gull *Larus californicus* **CAGU** L 21in

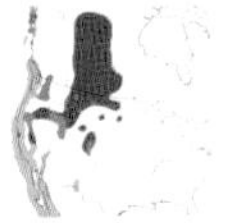

Uncommon breeder. Rare but regular to the E. Easily overlooked. **ID**: 4-yr gull. Intermediate between RBGU and HERG in size and shape. Focus on mantle, bill, and leg color. Long parallel-sided bill, thinner than HERG. Long-winged. Ad/3rd-yr: slightly darker mantle than HERG and yellow/green/gray legs. Bill with red and black spot (winter). Dark iris. Extensive black on wingtips. 1st/2nd-yr: uniformly checkered warm brown (like some HERG). Focus on size, shape, and upperwing with uniform dark greater coverts and inner primaries. 2-toned bill pattern consistent. Gets darker mantle in 2nd year.

Mew Gull *Larus canus* **MEGU** L 16in

Very rare visitor. 2 distinct ssp: Mew (*brachyrhynchus*) from the W Coast and Common (*canus*) from Europe. **ID**: 3-yr gull. Both smaller than similar RBGU with shorter and thinner (more pointed) bill. Mantle darker gray. Larger mirrors. Compared with Mew, adult Common has crisper head markings, typically greener bare parts, bill with poorly defined ring (usually absent in MEGU). Wingtips more extensively black with less white. 1st-year: Common also has crisper head markings, more defined tail pattern, white rump, and more contrastingly marked plumage.

Herring Gull *Larus argentatus* **HERG** L 25in

The common, large pink-legged 'default' gull. Scarcer to the W. Adapted to man with largest concentrations (1000s) at landfills in winter, roosting on reservoirs. 'American' Herring Gull probably full species with differing populations within. Hybridizes occasionally. **ID**: 4-yr gull. Remarkably variable. Sometimes appears confusing: focus on size, shape, bill proportions, mantle, and leg color in ad/3rd-yr. Younger imm: uniformly dull pattern of color, size, and bill. Focus on wing and tail patterns to differentiate from other species. Ad br: pale gray mantle. Pink legs (variable). Pale iris (usually). Deep yellow bill with red and often black spots. Ad nonbr: streaked/mottled head and neck. Bill paler. 3rd-yr: as adult with some dark in coverts, tail, and bill. Juv: variably brown (some grayer) with pale-fringed upperparts. N. populations larger, grayer, and molt later. Dark bill and iris. 1st-w: scapulars and mantle have complex dark centers and extensive light gray edges. Becomes increasingly bedraggled into summer, some bleach very pale, some wear confusingly dark. 2nd-w: newer feathers with complex patterns, variegated look, superficially grungy. Gray mantle and other adult-like feathers molted through year.

Black-tailed Gull *Larus crassirostris* **BTGU** L 19in

Very rare Asian visitor. Surprisingly, almost annual (mostly adults). **ID**: 4-yr gull. Unusually, all ages have broad subterminal black tail band bordered by white. Medium-sized, long-winged, and long, bicolored bill with unique shape. Slate-gray mantle, similar to LBBG. Yellow/green legs. Features in combination make ID straightforward. Older imm: shares many features of adult.

Yellow-legged Gull *Larus michahellis* **YLGU** L 25in

Very rare European visitor. **ID**: 4-yr gull. Compared with HERG, older birds show slightly darker mantle, yellow legs, extensive black square-ended wingtips, orange-red orbital ring, larger bill, pale head (only a few streaks around eye in winter). 1st-yr: very similar to large-billed LBBG. Hybrid HERG x LBBG regular and very similar. Usually slighter, thinner bill, smaller streaked head, pinker legs.

Slaty-backed Gull *Larus schistisagus* **SBGU** L 25in

Very rare. Adults increasingly reported. **ID**: 4-yr gull. Variable slate-gray upperparts, similar or darker than LBBG. Compared with LBBG is bulkier, has long parallel-sided pale yellow bill, deep pink legs, and broad white trailing edge to wings. The clincher is the 'string of pearls:' white separating the gray from the black wingtips of middle primaries. 1st/2nd yrs are pale like THGU.

Kelp Gull *Larus dominicanus* **KEGU** L 24in

Very rare. **ID**: 4-yr gull. Has bred in LA, where hybridized with HERG. Only older birds safely identified by combination of almost black back and yellow-green legs. Look for wingtip pattern (usually 1 mirror on outer primary) and bill size/shape. Younger imm: similar to a big LBBG with a large bill.

Great Black-backed Gull *Larus marinus* GBBG L 30in

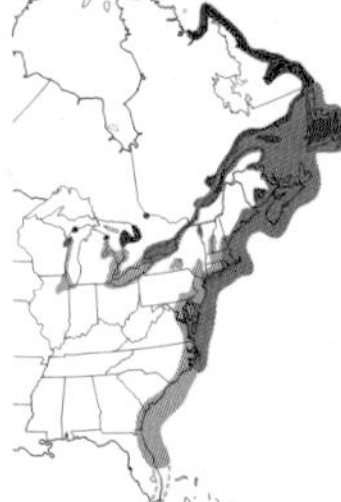

Very common on n. Atlantic coast, less numerous s.; often seen lounging on beaches. Scarcer inland. Survives frozen N by adopting versatile diet (using huge bill). Scavenges but also kills live prey, including birds such as AMCO. **ID**: 4-yr gull. A real tank. Our largest gull, and bulky with a killer-like thick bill. Darkest-backed common gull with short wings. Dark colors add to mean appearance, but overall impression is clean and bold. Younger birds' contrasting plumage is neat and fairly distinctive. If in doubt, look at that bill. Ad br: dark gray and white. Bright yellow bill with red spot. Pink legs. Ad nonbr: a few dusky marks on head but still essentially white-headed. Some have black bill spot next to red. 3rd-yr: variable number of imm feathers in wing, variable black on bill and, generally, some black in tail. Juv to 2nd-s: superficially similar with boldly checkered upperparts and well-defined markings on underparts. Pale-headed. Strongly patterned tail. Juv: mantle and scapulars show fairly uniform dark centers and broad pale fringes. New feathers, molted through first winter, have more complicated internal markings. Bill typically mostly dark until 2nd-s, when also starts to get dark adult-like mantle.

Lesser Black-backed Gull *Larus fuscus* **LBBG** L 21in

Scarce but increasing, particularly on beaches, garbage dumps and in grass fields. **ID**: 4-yr gull. Slighter, shorter-legged and longer-winged than HERG. Usually slim-necked, small-headed, and thin-billed. Ad/3rd-yr. distinctive slate-gray upperparts (between HERG and GBBG) and yellow legs. Red orbital ring. Yellow iris. Extensive black on wingtips, usually 1 mirror. Head heavily streaked in nonbr plumage. 2nd/3rd years have largely black bill with yellow tip (scarce in other gulls) or black in center of bill. 1st-/2nd-yr. like a miniature GBBG. Dark upperparts with dark area around eye. Underparts have distinct blotches, HERG has different pattern. Often easy to spot in flight—uniform upperwing pattern, dark bases to all greater coverts, dark underwing, contrasting with white 'rump' and dark tail band. Bill black with some pale after a year. Legs pink until 3rd year. Most birds are of the n. European race *graellsii*, a few are darker-mantled like the Baltic race *intermedius*.

Heermann's Gull *Larus heermanni* **HEEG** L 19in

Very rare, distinctive 4-yr Pacific gull. Medium-sized with variably dark appearance, except for breeding adult, which has striking white head.

Glaucous Gull *Larus hyperboreus* **GLGU** L 27in

Hardy n. species, fairly common in Canada, scarce winter visitor s. of border (mostly 1st-yrs). **ID**: 4-yr gull. The large 'white-winged' gull. Compared with ICGU, the other gull with white wingtips, is larger with longer parallel-sided bill and flatter head giving fierce impression. Front heavy, with shorter wing point and primary projection. Adult: white primaries contrast with pale gray upperparts. Striking in flight. Lightly streaked head in winter. Orange/pink orbital ring. 3rd-yr: as adult, some imm feathers and dark on bill. Juv/1st-yr: striking 2-toned pink bill with well-defined black tip. Finely patterned creamy buff fading to white through winter. Eye appears small on large head. Dark iris. 2nd-yr: like 1st-yr but has uniform gray mantle, pale iris, and bicolored bill. Hybridizes with HERG ('Nelsons Gull'). The latter has intermediate features.

Glaucous-winged Gull *Larus glaucescens* **GWGU** L 26in

Very rare Pacific gull. Hybridizes frequently. A number of potential ID pitfalls. Focus on size and large bill in combination with adult wingtip pattern. Finely marked head often has hooded look (nonbr). 1st/2nd-yr: uniform tea-brown (some grayer) appearance with black bill.

Iceland Gull *Larus glaucoides* **ICGU** L 22in

Uncommon in winter in the NE. Contentious taxonomy. Race *kumlieni* perhaps originates from nominate race *glaucoides* (has white wingtips) and THGU mix. This explains incredible variation in wingtip pattern. **ID**: 4-yr gull. Averages smaller than HERG. Short slim bill, potbelly, short-necked and dome-headed. Long primary projection and wing point. 'Gentle' or dove-like appearance. Short legs. Ad: wingtip pattern from all-white to dark gray (not black) similar to THGU. Iris pale yellow to brown. Deep red orbital ring. Pale underwing stands out. 1st-yr: dark bill. Very similar to THGU.

Thayer's Gull *Larus thayeri* **THGU** L 23in

Scarce in winter to the S. Rare in the E. **ID**: 4-year gull. Size and shape as largest ICGU. Id from the latter with great caution: Adult: hooded appearance (nonbr), most (80%) have dark iris, deep pink orbital ring, extensive blackish wingtips (with white 'tongues'), slightly darker back. From HERG by size and shape, bill size, slightly darker mantle, dark iris, wingtip pattern (whitish underwing usually most striking). 1st-yr: like ICGU but averages darker brown, dark-centered tertials and primaries (darkest part of bird), solid tail band. Dark ICGU and small pale-winged HERG have overlapping features.

WALKING WATERBIRDS

An incredibly diverse and interesting group of birds, shorebirds, or waders as they are known outside North America, are popular for 2 reasons: they are fairly common in suitable habitat, usually wet areas but also certain types of fields and beaches; and a number of different species are often seen together, giving repeated opportunities for comparison.

The majority of species breed on inaccessible tundra, so they are seen most frequently during migration or on their wintering grounds much farther south. Turnover at these spots can be high, so repeated visits will invariably yield new birds. Looking out onto a marsh at hundreds of birds can initially seem confusing and even overwhelming. However, with decent views (a telescope always helps), and prolonged careful observation, people are often surprised by how much can be learned, and what may first appear as a lot of 'little brown jobs' can end up as 14 different species in an intriguing mix of plumages. Size and shape, as with all birds, are crucial in getting to grips with identification.

It can be difficult to get close to birds on a mudflat or marsh area, and judging size is not always easy. Remember, a bird's size is hard to measure precisely in the field, and the real value of size is that it allows comparison between individuals. Careful observation of relative differences in size is very important. After size, look at shape. Important factors are bill length and shape, length of the wing relative to the tip of the tail, and leg length. Behavior and feeding styles are also great clues in successful ID.

In some cases, birds have longer legs for wading in deeper water, and thus they need longer bills with which to feed. On the other hand, birds such as plovers, are found on mud, sand, or short grass. They can run to their prey and pick it up in a deliberate fashion with their short bills (run-and-pick style). Color is useful, but many birds tend to be shades of gray and brown, creating real ID challenges. More pertinent are the patterns of color—the areas of light and dark and the contrast between them. Learning to think in black and white can be helpful here.

Probability can also play a major role. Often we identify a distant flock to species largely because of the abundance of birds. For example, a thousand small 'peep' crawling over a marsh in Massachussetts will almost certainly comprise mainly Semipalmated Sandpipers. The slightly smaller but darker, shorter, and more rounded ones will usually be Least Sandpipers; and the ones that are 20% bigger with longer wings are probably White-rumped or Pectoral Sandpipers. Familiarity with the area and a full appreciation of probability of occurrence is a huge help!

In many cases, it is important to age or know the specific plumage of the bird before we can identify it. As a broad generalization, juvs look very tidy, often with lots of buff coloring and neatly patterned upperparts (many described as 'scaly'), and breasts with a wash or neatly patterned streaking. Adults often show a combination of breeding and nonbr feathers, lacking the neat pattern of juvs. Adult shorebirds typically molt twice a year: once in spring to create breeding plumage and once in fall to create nonbr plumage. Most juvs also molt into nonbr plumage during fall and superficially look like adults but can usually be aged by retained juv wing coverts, which they keep through the winter. Some species, such as Semipalmated Sandpiper and Long-billed Dowitcher, molt into an adult-like plumage in spring. Others, such as Western Sandpiper and Short-billed Dowitcher, typically appear as nonbr birds with a number of breeding-type feathers mixed in. However, it's variable, with some in complete nonbr and others in full breeding plumage. Many immatures stay on the wintering grounds, some move north but very few reach the breeding grounds. At the end of the summer they molt into nonbr plumage.

A number of shorebirds travel amazing distances from the high Arctic to southern South America, often flying for days without stopping. Recent research has shown that they essentially shut down organs to conserve energy. The importance of birds being able to feed effectively on or before migration cannot be underestimated. For example, in spring on the Delaware Bayshore, shorebirds need to double their body weight within a week to 10 days. Birds that travel long distances all have one feature in common: very long wings. This can be seen on sitting birds and is an important feature in ID for certain species, for example, White-rumped, Pectoral, and Baird's Sandpipers. These wing feathers are never molted on migration (a hole in any wing is never a good thing!). Birds that winter locally, or do not have to travel so far, generally have shorter wings, giving them more agility, and they often molt while they are on migration. This is useful to know. For example, Least and Western Sandpipers will often have 'gaps' in their wings on migration since they don't have far to travel, whereas Semipalmated Sandpiper always has a complete set of flight feathers to allow it to make the journey to the Caribbean and/or northern South America.

In this book I do not show or mention where a bird's wintering grounds are if they are outside North America. It should be assumed that these birds winter to the south of the mapped region in the Caribbean, South, or Central America.

Shorebirds are often seen in flight, frequently in mixed flocks. A raptor flying over often flushes everything. This can be frustrating but also provides you with a different set of ID playing cards. This is when shorebirds are most vocal, and many have distinctive calls. Tail patterns and wing bars are good features to look for. Sometimes large flocks will divide and sort by species. The one out on its own is often the 'goodie' that was missed on the ground.

The term 'peep' is frequently used. There is some ambiguity as to which species are included in this term. For the purposes of this book, the word covers Western, Semipalmated, and Least Sandpipers—and the stints.

Herons and egrets are a common sight, particularly in coastal marshes. They are usually conspicuous due to their large size and tendency to feed in accessible, open areas. Their beauty always draws admirers. In the past, demand for their ornate plumes resulted in large declines in populations. 'Heron' and 'egret' are loose terms, with some overlap. Egrets tend to be white and slimmer. Herons, egrets, ibis, and spoonbills nest in mixed colonies, which are often large, noisy, and smelly. Nests may be on the ground, in reedbeds, or bushes, but most commonly in trees. These sites are generally used for a number of years and are then abandoned, often for no apparent reason. Birds spear fish and small insects, but will catch other prey such as frogs and crabs. Prey is usually eaten whole, and it is not uncommon to see birds with swollen 'cobra-like' necks as they struggle to swallow their catch.

Molt timing in herons tends to be variable, as do the resulting plumages. Ageing these birds is often difficult. For a few weeks, at the peak of breeding season, when birds are courting ('high breeding'), bare parts, particularly legs and lores, change to brighter colors, usually reds and pinks. Herons tend to prefer warmer climates; hence there is a more southerly bias to their distribution. These birds breed earlier than more northerly ones and molt accordingly.

Rails live in marshes. They are primarily nocturnal but are often seen in the daytime. Adverse weather, particularly flooding but also freezes, will make them easier to find. They are often vocal but usually stay frustratingly well hidden, occasionally coming to edges, where they remain very wary. The term "thin as a rail" comes from their slim profile, a body shape that allows them to fit between reed stems. They frequently run for cover and fly as the last resort. They sometimes swim to cross gaps and even go under water to escape predators. Rail chicks start walking the marshes with mum when only a few days old—they are small and black. They lose the black feathers as they grow to full size in the following weeks. By Sep/Oct most juvs have molted into adult-like plumage.

ad. transitional

ad. nonbr.

juv.

ad. ♀ br. /transitional

ad. ♂ br.

Black-bellied Plover *Pluvialis squatarola* **BBPL** L 11.5in

Common coastal migrant and winter resident. Scarcer than AGPL inland. Often alone or in flocks, using a wide variety of habitats. As with other plovers, has distinctive run-stop-and pick feeding style. **ID**: Easily the largest and heaviest of the plover group with muscular neck and large body. A big head and large bill give it a distinctive 'not-to-be-messed-with' appearance. Bill size rules out all other plovers. In flight, diagnostic black armpits, white rump, and far-carrying *clu-wee* call tend to be a giveaway. Ad ♂ br: boldly spangled upperparts with gray crown diagnostic at distance. White vent contrasts strongly with black underparts. This contrast, easy to see at a distance, is lacking in golden-plovers. Ad ♀ br: subdued browner and less striking version of ♂, plumage most often confused with AGPL. Some show intermediate characters. On migration, many pass through in a mottled combination of breeding and nonbr plumage. Nonbr/1st-s: fairly bland brown upperparts with pale underparts and brown wash heaviest on breast sides. Juv: as nonbr but with more boldly fringed upperparts and distinct streaking on underparts.

ad. ♂ br.

ad. ♀ br.

juv.

ad. transitional

American Golden-Plover *Pluvialis dominica* **AMGP** L 10.5in

Most common on spring migration through e. plains. Wider distribution on fall migration, many migrating offshore. Usually in flocks. Typical plover feeding style. **ID**: Noticeably smaller and more petite than BBPL with smaller head, fine bill, and attenuated rear end created by long wings. In flight, is slim with narrow wings. Weak wing bars and uniform underwing, lacking black armpits and strong contrast of BBPL. Ad br: black underparts usually to tail tip, less so in ♀. White neck line bulges at shoulder and stops abruptly. ♀: duller and browner than ♂—many intermediate. On migration, many are in transitional plumage. Neck line often extends as barring down flanks but bulge still noticeable. Nonbr/transitional: at distance, striking white supercilium sets off dark cap—stands out among BBPL. Crisply marked gray upperparts with gray-fringed pale underparts. Juv: similar with smaller spotting on upperparts and uniform pattern on underparts. Most are gray-brown but some golden, brightest on mantle. Formerly conspecific with PGPL. AMGP has longer wing point, shorter tertials, and therefore longer primary projection (typically 4 pp's showing, 2/3 in PGPL). See PGPL for other differences.

Pacific Golden-Plover *Pluvialis fulva* **PAGP** L 10.25in

Very rare. Very similar to AMGP (formerly conspecific). On average, bigger-headed, larger-billed, longer-legged, but overlap in all features. Focus on wing formula to confirm ID (see AMGP). Many recall BBPL in structure, but some look tall and skinny. Ad br: focus on white extending down flanks without obvious bulge. Juv/nonbr: more golden (see AGPL) with subtly different breast pattern.

European Golden-Plover *Pluvialis apricaria* **EUGP** L 10.5in

Very rare European visitor to the NE. Compared with AMGP is noticeably more squat, finer-billed, shorter-winged, and bigger-headed with brighter golden plumage, similar to PAGP. Smaller spots on upperparts. White line extends down flanks in breeding plumage. Diagnostic white armpits. Softer whistled call, a *tooo*.

Northern Lapwing *Vanellus vanellus* **NOLA** L 12.5in

Fantastic and unusual plover but very rare European visitor, primarily to the NE. Locally called 'Peewit' after call. Typically found in pastures, often with animals. Very distinctive appearance with striking rounded wings and color pattern in flight. All ages have similar bold plumage and crested 'professorial' look.

Lesser Sand-Plover *Charadrius mongolus* **LSAP** L 7in

Asian species. A few records. Structure similar to slightly smaller and shorter-billed SEPL. Breeding plumage striking with burnt orange across breast. Male has more black in face. Worn/molting birds show ghost pattern, variably buff breast, and contrasting pale throat. Juv: brown above with broad buff feather fringes and buff breast patches. Recent FL record of similar Greater Sand-Plover.

ad. ♀ with chick

ad. ♂ br.

juv.

Killdeer *Charadrius vociferus* **KILL** L 10.5in

The familiar plover. Widespread and hardy. Found living and nesting just about anywhere—including stone driveways, parking lots, playing fields, supermarket roofs, farm fields, sod farms, and muddy edges to ponds, often near people. Nest is a shallow scrape in the ground (easy to tread on). Plovers and other birds often feign a broken wing to distract from eggs/chicks. Very loud. Named after call, though not a good likeness. **ID**: Large, tall, slim, and elegant with a very long tail—strikingly different from all others. Unique double breastband. Bold tail pattern. Migrants often travel in tight flocks but spread out to feed. Slim, angular shape striking in flight. Bold white underparts striking at distance. Ad ♂: more extensive black in face and brighter red orbital ring than ♀ but differences subtle and best worked out with pairs in view. Variation in upperpart color from bright rufous to dull brown. Pale-fringed upperparts make ageing from juv difficult. Juv: downy young have one breastband. Quickly becomes more adult-like but has narrower fringes. Darker and broader-fringed coverts and scapulars grown in through winter. Soft filamentous tail feathers droop and show excessive wear.

Mountain Plover *Charadrius montanus* **MOPL** L 9in

Breeds on short-grass prairie. Rare winter visitor and migrant to bare earth fields and burnt stubble. Winters in tight flocks. Distinctive shape—tall and slim with fine bill. Fairly nondescript in all plumages with fawn-brown upperparts and white underparts with buff wash on breast. Dull pale legs. Ad br: shows dark forehead and lores, slightly stronger in male. Nonbr and juv: very similar with broad orange fringes like KILL. Shape and bland colors unique in N America.

Wilson's Plover *Charadrius wilsonia* **WIPL** L 7.75in

Uncommon resident on beaches. Frequently wanders out of range. Often alone, though sometimes mixes with other plovers. **ID**: Largest 'collared' plover other than KILL. Very large bill, long legs, angular head, and squinty-eyed appearance give it a 'hard' look. Front heavy and often seems to be hunkered down, even when moving. Pale legs variable in color. Ad ♂ br: molts into breeding plumage very early in year. Black breastband often complete, varies in thickness. Lores and forecrown also black. Ear coverts often warm brown. Black replaced by mostly brown in ♀. Nonbr: lacks black. Juv/1st-yr: Very similar to nonbr with pale-fringed coverts. Beware of similarly patterned 'downy' KILL.

juv.

♀/transitional

ad. nonbr.

ad. ♂ br.

Semipalmated Plover *Charadrius semipalmatus* **SEPL** L 7.25in

The common small plover, particularly in coastal areas. Occurs widely, scattered over beaches, mudflats, and fields. Frequently numbers in the hundreds at certain roost areas where quite happy to mix with other shorebirds, and is reasonably approachable. Palmations (webs) between toes. Sometimes foot-patters to bring food to surface. Typical plover run-stop-pluck feeding style. Often aggressive towards other birds. Vocal: sharp *che-WIT*. **ID**: Nicely proportioned. Much darker than similar-sized or smaller plovers with bolder breastband. Strong contrast with white underparts makes it stand out, even at long range. All ages and sexes have same basic pattern but with a lot of variation particularly on breast and head. Leg color variable yellow-orange. Bill black with orange base. Ad br ♂: most extensive black on head and breast. Many are as ♀/transitional with brown mixed in. Nonbr: browner with less extensive breastband. Juv: as nonbr with well-defined pale fringes to upperparts. Focus on upperpart color and head pattern when small plovers are in flight.

Common Ringed Plover *Charadrius hiaticula* **CRPL** L 7.5in

Very rare breeder on Baffin Island. Few records elsewhere. Very similar to SEPL. Usually found by call: a mellow *tooe*. Try to confirm ID with lack of palmations (webbing) between toes. Several subtle differences; the best is face pattern, white forehead extending to under the eye with longer dark parallel-sided stripe from lores onto cheek. In fall they are very worn with patchy pale gray and cold brown upperparts. Also darker orbital ring, longer, thinner, and more pointed bill with smaller black tip. Juv: white line between gape and dark lores.

ad. ♀

ad. ♂

nonbr.

Snowy Plover *Charadrius alexandrinus* **SNPL** L 6.25in

Uncommon. Separate inland and coastal populations. Dry mudflats, salt pans, and sandy beaches. Often in small groups. Runs faster and farther than other plovers. **ID**: The smallest plover with a distinctive shape. Fat body on long legs. Big head often with exaggerated steep forehead and square-headed look, too big for such a thin bill. White forehead, pale lores, and small distinct breast patches enhance clean pale appearance. Dull pale gray or pinkish legs. Ad br: ♂ has dark forehead and breast patches, typically browner in ♀. Nonbr: lacks black. Juv/1st-w: pale fringes to coverts soon wear off.

Piping Plover *Charadrius melodus* **PIPL** L 7.25in

Uncommon. Globally endangered. Different inland and coastal populations. Inland birds move to coast in winter where 2 populations often mix. Found on sandy beaches but stays away from water except to feed. Can be very inconspicuous. Competes with tourists for beach space with resulting problems. Nests often protected from predation by wired cages. **ID**: Strikingly pale. Large domed head on neckless fat round body—a very distinctive shape. Short orange legs. Black eye stands out on pale face. Stubby orange bill is black-tipped in summer, all dark in winter. Inland race has slightly darker upperparts, more complete breastband, and darker lores. Ad ♂ br: black forehead and collar. Ad ♀ br: black areas of ♂ mostly gray. Faded and very worn by end of summer. Nonbr: a more uniform pale gray-brown with little or no black; bill changes color. Juv/1st-w: as nonbr with pale-fringed upperparts that quickly wear away. Strikingly pale in flight with white wing bar and bold black tail band. Call: a soft whistled *peep*. SNPL has a different shape, dull-colored legs, and differently colored upperparts. SEPL is always strikingly darker.

juv.

ad.

American Oystercatcher *Haematopus palliatus* **AMOY** L 17.5in

Widespread coastal breeder on beaches. Winters in small flocks, often on mudflats and intercoastal waterways, wherever shellfish are plentiful. Chisels or hammers shells to open them. Often seen flying around in pairs or small groups in courtship display. Very noisy high-pitched whistling *kweep*, a familiar sound of the beach, frequently repeated, sometimes reaching a crescendo. Breeding pairs are very aggressive, regularly chasing other couples on the ground or in flight down the beach. Nest is a shallow scrape. Young birds take over 2 months from hatching to independence, though often stay with parents through winter. **ID**: Unmistakable. Large, heavy-set and potbellied. The beady orange eye seems to stare at you. Long, thick, sword-like red bill, visible from distance. Pied plumage striking in flight with bright white belly. One plumage year-round for both sexes. Juv: black-tipped bill, dark eye with hard-to-see pale-tipped coverts.

Black-necked Stilt *Himantopus mexicanus* **BNST** L 14in

Locally common in suitable shallow ponds, where freely mixes with other species. Often wanders out of range, particularly in spring. Usually found in loosely scattered groups, often with AMAV. Daintily picks food from knee-deep water. **ID**: Nothing belongs on such long skinny legs—the 'stilts.' Sometimes rests by sitting on lower leg. Needle-like bill. "Could be made of glass." Small head, slim neck, chunky body, long wings with attenuated rear end. Appears 2-toned dark and white. In flight, very slender, dragging long pink legs behind. Often gives distinctive short, sharp *yip* call notes (hence nickname of 'Marsh Poodle'). Ad ♂: black upperparts. Ad ♀: brown mantle and scapulars, often appears blackish at distance. Juv: pale-fringed brown upperparts. Becomes more adult-like through winter but with diffuse dark on neck and head.

juv.

ad. br.

ad. nonbr.

American Avocet *Recurvirostra americana* **AMAV** L 18in

As BNST, locally common in suitable shallow ponds, marshes and mudflats. Regular out of range. Flies around in tight flocks, which can number in the hundreds. Seems to crouch low, body at 45 degrees, to keep bill horizontal in distinctive feeding style, swishing bill backward and forward as it marches through the water. Sometimes picks from surface. Often seen standing in long spread-out lines, many on one leg with head tucked in while sleeping. Typical call is a whistled *bleet*. **ID**: Instantly recognizable by unique upturned bill. Sometimes ♀'s bill is more strongly upturned than ♂'s but difficult to judge. Large bird with sturdy body and long neck and legs, giving an elegant look; large flocks are particularly striking. All plumages have dark upper scapulars and coverts on a white background. Adult: marmalade-orange neck and head. Nonbr: head and neck become pale gray. Juv: very similar to nonbr but with very thin, pale fringes to scapulars and coverts—often very difficult to see. Cinnamon wash to head.

Eurasian Curlew *Numenius arquata* **EUCU** L 21.5in

Very rare in fall and winter. Closest to LBCU but much colder brown with heavier streaking on the underparts and a different underwing pattern. Bill long and downcurved. The bold tail pattern puts ID beyond doubt! Beware of WHIM, particularly European race. Latter is significantly smaller with boldly striped head.

Spotted Redshank *Tringa erythropus* **SPRE** L 12.5in

Very rare, mostly late fall or winter with GRYE. Always id by shape, bold orange-red legs, and black bill with red at base of lower mandible. GRYE-size with fat, rounded body, long legs, straight bill with droop at tip, and small head. Dowitcher-like tail pattern. Breeding plumage all-black including legs. Nonbr: similar to GRYE. Beware of GRYE with bright orange-tinged legs. Loud *chewit* call.

Common Redshank *Tringa totanus* **COMR** L 11in

Very rare to the NE. Id by bright orange-red legs—striking at all times. Bill usually orange-based. LEYE-size but more compact. Relatively nondescript apart from legs, until it flies, revealing bold upperpart pattern with striking white trailing edge to wings. All plumages have brown upperparts, diffusely streaked underparts, and prominent eyering with indistinct eyestripe and supercilium.

Wood Sandpiper *Tringa glareola* **WOSA** L 8in

Very rare. Similar to a squat version of short-legged, short-billed LEYE. It can also recall a boldly marked SOSA. All plumages have distinct white supercilium that sets off dark cap. Upperparts are dark brown with obvious large pale spots. Diffusely streaked breast. Legs yellow-green. In flight, brown-winged with a white rump and strongly barred tail.

juv.

ad. br.

ad. nonbr.

Spotted Sandpiper *Actitis macularius* **SPSA** L 7.5in

Widespread on lakes, marshes, streams, beaches, rocky shores—possible just about anywhere. Frequently occurs alone, in drier habitats with no other shorebirds. Distinctive at distance by feeding style of 'crouch and run' as it hunts insects. It will stand with head close to ground, tail in the air, constantly bobbing, and then suddenly scurry to grab its prey. Bobs even when standing still. Equally recognizable flight style: very fast, shallow flickering wingbeats followed by short glides. Early fall migrant, with birds heading s. by late Jun. Flight call: a clear *peet-weet-weet*. **ID**: Short neck, chunky rounded body, and long tail on short yellow-green legs. Confused with SOSA but moves differently, is smaller, fatter-bodied with shorter neck and legs. Ad br: boldly spotted underparts. Brown upperparts have black internal bars. Nonbr: plumage sometimes has a few spots on rear flanks, otherwise pale underparts with brown breast sides. Pale-tipped coverts. Juv: boldly patterned with pale tips to feathers and dark subterminal markings to all upperparts. This pattern becomes heavily worn through winter.

juv.

ad. br.

ad. worn/nonbr.

Solitary Sandpiper *Tringa solitaria* **SOSA** L 8.5in

Widespread and fairly common in a variety of wet muddy areas—muddy overgrown ponds, flooded fields, 'duck' ponds, farm slop ponds, and any water, puddle-sized or larger. Breeds in forested bogs. Often 'solitary,' in part due to choice of feeding places. Always looks nervous, jerking head up and down, and walks in very deliberate manner, picking insects off surface. When flushed, is usually vocal, rising rapidly with exaggerated flicking wingbeats. Call similar to SPSA but higher-pitched, faster, and more emphatic. **ID**: Between SPSA and LEYE in size and proportions. Distinctive shape with fat body, attenuated rear end, small head, and medium-length bill. Upright posture when not feeding. Unique color pattern in flight: uniform upperparts with dark-centered barred tail and very dark underwing. All plumages similar; dark brown, spotted upperparts, heavily streaked breast on white underparts. Striking spectacled look with bold eyering and white supraloral. Green legs. Ad br: upperparts have distinct white spots. These wear off through summer/fall. Juv: fresh plumage with obvious white spots, otherwise similar to adult.

1st-w.

juv.

ad. transitional

ad. br.

Lesser Yellowlegs *Tringa flavipes* **LEYE** L 10.5in

Common and widespread in all types of wetland. Graceful and elegant appearance with 'soft' demeanor. Quietly picks food from surface, body at 45 degrees. Occasionally runs around crazily like GRYE. Often in loosely scattered groups, occasionally with GRYE or SPTS. Flies in compact flocks. Frequent *tu* call given singly or repeated several times in rapid succession. **ID**: Smaller, more rounded body than GRYE (it is not worth cooking!) with proportionally longer legs. Always looks dainty. Occasionally shows bulging sternum of GRYE. Best ID feature is shorter, narrower-based bill that looks straight. Has a more rounded head, and a less distinct eyering and more muted face pattern than GRYE, though GRYE's supercilium tends to be more pronounced. In flight, id by size and structure. Plumages similar to GRYE but differ as follows: Ad br: finer black streaking restricted to breast. Upperparts less black. By fall migration, upperparts are a mixture of heavily worn breeding feathers and new nonbr. Juv: aged by well-defined spots to brown upperparts and uniform breast streaking. Looks subtly paler with less distinct breast markings and less 'notched' upperparts than GRYE.

nonbr.

ad. br.

juv.

ad. transitional

Greater Yellowlegs *Tringa melanoleuca* **GRYE** L 14in

Common and widespread in all types of wetland. Feeds in loose flocks, sometimes with LEYE and other species. Collects to roost in tight packs, head often tucked in. Runs around crazily, seemingly in circles, trying to catch up with small fish. Looks from side to side—nothing is safe! Also carefully picks from surface. Frequently gives distinctive, loud whistled *tew tew tew* call—quite easily imitated. **ID**: 20% larger than LEYE. Bulky body, usually with angular bulging sternum. The large body size (it has enough meat on it to cook!) is arguably the easiest way to separate from LEYE. Proportionally small-headed and thin-necked, frequently with a kink in neck. Noticeably longer-billed and bill slightly upturned compared to LEYE with broader base, finer tip. and usually pale at the base. Striking bright yellow legs. Bold eyering, coarse head streaking, and square-headed look. Ad br: feathers black or gray. Extensive dark flank bars, lacking in LEYE. By mid-summer much browner with fewer breast markings. Nonbr: complex feather markings on paler gray-brown upperparts. Juv: similar to nonbr but browner with obvious notches to upperparts. Breast tends to be more crisply streaked than LEYE's.

ad. transitional

juv.

nonbr.

ad. br.

Western Willet

juv.

1st-w.

ad. br.

Eastern Willet

Willet *Tringa semipalmata* **WILL** L 15in

Two distinct populations: prairie breeders that winter on s. beaches, and coastal marsh breeders that always winter s. of the US. A familiar sight on tourist beaches, competing with kids for sand crabs. Runs in and out of waves and probes deeply in mud and sand. At other times walks elegantly, standing tall with a stiff gait. Very early fall migrant. Juv e. birds leave breeding grounds the minute they can fly. Very noisy, raucous *wee-willet* call, often repeated several times. **ID**: A large, thick-set but well-proportioned shorebird with a long thick bill. Relatively nondescript brown or gray but morphs into a different beast in flight with striking black-and-white wings. Prairie birds are 15% larger and more elegant and godwit-like, due to longer legs, slimmer neck, and a bill that is narrow-based, longer, and straight or slightly upturned. E. birds have very broad-based bills, which usually appear downturned and blunt-tipped. Most have a much stronger loral pattern. E. birds are also darker and browner in all plumages. Ad br: coarsely marked brown and gray. Nonbr: uniform gray upperparts, pale underparts. (e. birds rarely seen in this plumage). Juv: upperparts have complex internal markings.

ad. nonbr.

ad. br.

juv.

Upland Sandpiper *Bartramia longicauda* UPSA L 12in

Uncommon and rapidly declining summer visitor to prairie and grasslands On migration, can be seen on bare earth and sod farms, always keeping to itself. Ungainly and almost freakish appearance due to gait, proportions, and odd head. Struts, but you usually only see its funky head raised above grass. Walks steadily but purposefully, head rocking backward and forward before daintily picking something from the ground. On breeding grounds, frequently perches on posts, where it raises wings after landing. Often, particularly on migration, heard and not seen as it flies overhead giving rolling, bubbly whistle that is very similar to CAWR call. Distinctive long and drawn-out descending whistled call on breeding grounds. Related to curlews but somehow feels more like a roadrunner. **ID**: Fat body, very long tail, skinny neck with oversized eye on small square head. Short, yellow-based bill and yellow legs. All plumages similar. Adult: warmer buff tone than juv. Black subterminal markings to upperparts. Juv: aged by well-defined pale fringes and dark subterminal markings on upperparts and breast sides. In flight, looks slim with a long tail.

ad. Eurasian Whimbrel

ad. Eurasian Whimbrel

ad.

Whimbrel *Numenius phaeopus* **WHIM** L 17.5in

Fairly common n. breeder and coastal migrant. Scarce elsewhere. Usually occurs in flocks in the double figures on marshes and tidal areas, but also found on beaches and rocky coasts. Walks steadily but purposefully. Rarely mixes with other species except at tidal roosts. Far-carrying whistled call: *pi pi pi pi*. **ID**: Striking downcurved bill and striped head pattern. Otherwise, a large and sturdily built, well-proportioned brown shorebird. In flight, has long pointed wings, uniform brown above, and barred underwing but no strong contrast anywhere—the bill is noticeable at fairly long range. Ad: plumage same year-round. Breeding birds often have all-black bill, pink-based the rest of the year. Juv: very similar to adult. Best aged by fresh plumage with larger pale notches to upperparts and fine streaking in underparts. European race annual on E coast in WHIM flocks. White 'V' extends up back. Underparts, particularly wing lining, whiter. Also colder brown overall. More extensive streaking shows high contrast on white background.

Long-billed Curlew *Numenius americanus* **LBCU** L 23in

The largest shorebird. Uncommon breeder on prairie grassland. Small numbers winter on beaches, tidal mudflats, and fields. Feeds on crustaceans by sticking incredibly long bill deep into ground. **ID**: Diagnostic bill. All plumages strongly buff. Fairly uniform head and underparts with fine streaks on breast. Uniform upperparts and striking cinnamon underwing in flight. Adult: complex feather markings. Juv: more boldly marked with broad buff fringes to dark-centered coverts. WHIM only other bird with downcurved bill, is much shorter-billed, smaller, and duller. Sleeping MAGO has dark legs.

Black-tailed Godwit *Limosa limosa* **BTGD** L 16.5in

Very rare to the E coast and Great Lakes. Large long-legged godwit with long broad-based bill. Most striking in flight with white underwing, upperwing bar, and rump; black tail. Breeding birds are largely burnt orange with paler barred belly. Nonbr: pale with uniform gray upperparts. Juv: a lovely orange-buff. Bold pattern in flight should remove any doubt!

Bar-tailed Godwit *Limosa lapponica* **BTGO** L 16in

Very rare to the E coast. Small compact godwit with short legs and neck. Bill narrow-based but quite long. 'Bar-tail' and white 'V' up back are the easy ID clinchers. Nonbr/juv: evenly patterned upperparts and pale underparts. Breeding ♂ rich chestnut, ♀ much paler. European and AK races possible. AK birds (juv shown) have heavily marked underwing.

juv.

ad. nonbr.

ad. ♂ br.

ad. ♀ br.

Hudsonian Godwit *Limosa haemastica* **HUGO** L 15.5in

Uncommon n. breeder in marshes and spruce bogs. Easiest to see as it migrates n. through e. plains in spring. Most migrate in flocks far offshore in fall. Smaller numbers down the E coast. Found in a variety of wetlands. Frequently seen steadily probing for food in knee-deep water with other shorebirds. Small size makes it easier to overlook than other godwits. **ID**: Often seen in flight when striking black underwing diagnostic. Black tail, white rump, and wing bar all make for easy ID; not much larger than GRYE. Fat body and very long wings—perfect for long flights to Argentina. All plumages show supercilium, boldest in front of eye. Narrow-based bill, larger in ♀. Ad ♂ br: bold rusty underparts with dark brown upperparts, often with paler-headed appearance and orange-based bill. Ad ♀ br: lighter than ♂ with more barring and pale areas. In fall, most migrants are a mixture of breeding and uniform pale gray nonbr plumage with pink-based bill. Juv: buff-tinged breast and subterminal anchors and bars to pale-fringed upperparts.

ad. br.

nonbr.

Marbled Godwit *Limosa fedoa* **MAGO** L 18in

Fairly common prairie breeder. Wintering and migrant birds use beaches, mudflats, wetlands, and fields, often in groups. Walks steadily. Feeds by probing deeply for worms and crustaceans, often with bait fishermen for company. Roosts in flocks with smaller WILL, head tucked in. **ID**: The largest godwit. Long broad-based bill is strikingly pink at base. Shows some buff in all plumages and a striking cinnamon underwing. Plumage almost identical in LBCU, but not the bill or gray legs! As in all godwits, ♀ is slightly larger and longer-billed than ♂. Ad br: least seen plumage and often confused with HUGO. Paler, with more extensive fine barring on underparts and with more complex markings on upperparts than HUGO. Nonbr: extensively orange-buff, upperparts with complex internal feather markings. Juv: very similar to nonbr with paler head and larger dark feather centers to upperparts. Molts out of juv plumage very quickly.

Ruddy Turnstone *Arenaria interpres* **RUTU** L 9.5in

Common in coastal areas. Scarce migrant inland. Typically in small groups, but occasionally in large flocks on migration. Prefers rocky shores and jetties, but also frequents beaches. A bird with real character: the 'sausage dog' of shorebirds. Walks quickly, seeming to sniff around like a terrier, head close to ground, looking for anything edible. Uses bill to turn over rocks and debris at wrack line to find hidden goodies. Also digs holes in beach, sand thrown everywhere. **ID**: Pocket battleship: small with very short legs, thick-set long body. Short chisel-shaped bill for tipping over objects. Plumage boldest in breeding ♂ and dullest in nonbr, in latter plumage has the same basic pattern with complex bib and head pattern. Bold upperparts most striking in flight. Ad ♂ br: superbly marked with bold, well-defined colors. Ad ♀ br: a subdued and messier version of ♂ with brown in cap and upperparts. Ad nonbr: most uniform plumage, generally duller brown. Sexes similar. Juv: uniform upperparts with well-defined white or buff fringes. Plumages show a lot of variation.

Red Knot *Calidris canutus* **REKN** L 10.5in

High-Arctic breeder; suffering significant population declines. Winter and migrant visitor to coastal areas. Scarce inland. Migrants and wintering birds prefer beaches and tidal mudflats. Roosts often number in the hundreds. Traditional stopovers on migration in late spring include Delaware Bay, where feeds on horsheshoe crabs, which have suffered a 90% decline in 20 years. Moves slowly, usually picking food off surface. Call a nasal *whet whet*. **ID**: Unique shape. Built to travel long distances. Large body (fuel tank) with long pointed wings. Short legs; many similarities with RUTU. A horizontal bird! Stout, fairly short straight bill. Sexes similar. Ad br: extensive variation. Rufous underparts with contrasting white vent. Upperparts coarsely spangled. Ad nonbr: pale gray upperparts, breast streaking/barring extending onto flanks. The pale supercilium stands out at a distance. Similarly marked SBDO is much longer-billed and -legged. Juv: as nonbr but with dark subterminal lines and white fringes to upperpart feathers. Fresh juvs can show buff wash on breast. 1st-s: most as nonbr or with some orange.

Sanderling *Calidris alba* **SAND** L 8in

Common migrant and winter visitor to beaches. In the Interior also on alkaline lakes. One of my favorites, always on the run. It's the 'army' on the beach that chases waves into the ocean, only to be chased back out, then return for more. Its short legs move as fast as they can. Roosts in tight flocks higher on the beach. Feisty, squabbling with others. In spring, often found with other 'Arctic' migrants—REKN and RUTU. Moves more carefully, picking food from surface or digging holes for buried goodies. The boldest white wing bar in flight with clean white underwing. Often calls, a liquid *qweet*. **ID**: Incredibly variable in coloring. A chunky bird with short black legs, thick neck and rounded head. All-black straight bill. Sexes similar. Ad br: breast, head, and upperparts range from dull brown to bright chestnut to dark or pale gray and black. Colors fade through summer, often becoming pale orange and worn combinations of white through black. Molts late summer. Nonbr: striking white underparts, very pale gray upperparts. Dark bend of wing often striking. Large black eye stands out on white face. Juv: pristine spangled upperparts, often buffy when fresh; these soon wear to white.

Dunlin *Calidris alpina* **DUNL** L 8.5in

The most abundant wintering shorebird in many coastal areas, sometimes taking over beaches and mudflats, also marshes, particularly on migration. Uncommon migrant inland. Migrates late in fall. Feeding flocks spread out, walking steadily and picking food from surface. Roosting birds bunch up, often into large chattering flocks. Call: drawn-out, buzzy *jeesp*. **ID**: Larger than most of the small shorebirds. Short black legs, fat neckless body with noticeably long and slightly downcurved black bill. Often looks hunchbacked. The key 'comparison' bird, so important to know size, shape, and characteristic bill. Often found with same-size SAND. Ad br: striking rufous upperparts, pale head and underparts with bold black belly patch. Ad nonbr: uniform gray upperparts and breast with indistinct supercilium. Juv: rufous and buff-fringed upperparts with diffuse streaking on underparts (unusual in juv shorebirds). Most birds have molted many or most juv feathers before leaving breeding grounds, so appear largely as nonbr with pale-fringed coverts. Structurally similar WESA is smaller and in nonbr plumage is paler above and lacks darker breast.

Curlew Sandpiper *Calidris ferruginea* **CUSA** L 8.5in

Rare but annual visitor in spring and fall, mainly to the E Coast. Slightly larger than DUNL with longer legs, wings, neck, and more down-curved, fine-tipped bill. White rump and striking rufous underparts. Many birds a combination of pale gray nonbr and some breeding feathers. Juv: scaly, clean underparts and prominent supercilium.

Purple Sandpiper *Calidris maritima* **PUSA** L 9in

Localized winter visitor to rocky shores and jetties, often with RUTU. Late spring migrant. Its habits, shape, and dark overall appearance distinctive. Usually in small flocks, clambering over rocks in search of mussels. Jumps from rock to rock, dodging waves. **ID**: The fattest shorebird. Smoky gray upperparts and diffusely streaked underparts; blends into dark rocks. Purple iridescence often hard to see. Typically orange legs and bill base create strong contrast. 1st-w: as nonbr with white-edged coverts and tertials. Many molt into breeding plumage on wintering grounds. Boldly patterned in flight.

ad. ♂ br.

ad. br.

juv.

ad. fall

Pectoral Sandpiper *Calidris melanotos* **PESA** L 8.75in

Common migrant throughout, scarce early spring migrant in the E. Occurs on wetlands, sod farms, and in wet agricultural areas, often sneaking through tangled vegetation. Sometimes in single-species flocks but also mixes in with other shorebirds. Walks around steadily picking at surface, head rocking slightly as it moves. Call: BASA-like trill but deeper and throatier. Also gives a deeper grunting *drrrp*. **ID**: Much larger than peep, and slightly larger than similar looking WRSA/BASA. Chest is almost disproportionately large and bulging, yet overall bird is attenuated with long wings. Sometimes this can make it appear small-headed. All plumage patterns similar. Focus on strongly marked breast and well-defined pectoral band—striking even at long range. ♂ is 25% larger than ♀. Ad br: rufous-fringed scapulars, ear coverts, and crown with orange-based bill. Ad ♀ br: duller than ♂, darker-billed with sparser breast streaking. Worn late summer/fall birds are much duller and lack rufous tones. Juv: similar to adult but with fresh plumage and warmer tones plus crisper edging to upperpart feathers. In flight, is dark with wing bars and white outer tail—the pectoral band still stands out!

ad. transitional

juv.

nonbr./1st-s.

ad. br.

White-rumped Sandpiper *Calidris fuscicollis* **WRSA** L 7.5in

Fairly common late migrant spring and fall. Most go through the Great Plains in spring, E Coast in fall. Often with SESA and other small shorebirds, sometimes in single-species flocks. Found in a variety of wetlands, often in water, but just as happy in grass and muddy borders. Frequently heard before seen: an unusual short high but nasal *tzeek*. Diagnostic white rump not always easy to see in flight. **ID**: Looks just like SESA but is 20% larger, has a striking tapered rear end due to very long primaries. The latter stand out as a black wedge, even at distance. Size and structure as BASA but with flatter crown and thicker neck. BASA has a buffier breastband and differently patterned upperparts. Orange base to bill is diagnostic but difficult to see. Supercilium distinctively straight, often upturned behind eye. Ad br: neatly streaked underparts, extensive gray fringes to upperparts in early spring become rufous. Fall birds, like SESA, an abstract mixture of gray (nonbr) and black (br) feathers. A few reach nonbr in late fall. 1st-s: vary from nonbr to breeding. Juv: commonest in Oct, very scaly upperparts with contrastingly rufous upper scapulars, often with molted nonbr gray scapulars mixed in.

ad. transitional

ad. br.

juv.

Baird's Sandpiper *Calidris bairdii* **BASA** L 7.5in

Common migrant through the Great Plains, where sometimes the commonest shorebird. Scarce fall migrant in the E. Found in wetlands, preferring dry ground and short-grass areas at edges; also sod farms. Call: a *trrrrt*—softer than PESA. **ID**: Size and structure like WRSA with steep forehead and small square-headed appearance on thinner neck. Very long-winged. Eye often stands out on bland face. Slightly downcurved bill is all black with narrower base and finer tip. Buff breast and face with diffuse streaking are always strongly indicative of this species. Ad br: spring migrant upperparts have mostly gray nonbr feathers with black-centered breeding scapulars. On breeding grounds, full breeding plumage is dark-streaked breast and upperpart feathers with bold black centers and silver fringes. Fall birds are much plainer with a mixture of worn breeding and buff nonbr feathers. Nonbr: birds have browner tones to breast and upperparts than similar species. Juv: overall impression is of a strongly buff bird with broad pale fringes to upperparts giving a distinctly scaly look. In flight, lacks strong contrast, but note very long and pointed wings—for flying to South America!

1st-s.

juv.

ad. transitional

nonbr.

ad. br.

Western Sandpiper *Calidris mauri* **WESA** L 6.5in

W. peep. Uncommon fall coastal migrant in the N, commoner farther s., where also winters. Almost no spring records in the NE. Scarce inland. Habitat preferences as SESA, also drier habitats. Typically winters on beaches and wetlands. Call: sneezy qualities of WRSA. **ID**: Slightly larger than SESA with DUNL-like shape: fat round body, a neck on steroids with big head and steep forehead. With practice, most birds are identifiable by shape alone. Bill longer than SESA, but much overlap, broad-based, decurved at end with fine tip, irrespective of length. Stands more upright at rest. Ad br: strong contrast between rufous scapulars, cap, ear coverts, and rest of upperparts. Well-defined arrows on underparts. SESA can appear rufous but color is mostly found on the fringes, not bases of feathers. By late summer upperparts mostly nonbr. Pale appearance with some rufous in upperparts and arrows on underparts. Nonbr: common by late Aug—pale gray with narrow white fringes. Wing molt is diagnostic from SESA (latter finishes molt in South America and never molts any flight feathers). Pale gray fades browner through winter. Ist-s: as nonbr or with some adult-type scapulars and breast streaking.

juv.

ad. transitional

ad. transitional

ad. br.

Semipalmated Sandpiper *Calidris pusilla* **SESA** L 6.25in

Very common migrant. A few winter records in s. FL. The common peep, swarming over mudflats and other wetland areas, often with tail pointed in the air. Occasionally on fields. Strongly territorial, fighting for space. Call: a sharp *jeet*. **ID**: Between WESA and LESA in size. Similar to WESA but noticeably slimmer-bodied. Bill typically shorter but more tubular and thicker-tipped. Some long-billed but shape is consistently different from WESA. Both have quite long legs and partial webs (palmations), most noticeable between outer toes. Ad br: upperparts gray-brown with dark centers, though some quite rufous (most obvious around ear coverts, crown and scapular edges). By fall, upperparts are a combination of evenly spread gray (nonbr) and black (worn) feathers. Finishes molt to nonbr on wintering grounds. 1st-s: as adult (retains a few juv coverts but only viewable in hand—see WESA). Juv: uniform scaly gray-brown upperparts with diffuse internal feather markings. Early fall birds can have light buff on breast. Compared with WESA has darker head markings with capped appearance and strong contrast between rufous upper and gray lower scapulars.

Least Sandpiper *Calidris minutilla* **LESA** L 6in

Common migrant and winter visitor. Found in muddy and marshy areas but tends to stay in drier spots with vegetation. Also on beaches, fields, sod farms, small ponds, and a variety of habitats. The only peep with green legs. Feeds crouched over on bent legs, slowly sweeping from side to side as it picks food from the surface. In flight, easily recognizable by short angled-back wings, dark broad wing linings and rocking motion with shallow wingbeats. Distinctive *kreep* call. **ID**: The smallest peep. Distinctly short, fat, and chunky with short neck and small head; these features should be the basis of all ID. Narrow-based bill, noticeably downcurved and with fine tip. In all plumages, the peep with the darkest upperparts and breast. Ad br: dark-streaked breast on dull background. Broad pale fringes are worn away by late summer, making upperparts even darker. Ad nonbr: darker and browner tones than other peeps in similar plumage. Upperpart feather centers noticeably darker than fringes. Complete or near-complete breastband. Juv: upperparts richly colored (first juvs often mistaken for stints) with white tips to scapulars forming mantle 'V.' Breast streaked and darker than in other juv peeps.

Red-necked Stint *Calidris ruficollis* **RNST** L 6.25in

Annual, mostly Jul/Aug. Similar to SESA with subtly shorter legs, steeper forehead, shorter neck, and longer rear end. Pale-headed, rufous scapulars, red neck always darkest on throat. Throat bordered by dark spots, wearing to orange, and disappears through fall. All birds recorded to date have had some element of breeding plumage. Try to find the first juv!—they have unwebbed toes like LIST.

Little Stint *Calidris minuta* **LIST** L 6in

Almost annual, mostly Jul/Aug. Long-legged, potbellied, and small-headed with fine slightly decurved bill. Breeding plumage more orange than red, strongest on ear coverts and palest around chin/throat. Spots mixed in orange on throat. Rufous upperparts include tertials and coverts. Juv: very boldly marked upperparts and is pale-headed with split supercilium and white forehead.

Sharp-tailed Sandpiper *Calidris acuminata* **SPTS** L 8.5in

Very rare fall visitor. Adult usually in late summer, juv in fall. Similar to a small-headed and small-billed PESA with a chestnut cap, bolder eyering, and different breast pattern in all ages. Adult: dull buff breast with varying amounts of bars and chevrons mostly on breast sides and flanks. Never streaked, and with well-defined breastband like PESA. Juv: strikingly orange-buff with fine streaks across upper breast. Can recall BBSA at distance. The cap is always striking.

Buff-breasted Sandpiper *Tryngites subruficollis* **BBSA** L 8.25in

Uncommon migrant to the Great Plains. Fall migrant to the E Coast and offshore (mostly juvs). Found in farm fields, sod farms, and sandy areas, often in flocks. Migrants in spring display at staging areas, pumping out breast, spreading wings. White underwing with 'comma' is best way to find them. Moves with jerky strut. **ID**: Gentle appearance, large eye on bland face. Square head, short, thin bill, slim neck, and attenuated rear end. All plumages buff. Ad br: broad buff fringes to upperparts. By fall, a few nonbr scapulars, with broader and darker fringes, are molted in. Juv: neat, scaly upperparts.

Ruff *Philomachus pugnax* **RUFF** L 11in

Rare migrant, possibly breeds. Structurally like a large, ungainly BBSA with a longer bill. Looks disproportionately large-bodied with long legs but small head. Stands tall. ♀ (Reeve): 20% smaller than ♂ and smaller-bodied. All have yellow or orange legs. In flight, has bold white 'V' on rump. Ad br ♂: has stunning breeding plumage, a variable mixture of white, black, and tan. Neck feathers ('ruff') molted by early fall, otherwise still boldly marked. ♀ has mottled breast. Ad nonbr: similar to breeding ♀ but duller. Often has pale feathering at base of bill. Juv: buff, paler on belly. Upperparts scaly with bold buff fringes. Size and shape very different from BBSA.

Lesser Yellowlegs

ad. br. transitional

1st-w.

ad. br.

juv./1st-w.

juv.

Stilt Sandpiper *Calidris himantopus* **STSA** L 8.5in

Fairly common migrant throughout on freshwater pools. Scarce spring migrant in the E, and winter visitor. Usually found knee-deep in water, often mixed in with dowitchers. Very long-legged so tips like an oil-derrick to feed. Bill often held perpendicular to water as it peers for food. Feeds in the same spot like dowitchers, never tearing around like yellowlegs. **ID**: Very elegant. Fat body with long green legs, slim neck, and longish, gently decurved bill on pointed head. Ad br: appears dark with heavily barred underparts and often striking chestnut ear coverts and cap—very dapper. Early fall migrants are often a combination of worn breeding plumage and freshly molted gray nonbr feathers. Late fall migrants are usually in mostly nonbr plumage. Nonbr: upperparts pale gray with long, white supercilium. From dowitchers by smaller size, disproportionately small head with bold supercilium, downcurved bill, and duller plumage.

Long-billed Dowitcher *Limnodromus scolopaceus* **LBDO** L 11.5in

Uncommon and local. Migrants have strong preference for freshwater pools; rarely tidal mudflats. Wintering birds often found on mudflats. Feeds knee-deep in water, head down, rapidly probing, 'sewing-machine' style. Single *kik* call, given standing or in flight, diagnostic. Calls, also given in succession, are often misidentified as repeated *tew* call of SBDO. **ID**: Many have diagnostically long bills that usually appear straight with narrow base and fine tip—SBDO shorter-billed with broad base and blunt downcurved tip. Often with SBDO. LBDO are the larger and rounder-bodied ones, often appearing hunch-backed and are found in deeper water using their longer legs. Ad br: darker orange underparts with white-edged black bars, heaviest on upper breast. Upperparts blacker with white tips (SBDO browner with buff fringes). Paler by fall but with a few horizontal breast bars (spots on similar *hendersoni* SBDO). Ad nonbr: upperparts darker, feather centers fading paler to edge (SBDO paler gray with white fringe). More uniform dark breast and flanks. Juv: easy to overlook as nonbr but has narrow rufous fringes to tertials and scapulars—lacks bold patterns of juv SBDO. No 1st-s. plumage—appears as adult.

ad. nonbr.

1st-w.

juv.

ad. br. *griseus*

ad. br. *hendersoni*

Short-billed Dowitcher *Limnodromus griseus* **SBDO** L 11in

Very common migrant on coast, uncommon inland. Uncommon winter visitor. 2 races: *hendersoni* (prairie population) breeds w. of Hudson Bay, uncommon through plains. Fall migrant and winter visitor to the E, *griseus* (Atlantic), breeds e. of Hudson Bay, the common migrant 'dow' in the E. The common dowitcher, particularly on tidal mudflats. Habits and feeding as LBDO. **ID**: Prairie birds average larger, longer-legged and -billed: midway in many respects between *griseus* and LBDO. Striking tail pattern appears paler than LBDO, lacking contrast between dark tail and whiter back. Ad br: considerable overlap between races. *Hendersoni* has orange underparts, *griseus* is more heavily barred and less orange. Worn and paler by late summer. *hendersoni* is orange with spots down the breast sides and flanks (see LBDO). Nonbr: pale gray upperparts with well-defined pale fringes (wear duller through winter). Less solidly marked underparts, particularly on flanks, than LBDO. Juv: often strikingly orange underparts and complex fringes to upperparts (variation and all wear duller). 1st-s: mostly as nonbr with a few breeding feathers—any summer dowitcher in nonbr plumage is this species.

Wilson's Snipe *Gallinago delicata* **WISN** L 10.5in

Widespread and common breeder, migrant, and winter visitor. Found in wet bogs, fields, and marshes. Usually not seen until flushed. Explodes from underfoot with rasping *scaarp*, flying off in zigzag pattern. Usually drops back into cover, runs off, and disappears. Occasionally confiding, often when hungry or where lots of worms. Sits on posts and other perches on breeding grounds. Often in small groups if habitat suitable and occasionally flies in small clusters. Dramatic and familiar display on breeding grounds: flies high into sky making 'whurring' sound (winnowing), circling territory before dropping to ground. Heavily barred flanks and underwing contrast with white belly—striking in flight. Often seen crouched among tussocks, cryptic plumage makes it difficult to see. Pumps body up and down when feeding. **ID**: Structurally most like a short-legged dowitcher but more often confused with AMWO because of behavior and cryptic plumage. All plumages similarly dark brown with striped back and head. Molts in spring and late summer, so fall migrants are in fresh plumage and look like juvs.

American Woodcock *Scolopax minor* **AMWO** L 11in

Supposedly the commonest shorebird but you would never know it. Found in damp forest. A loner that blends into forest floor. Usually found by listening for sound as it tries to walk away. Often flushed. A dark, stocky bird with broad wings that whistle in flight. Chiefly nocturnal, coming to forest edge at dusk to feed in adjacent open areas. On breeding grounds displaying birds give a CONI-like *beent*. Flies high around territory in courtship display, usually seen as a silhouette, with steady but fluttery wingbeats—the easiest way to find one. Feeds mostly by probing for worms. When ground is frozen, is pushed out of woods to thawed sunny edges. Rocking motion as it walks, often pausing as if to take a breath. Sometimes spreads tail to defend feeding territory. **ID**: Weird-looking with large eye placed high on oddly domed head. Muscular neck on fat body. Tends to look very dark at first glance, but orange underparts striking with good views. All plumages similar.

juv.

ad. nonbr.

ad. ♀ br.

ad. ♂ br.

Wilson's Phalarope *Phalaropus tricolor* **WIPH** L 9.25in

One of the commonest shorebirds in the prairies, uncommon in the E. Often in small flocks, swimming and constantly picking insects from the surface. Moves from side to side, grabbing an insect from behind itself. Like other phalaropes, will sometimes spin, creating a vortex to bring food to surface. Walks more frequently than other phalaropes. Legs dark in summer, yellow in winter. Looks awkward with fat body on short legs, tail tipped up, and head held close to the ground, like PESA. In flight, shape and white rump striking, but jerky shallow wingbeats sometimes more of a giveaway at distance. Call: a grunt! **ID**: Largest phalarope with noticeably longest and slimmest neck. Head pointed due to shallower forehead. Incredibly fine needle-like bill most striking feature; looks like it will snap! Ad ♂ br: fairly pale and nondescript. Ad ♀: spiffy. At distance, dark lines from bill down neck divided by pale on crown and nape (unique). All birds molt to nonbr in late summer—most fall migrants in partial or complete very pale nonbr plumage. Juv: dark upperparts with buff fringes and wash to breast. Molts to nonbr early, so rarely seen in full juv plumage. Other shorebirds do occasionally swim.

ad. ♀ br.

ad. ♂ br.

1st- w.

ad. transitional/
nonbr.

Red Phalarope *Phalaropus fulicarius* REPH L 8.5in

Uncommon winter visitor offshore; rare on migration everywhere else. Migrants not usually seen before mid-Sep. Most easily seen on breeding grounds or from boats, feeding on weed lines. Occasionally blown onshore during storms. Usually in small flocks, often with similar RNPH. **ID**: Larger than RNPH. Focus on noticeably thicker bill with pale area at base (often difficult to see). Larger size exaggerated as it sits higher in water, also thicker neck and squarer-headed look. Molts early in fall and fairly late in spring, so wintering birds and migrants are mostly in nonbr plumage. Upperparts uniform pale gray, with narrow pale fringes. Often more extensive white on crown and paler gray nape than RNPH, giving paler-headed look. Ad ♂ br: as with other phalaropes, ♀ brighter. Both striking, but if in doubt confirm with bill shape and color. Juv: black upperparts with broad orange-buff fringes. Strongly orange-buff breast. Molts early to nonbr, when aged by retained buff-fringed juv coverts and tertials. Molting adults have old dark-fringed tertials. In flight has broader wing bar and whiter underwing than RNPH. Beware of SAND, which has even bolder wing bar.

ad. nonbr.

juv.

ad. ♂ br.

ad. ♀ br.

Red-necked Phalarope *Phalaropus lobatus* **RNPH** L 7.75in

Widespread but scarce land migrant through region on wetlands. Common offshore, particularly along weed lines. Behavior as other phalaropes. Usually in flocks offshore. Gives sharp *chet* calls when flushed. **ID**: The smallest phalarope. Small square head and thin neck add to petite feel of this species. Needle-like bill much shorter than WIPH. Most birds identifiable on these features. Ad ♀ br: a beauty, marked with white eyelash as if to emphasize striking head. Ad ♂: a washed-out version of ♀ and more like ♂ WIPH. Spring migrants often in breeding plumage. Fall migrants are usually in molt, so have a mixture of breeding and nonbr feathers with upperparts that often look stripy. Ad nonbr: pale gray but with broad white feather fringes. They appear darker and striped rather than uniformly pale as in larger REPH; also apparent in flight, with darker wings and narrower wing bar on upper- and underwing, all adding to darker and more contrasting appearance. Juv: strikingly dark above with buff fringes. Black cap and ear coverts. Molts on migration so often a mix of gray and black above.

American Bittern *Botaurus lentiginosus* **AMBI** L 28in

Uncommon and widespread in reedbeds. In winter also found in coastal marshes. Very secretive, usually staying well hidden. Occasionally comes to reedbed edge but often 'freezes,' sometimes with head pointing skyward, "like a reed," making it easy to overlook. Walks deliberately, foot kicked high before slowly moving forward. Sometimes has such confidence in its camouflage that it allows close approach, looking at you with weird facial expression. Usually calls: a night-heron-like *squak* but higher-pitched. Superb breeding-ground booming call: a gulping *gunk a gunk*, usually repeated several times—the sound of n. marshes! **ID**: Fat body and long neck that, when stretched, gets progressively narrower. Dark malar disappears with change in posture; boldly striped breast. Ad ♂: averages longer-billed. Malar averages bolder, and has intricately marked upperparts, slightly coarser in adults than juv but differences difficult to judge. Imm night-herons smaller, shorter-necked, and smaller-billed, not so brown, and lack black malar stripe. In flight, AMBI's flight feathers contrast with upperparts. Shape very similar to much smaller GRHE; in silhouette, AMBI has slower wingbeats, longer bill, and huge trailing feet.

juv.

ad. ♀ br.

ad. ♂ br.

Least Bittern *Ixobrychus exilis* **LEBI** L 13in

Uncommon and local in reedbeds. Very secretive and easy to overlook. Sometimes stands at muddy edges, tending to look very small next to large reeds. Frequently perches on reeds, legs often splayed for balance. Slowly moves head and neck toward prey before incredibly fast grab with bill. More often heard than seen. Gives raspy *kek-kek-kek-kek* call. Also a soft BBCU-like *coo coo coo* (day or night). Most often seen in flight moving from one patch of reeds to another. Size, shape, and bold color patterns make this bird a straightforward ID. **ID**: Very small. Appears fat but can also look skinny when neck stretched. A lightweight bird. All plumages same 2-toned pattern: warmly colored with dark cap and back and with white lines. Ad ♂: black cap and back. Brown replaces black in ♀. Juv: similar to ♀ with bolder breast streaking and pale fringes to coverts. Back feathers are molted during winter—black for 1st-w ♂. Often confused with larger and chunkier GRHE. LEBI is always paler and buffy, especially on the wings.

1st-s.

juv.

ad. br.

Green Heron *Butorides virescens* GRHE L 18in

Widespread, but common nowhere. Usually found alone, sitting quietly in streams or at ponds, though frequently sits higher in trees. Also in marshes, especially winter. Very easy to overlook. Stands with body at 45 degrees, suddenly lowers its head, walks a few paces, leaning forward, before quickly snatching its prey. Most visible flying between feeding areas or when flushed, when it often gives a surprisingly loud and piercing *skeu*. Tends to be active at twilight, when in silhouette looks like a small AMBI with faster wingbeats and smaller feet. **ID**: A dark, short, and compact heron that usually appears neckless. Even though small, the large bill and glassy stare suggest it is not to be tangled with. Ad br: black cap, purplish brown neck, and line down breast. Turquoise upperparts with narrow white fringes. Bill and lores black, legs orange. Ad nonbr: legs, lores, and bill, particularly lower mandible, turn yellow. Juv: extended breeding season, so this plumage seen much of the year. Extensively streaked neck and broad buff tips to coverts. Adult-type feathers are grown in through winter. 1st-s: still retains some juv feathers, most easily seen on neck. This species shows a lot of variation in imm plumages.

Yellow-crowned Night-Heron *Nyctanassa violacea* **YCNH** L 24in

Scarcer than its cousin, and found in a wide range of wetland habitats, but often in swamps and more heavily wooded areas. Roosts and breeds with BCNH and other herons but also nests in small isolated colonies of a few pairs in wooded areas away from water. Feeds mainly on crabs, usually on falling tides. Call: similar to BCNH but higher-pitched. **ID**: Taller and slimmer than BCNH with longer legs and neck. Pointed head accentuated by long chin. Bill all dark—BCNH often has yellow base to bill. Adult: zebra-stripe head pattern on gray. 'Yellow crown' often appears white. Legs pink in high-breeding plumage. 3rd–yr: as adult but some aged by brown in crown and upperparts. 2nd-yr: brown in crown stripe and has imm feathers in upperparts; sometimes dusky striping on underparts. Juv: from BCNH by grayer appearance, smaller spots on upperparts, more distinctly marked underparts—always confirm by shape! In flight, all plumages have contrasting dark trailing edge to wing. Structurally YCNH appears significantly slimmer, wings narrower and more pointed, with feet projecting much farther past tail.

1st-s.

2nd-/3rd-yr.

ad.

juv.

Black-crowned Night-Heron *Nycticorax nycticorax* **BCNH** L 25in

Fairly common on ponds, marshes, and in other wet areas. Colonial nester, often with other herons and egrets. Usually roosts in trees or reedbeds, but also in unexpected spots such as under bridges. As the name suggests, it is mostly nocturnal (also YCNH); birds fly out of roosting areas to feeding grounds at twilight. Loud *wock* a common sound at night in marshes. When feeding, crouches like smaller GRHE, occasionally walking a few paces hoping for better luck. Typically eats fish. Compact shape, striking in flight, emphasized by rounded wings. **ID**: Fairly big and stocky with short neck and big head. YCNH has a different shape. Plumages show much variation. Adult: black cap and back, gray wings, underparts vary from white to pale gray. Legs pink in high-breeding plumage. White head plume. 2nd-/3rd-yr: a subdued version of adult with grayer back and brown mixed in coverts. Bill often pale-based. Iris orange/red. Juv: dingy brown overall with blurry streaks on underparts and large triangular spots to wing coverts. Bill has yellow at base. 1st-s: variable with some birds still in worn juv plumage and others with dark caps and brown backs without spotting.

ad. br.

juv.

ad. nonbr.

Tricolored Heron *Egretta tricolor* **TRHE** L 26in

Uncommon to locally common heron in coastal marshes, usually with other herons and egrets. Very active, often runs around frantically before spearing fish. Frequently raises one or both wings while hunting, possibly to create shadows on the water for enhanced visibility. In flight, very slim and long-winged with diagnostic white belly and underwing coverts. Its curled-up neck sits much lower than the rest of the body, a good ID feature. **ID**: Strikingly slim with very long slim neck. Pointed head and very long bill. At distance, all plumages appear 2-toned. Ad nonbr: upperparts and neck subtle blues and purples, highlighted by white belly. Back plumes year-round. Yellow bill and lores. Ad br: plumes grow longer, bill and lores become blue, and legs darker. Juv: same basic color pattern as nonbr but with chestnut neck and fringes to coverts.

Tricolored Heron

1st-s.

ad.

juv.

Little Blue Heron *Egretta caerulea* LBHE L 24in

Fairly common bird of freshwater and coastal marshes, becoming scarcer inland. Usually seen alone or in pairs rather than in groups. SNEG-size but noticeably more compact and neater. Tends to feed quietly, standing still patiently or moving slowly and deliberately with head kept close to water. **ID**: From TRHE by compact structure and lack of white belly. Colors similar to significantly larger REEG. If in doubt focus on the bill pattern. Adult: at distance appears dark blue-gray, but close views show a beautiful purple head and neck contrasting with paler blue base to bill. Juv: uniquely all white. Easily confused with SNEG but best told by uniform yellow-green feet and legs. Other differences are duller yellow-green lores, bicolored and differently shaped bill. The neck is subtly shorter and thicker, the head appears more rounded. In flight, has dark wingtips; these can be hard to see. By late winter, new adult-type blue feathers are molted in, creating a patchwork. 1st-s: blue feathers start to outnumber white. As adult by second fall.

juv. dark morph

br. white morph

2nd-yr. dark morph

ad. nonbr. white morph

ad. dark morph

Reddish Egret *Egretta rufescens* **REEG** L 30in

Scarce s. egret. Occasionally wanders n. Found in coastal areas, particularly open saline pools. Often solitary, dashing around frantically on tiptoes; quickly stops and changes direction before stabbing or grabbing its quarry. Wings often raised, creating an umbrella. **ID**: Larger than other similar species, alert and erect posture enhances this impression. Large and slim but still sturdy with hefty bill and dark legs. Only GREG has a longer neck and is bigger. Shaggy medium-length plumes. Behavior often draws attention, but other species can behave similarly. Adults have strikingly bicolored bills. 2 morphs create different ID problems. Dark morph is commoner of the 2 forms. Similar to 20% smaller LBHE with brighter neck. Breeding birds have a striking 2-toned bill. Nonbr and 2nd year: bill is darker and plumage duller; dark lores and bill give a mean expression. Juv: a very dull and pallid version of adult with pale-fringed coverts. White morph: Ad br: striking 2-toned pink-based bill is not shared by any other white egret.

juv./1st-w.
juv.
ad. nonbr.
ad. high-br.
ad. br.

Snowy Egret *Egretta thula* **SNEG** L 24in

Widespread and common. The most abundant heron in fresh and salt water in most coastal areas. Alone or in groups, it is also the dominant species when feeding frenzies occur. Feeding behavior is varied. Will stand patiently waiting for movement, at other times pats its feet or bill in the water to create fish-attracting ripples, then suddenly stands erect and walks with purpose. Also stands motionless with bill half in the water. Occasionally it breaks into a sprint and tears around like REEG. Antsy, it often gives a gutteral *aargh*. **ID**: The black-billed egret with dark legs and yellow feet. Yellow lores. Leg, bill, and lores change color, so know the size and shape well. A smaller egret with very slender neck and dagger-like bill. Ad high-br: plumage held briefly in spring with bright red lores and long head and back plumes. Lores soon turn yellow, and plumes on head, neck, and back become worn through summer. Legs all black. Ad nonbr: yellow extends up back of legs, plumes reduced but some still present. Bill paler at base. Juv: very young birds can be confusingly dark-lored, pale-billed, rounder-headed, and lack plumes. Similar juv LBHE has unicolored greenish legs. Soon looks like ad nonbr, but lacks plumes.

Cattle Egret *Bubulcus ibis* **CAEG** L 20in

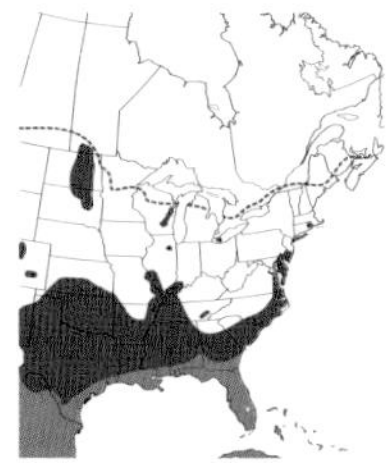

Common in the S, increasingly scarce farther n. Usually occurs in flocks. The odd egret out: small, hunchbacked, and dumpy, feeding around pastures, mostly on insects, as it struts with head rocking back and forth. Often found precariously near cattle and other livestock (even tractors) that kick up grasshoppers and other insects. Can occur in wet meadows, occasionally near other egrets. Chooses roosts and nest sites near water, sometimes with other egrets. Usually in groups that fly to roost sites in tight packs. Small size, flat undercarriage, and compact shape with short 'tucked-in' neck, separate it from other egrets. **ID**: The small orange-billed egret with short legs and neck. Proportions and shape at odds with other egrets. Only other egret with orange bill is GREG. Ad high-br: orange plumes on head, back, and breast. Vividly bright bill, lores, and red legs. Plumage held briefly in spring. Ad br: the bright bare part colors are replaced by the more familiar orange. Nonbr: completely white with yellow-orange bill and darker legs. Juv: as nonbr other than black bill, which soon becomes adult-like.

ad. high-br.

ad. br.

juv. /ad. nonbr.

Great Egret *Ardea alba* **GREG** L 39in

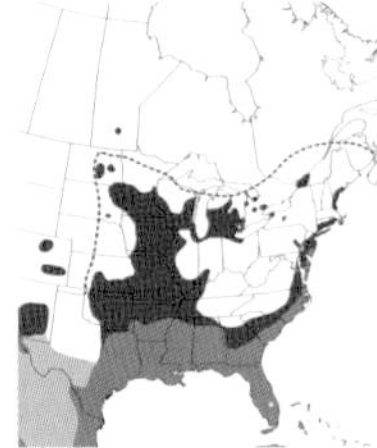

Widespread and common, particularly in coastal areas. The only egret in many inland areas. Found primarily near water but also in fields. Elegant, with snake-like long neck. Often seen standing, quietly waiting for prey, but sometimes stalks with imperial gait, or with neck at 45 degrees, scanning side to side. Doesn't run. In flight, has slow, deep wingbeats, more like GBHE than other egrets. The wings are long and slim, and the kinked neck droops a long way below the body. When agitated will fly briefly with neck held straight. Takeoff can be a struggle, and is often accompanied by a grunting *kaark*.

ID: By far the largest egret. A little smaller than GBHE. The only egret with orange bill, except for smaller CAEG. No other egret has such a long, thin neck, often held in an exaggerated 'S,' helping to id the most distant specks. Long legs enable it to wade in deep water. Ad high-br: plumage held briefly in spring with bright green lores and darker bill. Develops very long plumes on back called aigrettes (the name 'egret' comes from this French word). During summer, lores become yellow, bill paler yellow, and plumes shorter. Adult nonbr/juv: by Aug most lose plumes and look the same as juv.

3rd-yr.

2nd-yr.

juv.

ad.

Great Blue Heron *Ardea herodias* **GBHE** L 46in

The common and familiar large heron in most areas. A loner that usually feeds near water, often at night. Movements, if any, are usually slow and deliberate. Nests in colonies, with large stick nests in treetops (built in or returned to in late winter, so often easy to see). On migration, usually travels in flocks. Frequently calls in flight, a loud, slightly scary *raak*, mostly heard at night. **ID**: Largest and sturdiest heron with heavy bill and lots of variation in plumage. All have gray upperparts, black on crown, paler underparts. Most show black-and-brown bend to wing and brown thighs. 2 color forms. See Great White Heron (white form, p.201). Ad nonbr (dark form): sexes similar. 4–5 years to full adulthood. Aged by adult coverts and white crown stripe. Legs dark, bill yellow. Ad br: soft-part coloring brighter and variable. Bill deep yellow or orange, legs can be green, yellow, red, or black, and lores purple-/blue-toned. 3rd-/4th-yr: similar to adult with gray in crown. 2nd-yr: gray in crown, usually some old juv feathers. Juv: plumage held for most of first year. Overall dingier and more uniformly marked than older birds. Upperpart feathers pale-fringed, usually cinnamon; dark cap, extensive streaking on neck.

Great Blue Heron (Great White Heron) *Ardea herodias* **GBHE** L 46in

White form of GRBH is sometimes considered a distinct species. Restricted to FL Keys, though vagrants have reached as far as the NE. Very easy to overlook as GREG but is larger with a substantially bigger bill and yellowish legs. Dark lores add to mean look. Compared to dark form of GBHE is 10% larger, bigger-billed, and has much shorter head plumes. Intermediate birds, known as 'Würdemann's Heron,' have characters essentially like a pale GBHE with a light head.

Little Egret *Egretta garzetta* **LIEG** L 25in

Very rare vagrant. Very similar to SNEG but with pale gray or duller yellow lores. Feet slightly paler yellow. Yellow rarely extends up back of leg and if so only a little but is often on the front too. Breeding birds have 2 plumes on back of head. Slightly larger and longer-necked, often with fiercer expression than SNEG due to different loral pattern. Bill slightly longer and narrower.

Greater Flamingo *Phoenicopterus ruber* **GREF** L 46in

Seen almost annually in the Everglades (thought to be wild birds). Picture clouded as commonly kept in captivity. This and other flamingo species sometimes escape from captivity. Remarkably long legs, funky bill, and head shape prove unmistakable. Adults are bright pink, imms paler with dark marks in the wing. Other flamingo species should be excluded.

Roseate Spoonbill *Platalea ajaja* **ROSP** L 32in

Uniqueness, elegance, and soft colors make this bird a real treat. Uncommon in the S in mangroves, swamps, brackish and freshwater wetlands. Occasionally wanders n. Colonial nester. Usually seen feeding in small groups, often with other herons. Walks steadily, often crouched, its spanking spatulate bill held nearly horizontally as it swishes from side to side through the water. Takes shrimp and other aquatic prey. Hard to overlook though worn and juv birds can look white rather than pink. In flight, neck and bill are held outstretched below the body, as though it is too heavy to carry. Usually flies in lines. Occasionally soars. **ID**: Bill shape always striking. Base of bill noticeably broad at face. Ad br: large, scaly black nape band. Plumage bright pink. Large ridges on gray bill. Orange tail and pink rump. 2nd-/3rd-year: similar to adult but averages duller, less black on nape, fewer ridges on bill, and dark wingtips noticeable in flight. Juv: pale pink and white with ridgeless pale bill. Pink tail and dark wingtips obvious in flight.

White Ibis *Eudocimus albus* **WHIB** L 25in

Common in se. coastal and freshwater wetlands, mangroves, beaches, lawns—even Disneyland! Often wanders out of range. Can be very approachable, tending to walk away with 'attitude' rather than fly. Feisty, often squabbling with others. Highly social, occurring in large flocks. Creates lines in flight, seemingly in sync, shallow flaps followed by short glides. **ID**: 2 distinct color patterns combined with shape and downcurved bill make ID straightforward--typical ibis shape with long downcurved bill, never all brown like other other regularly occurring ibis. Ad: all white with bright pink legs, bill, and face. In breeding season has stunning red tone. Juv: brown back and streaky neck, contrasts with white underparts. Young ibis are often smaller than parents with shorter bills. Unlike most birds, it can take several weeks for young to be fully grown after fledging. 1st-s: white adult-like feathers are molted in through first year, so many birds look pied. Some 2nd-yr birds can be aged by retained imm feathers.

Glossy Ibis *Plegadis falcinellus* **GLIB** L 23in

Fairly common in coastal marshes, ponds, and flooded fields. Probes mud or sweeps water for food. Usually occurs in flocks. Faithful to favored feeding and roosting places. Birds in saline areas regularly return to freshwater pools to bathe and drink. Small flocks often seen flying in evenly spaced lines, with intermittent flaps and glides. Range barely overlaps with similar WFIB. Both species regular in other's range (ratio 1: few hundred), happily mix, and sometimes hybridize, offspring showing mixed characteristics. Increasing frequency of out-of-range birds could be greater observer awareness or genuine range expansion. **ID**: Dark in all plumages with dark eye and loral skin bordered by pale blue. Ad br: light blue (usually appears white) border to lores. Iridescent green-and-purple coverts. Scapulars, mantle, and neck rich brown. Nonbr: neck streaked, reduced iridescence and generally duller. Juv: very young birds smaller, lack white border to lores, shorter-billed, often with white patches on head, neck, and bill. Becomes more like ad nonbr through fall with smaller rounder wing feathers, duller green wings, and purple brown-back with just a hint of iridescence. 1st-s: as nonbr.

1st-yr.

ad. nonbr.

ad. br.

ad. nonbr.

White-faced Ibis *Plegadis chihi* **WFIB** L 23in

W. counterpart of GLIB. Uncommon in shallow wetlands, flooded pastures, and agricultural fields. Tends to breed on islands with bushes and reeds. Behavior essentially identical to GLIB, and vagrants always mixed together. **ID**: Red iris diagnostic in all but younger juvs. The facial skin varies in color (pink to red) but is also diagnostic. Averages slightly larger than GLIB. ♂ longer-billed than ♀. Ad br: deep pinkish facial (loral) skin with complete white surround. The surround appears even-widthed, however wide it is overall. On GLIB it always disappears behind the eye. Legs variably pink, brightest at knee—a good way to check through crowds of usually duller-legged GLIB. Nonbr: no white border to duller pink facial skin and red iris makes them harder to pick out. Upperparts duller and more green, head paler brown with white flecks and bare parts darker. Juv: as juv GLIB until iris becomes red and facial skin pink in fall. 1st-s: as nonbr. Given that it is often found with GLIB in summer, it is not surprising hybrids are being spotted more frequently. Facial features of hybrids should show characteristics of both species.

Limpkin *Aramus guarauna* **LIMP** L 26in

Scarce and local in wetlands and swamps with abundant snails. Weird looking, more akin to a massive kiwi than anything found in the US. Despite large size, presence is often first announced by a loud, wailing cry: *KeEEEuuur*. Largely crepuscular, seen flying to feeding grounds with deep wingbeats and exaggerated flicked upstroke. **ID**: Emu-like with long legs and massive body; small head and medium-length bill designed for snailing. All plumages chocolate-brown with white flecks. Head and neck more white than brown. Juv: averages duller-billed and has narrower white flecking.

Northern Jacana *Jacana spinosa* **NOJA** L 9.5in

Very rare s. visitor. Turns up in wetlands but also in more unexpected places such as golf courses. Known as the 'Lily Trotter;' ridiculously long toes help it do this. Yellow 'spur' at bend of wing. Brilliant yellow bill and frontal shield. Unique appearance suggests taxonomic status somewhere between shorebird, rail—and the unknown. Dark body and maroon wings come alive in flight, with brilliant yellow flight feathers. Nondescript juv plumage becomes more adult-like after a few months. Vocal in flight—raucous rail-like notes.

Wood Stork *Mycteria americana* **WOST** L 40in

S. species, locally common in swamps, ditches, and wetlands, particularly in shallow food-rich pools. Occasionally wanders out of range. Taxonomically a relative of vultures, it also looks a little odd, with bald, scaly head a clear similarity. Sometimes alone, but also in big flocks, particularly in drying water holes. Stands laconically, sometimes walking slowly to a nearby spot. Feeds by putting bill into water and 'feeling' for prey, often with bill open. Also foot-patters to stir things up. It will frequently hold a wing out to provide shade and better visibility. Sometimes sits in trees for long periods. Soars high on thermals with wings slightly bowed, flapping less as it gets higher and until nearly out of sight. Look closely at shape to distinguish from similarly patterned AWPE. **ID**: Massive with a club-like bill. Adult: extensive wrinkled gray bare skin on neck, smooth black plate on crown, and smooth pale forehead. Juv: yellow-brown with black-feathered neck and head. Bill paler with yellow tones. 2nd–4th-yr: bare skin grows gradually to replace feathered area. Molt pattern and replacement of retained juv feathers helpful to age accurately.

juv.

1st-w.

ad. br.
'rust-stained' Lesser

ad. Greater

Sandhill Crane *Grus canadensis* **SACR** L 46in (Greater), L 41in (Lesser)

Widespread but localized with several populations. Breeds in marshes and disperses to open fields, pasture, or arable land. Its bugling call carries miles. Well-known spectacular winter and staging gatherings can involve flocks in thousands. Platt River, NE, has around half a million birds in spring. Frisky courting birds dance up and down, wings open and bill pointing skyward. Walks methodically with neck tilted forward, and feeds by picking and probing. **ID**: Cranes are tall and uniquely shaped, with a heavy body and elongated wing feathers that hang like a feather duster. Lesser Sandhill Crane breeds in Arctic Canada and winters in TX and to the w. Averages smaller, shorter-billed, shorter-legged, and longer-winged. Greater Sandhill Crane—cen. Canada s. with resident populations in the SE—is larger in all aspects but shorter-winged, has paler primaries and a higher-pitched call. Adult: gray with red crown. Plumage often rusty in summer due to staining from soil (a process still not fully understood). Juv: buff-fringed coverts and little or no red on crown. Some 2nd-yrs can be aged by molt pattern and amount of red on crown.

Whooping Crane *Grus americana* **WHCR** L 52in

Very rare and endangered. Breeds in Wood Buffalo National Park in s. NT. Winters in coastal marshes around Aransas NWR, coastal TX. Birds are found spaced out in pairs or family parties of up to 5 birds. Migrants seen in between breeding and wintering grounds. Introduced populations in cen. FL and WI. **ID**: Egret-like, all white, but a massive crane with red crown and whiskers. Black wingtips stand out in flight. Juv: strikingly orange, buff, and white through first year. Adults stay with young through first winter. Sometimes confused with AWPE in flight.

Common Crane *Grus grus* **COMC** L 48in

Very rare European species. It is unclear whether our birds are true vagrants or escapes. Most 'legit' records are of birds found in huge SACR flocks in spring in the NE and surrounding states. Has bred with SACR in NJ. Gray like SACR but larger and heavier with black-and-white neck, red crown, and more extensive black in tail. Hybrids show intermediate characteristics.

ad. with young

juv.

Gulf Coast ad.

eastern ad.

Clapper Rail *Rallus longirostris* **CLRA** L 14.5in

Common rail in saltwater coastal marshes. Sometimes seen at low tide coming into muddy creeks but generally more often heard than seen; call is a series of clapping notes, accelerating at end. Clap your hands rythmically and birds will often respond. Also gives a series of repeated grunts. Sometimes occurs in unusual places after flooding. Tends to walk hen-like, always alert for predators, and will suddenly lean forward and run back into cover. **ID**: Very large and heavy with long bill and neck. All races are duller than KIRA. Much individual and racial variation. Gulf Coast birds have brighter underparts than Atlantic Coast populations. Adult: breast uniform buff to cinnamon contrasting with dark, white-barred belly and gray-and-white flanks. Upperparts brown with diffuse black streaks. Gray head not so strongly contrasting as VIRA. Gulf Coast birds are most similar to KIRA, but are not so boldly marked on flanks, are paler-chested, and have less contrasting dark centers to wing feathers. Juv: very dingy version of adult. Molts to adult-like plumage by fall.

ad.

1st-w./ad.

King Rail *Rallus elegans* **KIRA** L 15in

Uncommon in freshwater marshes and coastal marshes in fresh or brackish water before it turns too salty (where CLRA range ends). Most birds are identified from very similar CLRA by habitat. CLRA's call is slightly deeper, slower, and more evenly paced than KIRA's. Grunt call noticeably deeper. Hybrids supposedly occur where ranges overlap (usually brackish marshes), to make ID even more interesting! Behavior as CLRA. **ID**: Adult: told with difficulty from CLRA by more richly colored underparts lacking gray, and more contrastingly barred flanks and belly. Upperpart feathers have darker centers with bolder rufous edges to coverts. The lower cheeks are usually buff, so the face appears less gray. Typically a bolder and brighter bird! VIRA is much smaller and slimmer with a more boldly patterned face and brighter bill. Juv: a dingy version of adult. Like juv CLRA but usually more boldly patterned with darker upperparts and with brighter rufous fringes to coverts. Most birds have molted into adult-like plumage by fall.

ad. with chick

juv.

ad.

Virginia Rail *Rallus limicola* **VIRA** L 9.5in

Widespread and fairly common breeder in freshwater and coastal marshes. More localized in winter with majority in coastal marshes. After freezing weather sometimes shows up in unexpected places away from water. A small slim version of KIRA. Fidgety, even for a rail, always flicking its tail and making sudden jerky movements. Never relaxes; the neck is always extended as it looks around, its speed of foot often changes, and it will suddenly dash for cover, often seemingly for no good reason. Heard freqently: a rapid staccato *kadik kadik kadik*, and also gives a *wink wink wink* (a descending squeal, like a pig). **ID**: A little larger than SORA, much smaller than KIRA. Large-bodied but with longish legs and neck. Its usually alert posture adds to its slim appearance. Adult: like a small KIRA with long narrow-based bill, largely orange, and a gray face that contrasts with black cap. Looks dark at distance or in dense cover. Juv: initially very dark and dull. Extensive black on breast, and dark-billed. Molts into adult-like plumage by fall.

juv./1st-w.

ad./imm. ♂

ad. ♂ br.

Sora *Porzana carolina* **SORA** L 8.75in

Widespread and fairly commom rail in all but coastal marshes (rare). Sometimes walks around edges oblivious to passersby but at other times is very shy. A more methodical strut than other rails with rear end held in the air and tail costantly flicked like COMO. Light enough to walk on horizontal plant stems, carefully picking food off ground or water. Song is a *Koo whit* followed by a descending whinny. Also *so-ra*. Call: a sharp *kip*. In flight, has narrow white trailing edge to wing on otherwise all-dark upperparts. **ID**: Smaller than long-billed rails but larger than other short-billed rails. All plumages have brown upperparts with dark-centered feathers and lines of white. Barred flanks, buff undertail, and broad-based short bill. ♂ has more extensive black throat than ♀ in all plumages. Breeding birds (Jan–Aug) have more extensive black on throat and brighter yellow bill. In winter they are duller, and ageing and sexing birds is especially difficult at ths time. Juv: brown neck, breast, and cap. Black in cap, face, and throat, and gray in neck is molted in through fall—migrants show a range of variation.

ad. ♂

Black Rail *Laterallus jamaicensis* **BLRA** L 6in

Scarce and declining in coastal marshes. Isolated pockets inland in reedbeds and meadows. Winters in s. coastal marshes. Brutally difficult to see unless pushed out at highest tides. Stays hidden behind reed stems even when in densest cover. Often confused with all-dark chicks of other rails. Song: a fairly quiet *kikidoo*, usually given at night. Agitated ♂ gives growling *grrr*. **ID:** Tiny (sparrow-sized), squat, all-dark rail with white spots, red eye, and brown nape. Juv: brown iris. Wings all dark in flight.

juv./nonbr.

Yellow Rail *Coturnicops noveboracensis* **YERA** L 7.25in

Uncommon breeder in wet knee-high prairie bogs, grassy marshes, and wet meadows. Winters in coastal marshes, rice fields, and wet meadows. A real skulker. Most often seen in flight: broad white secondaries striking (beware, SORA has narrow white trailing edge). In summer, listen (usually at night) for morse-code call, easily imitated by tapping pebbles together. Often flushed when rice fields harvested. Probably widespread migrant in Oct and winter range farther n. than known. **ID:** Between BLRA and SORA in size. Adult: 2-toned yellow-and-black. Nonbr/juv: duller brown tones.

Purple Gallinule *Porphyrio martinica* **PUGA** L 13in

Uncommon in freshwater marshes and has an affinity for water lilies. Scarcer in winter. Regular out of range. Happier walking on floating vegetation and climbing reeds than swimming—very large feet perfect for this. Tail often held cocked, revealing all-white undertail coverts. Often found with similar COMO, AMCO, and PUSW. It is smaller, slimmer, and longer-winged than these. Color patterns, brilliant in adults, make ID straightforward. **ID**: Adult: Beautiful purple-blue underparts, turquoise on neck sides. Upperparts vary from green or turquoise to copper, partially dependent on light conditions. Sky-blue ornamental shield larger in ♂. Bill bright red with yellow tip. Juv: pale brown, often with olive tones, darker crown, green legs, and dull bill. Brown upperparts with green cast to wings. Yellow legs bright on otherwise bland bird. Molts into adult-like plumage through first year. By spring is a dull version of adult and with a small frontal shield.

Purple Swamphen *Porphyrio porphyrio* **PUSW** L 18in

Established population in Broward and nearby counties, FL. Large and fat-bodied but has small head. Combination of purple and blues like PUGA, but much bigger; dark red legs, frontal shield, and bill.

juv.

ad. ♂

♀/nonbr.

Common Moorhen *Gallinula chloropus* **COMO** L 14in

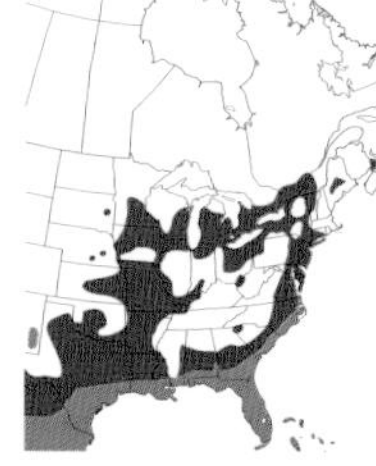

Common in freshwater marshes with reedbeds, often near humans. Usually stays in dense vegetation but sometimes ventures into open areas. Happily walks around water's edge or on floating vegetation but more often swims. Flicks tail while walking, and jerks head back and forth while swimming as if it will help it to go faster. Aggressive, often scrapping with other birds in its space. Small all-black chicks often cause confusion with other species. Noisy: a variety of screams, shrieks, and chatters. **ID**: Slim-necked, flat-backed, when swimming, with long tail. Serrated white flank line, and white undertail coverts have a black divide; green legs on otherwise dark bird. Ornamental shield (connected to bill) larger in ♂ and older birds. Adult: yellow-tipped red bill and frontal shield. Slate-gray underparts with white flecking in ♀, brown upperparts. Plumage averages duller (Jul–Dec). Juv: generally brown, grayer below with pale throat—dingy overall with brown bill that lacks 'oomph' of parents'! Frontal shield develops through first year and colors more solidly dark; adult-like after a year. Often found with larger and heavier AMCO. Latter is uniformly dark bodied with a pale bill.

juv.

ad.

American Coot *Fulica americana* **AMCO** L 15.5in

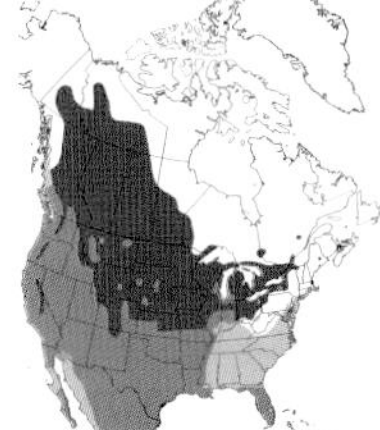

Widespread in wetlands, usually with open water, where upends or dives for submerged aquatic vegetation. Very common in some areas, particularly in the S, sometimes in many hundreds. Usually in loosely formed flocks but also forms dense lines. Mixes with ducks but is clearly the odd one out. Walks on lawns and golf courses with huge funky feet (akin to flippers). Also a regular for bread at the town park. Often flushed by boats—has a running takeoff (dark gray with a white trailing edge to wing) but is a strong flyer once it gets going. Can be very aggressive (see plate above). Has a variety of short nasal clucking call notes, similar to COMO but deeper. **ID**: Slate-gray with a white beacon on head. Dark colors often result in it being misidentified for COMO, but sturdy build and white bill make for straightfoward ID. Rust-red frontal shield (not connected to bill) larger in ♂ and older birds, smaller Jul–Dec. Adult: uniform slate-gray with white bill and red subterminal tip. Juv: medium-gray above, paler below with 'dirty' gray-white bill. By late fall, most birds are similar to adults.

UPLAND GAMEBIRDS

Gamebirds are chicken-like. They spend most of their lives on the ground. Yet, although most species are thought of as ground-dwelling, some are often happy in trees, eating buds and roosting, particularly in cold winter months.

They tend to be very hard to see under most conditions, and will often hunker down, their cryptic colors keeping them amazingly well hidden. When you get too close they 'explode' from the ground, no doubt scaring you more than them. Round-winged, they fly with a series of shallow flaps followed by glides, and always run back into cover as soon as they land. Most species are usually found in small flocks, called coveys, and many have distinctive calls that give away their whereabouts. All are sedentary and extremely hardy, quite happy to dig under snow for food and sometimes to roost.

Several species have spring leks where gatherings of males give incredibly boisterous displays to attract females. Some of these leks are open for public viewing and provide a fantastic spectacle. The sounds are equally remarkable.

Domestic Gamebirds

Common Peafowl (Peacock)

Helmeted Guineafowl

ad. ♂ western

displaying

ad. ♀ with chicks

ad. ♂ eastern

Wild Turkey *Meleagris gallopavo* **WITU** ♂ L 46in, ♀ L 37in

Fairly common despite being widely hunted, and is increasing in numbers and expanding range, often to more suburban areas. Typically found in forested areas with adjacent clearings. The wild ancestor of the domestic turkey. Typically wary, but can sometimes be quite tame, walking on roadsides, seemingly oblivious to traffic—surprising since it is hunted heavily. Often seen roaming around in flocks on field edges, walking to cover at its own pace when disturbed. Terrestrial but occasionally sits in trees, mostly to roost at night. In flight, has labored running takeoff before being high enough to glide with occasional flaps. Huge with pale rounded wings and black subterminal tail band. In courtship, tail is like a large Venetian fan and chest is puffed out. Familiar gobbled call, as in the movies. **ID**: Unmistakably massive body with with tiny, wrinkled head. ♂: body often appears black but in good light shows fantastic iridescent browns, greens, and reds. Head variably blue and red (wattle usually red) and has large black chest spur. ♀/imm: smaller, less iridescent, usually lacks chest spur and has grayer head. Birds in w. part of range show more extensive pale tips to tail feathers and primaries.

ad. ♂ European

ad. ♀

ad. ♂ Asian

Ring-necked Pheasant *Phasianus colchicus* RNEP L 21in

Introduced species found in open areas, particularly agricultural with nearby cover, ranging from low scrub to forest. Scarce in most areas but population fluctuates locally due to release from breeding programs, mostly catering to hunters. Struts around slowly and deliberately, often stopping to put head up into air to check surroundings. Frequently in loose-knit groups and often roosts in trees. Populations have variable appearance depending on origin. Crossbreeding makes subspecific ID almost impossible. In flight, bursts from the ground, wings making constant noise. Once away, it has intermittent short glide and wing noise. ♂'s call is a very loud, far-carrying *ko-ok.* **ID**: Very large with distinctive long tail in all plumages, shorter in ♀. Only GRSG has similarly long pointed tail. Ad ♂: unmistakable copper-toned stunner with bold red face. Birds resembling Asian race are commonest. Ad ♀: variable, from pale to dark brown. Fairly uniform and nondescript with spotted flanks. In flight, long pointed tail usually stands out.

Plain Chachalaca *Ortalis vetula* **PLCH** L 22in

Common and familiar bird in thorny forest in the Lower Rio Grande Valley, TX (introduced on Sapelo Is., GA). Normally secretive, many have become tame at feeding stations. Usually seen walking around on the ground in flocks, but quite happy eating fruit in trees. Hops on branches, often flies, crashing through trees en route. Frequently gives a scarily raucous call: a never-ending, up-and-down, *give-it-up* (*cha cha lac*), often given by several birds at a time—a real racket! **ID**: A slim, hen-like bird with a long white-tipped tail. Adult: fairly nondescript, gray head, olive breast and upperparts, dull yellow-orange below. Juv: similar but usually duller with fluffy undertail coverts.

ad. ♀ ad. ♂

Gray Partridge *Perdix perdix* **GRAP** L 12.5in

Introduced European species. Scarce and declining in open agricultural areas or in grass fields. Always in small groups. Very hardy, coveys often hide under fir trees and other windbreaks in winter. **ID**: 25% larger than NOBO. Often stands erect as if on guard. Compact with orange corners to very short tail. All plumages have gray neck and broad rufous flank bars. Ad ♂: striking dark belly patch with orange face. Ad ♀: lacks belly patch and has paler face divided by gray ear coverts. Juv: plumage held for 2–3 months, face and underparts dull and lightly streaked; no belly patch.

transitional (early summer) ♂

transitional (early summer)) ♀

Willow Ptarmigan *Lagopus lagopus* **WIPT** L 15in

Common in willow thickets and in low vegetation near treeline. Short-distance altitudinal migrant. Plumage matches environment. Unforgettable, comical call—a chuntering gargle of notes. Walks, occasionally standing on rocks. Flies to feeding areas—a few flaps followed by long glides. **ID**: Very similar to ROPT; separate by larger, broader-based bill, lack of black eyestripe (nonbr ♂) and by warmer red tones in breeding birds. Both sexes all white in winter with black outer tail. In spring, rufous head and neck. Ad ♂ br: all rufous with white belly. Ad ♀ br: coarsely barred. Warmer tones than ROPT.

Rock Ptarmigan *Lagopus muta* **ROPT** L 14in

Often in colder, higher, and more barren landscapes than WIPT. Stays white longer in early summer to blend in with habitat. Slightly smaller than WIPT with narrower-based bill. Nonbr: all white with black tail corners like WIPT. ♂: has black eyeline (WIPT doesn't) with red above. ♀ is all white. Ad ♀ br: yellow-brown rather than rufous of WIPT. Ad ♂ br: barred gray-brown head and neck, black eyestripe (most noticeable in winter). Colder yellow tones and smaller red eyebrow than WIPT. Call: a series of dry throaty *creeks*.

ad. ♂

ad. ♀

Greater Sage-Grouse *Centrocercus urophasianus* **GRSG** L 28in

Localized and uncommon in sagebrush plains. Often in flocks, some large out of breeding season. Even though the largest N American grouse, blends into surroundings and is difficult to see. Crouches when approached but happy to fly long distances. **ID**: Range, and hulking appearance with long tail, make ID straightforward. ♂: much larger than ♀ with dark throat, white breast, and extensive dark belly. ♀: underparts same color as upperparts but with small black belly patch (lacking in RNEP). Formerly conspecific with Gunnison Sage-Grouse found in CO.

Sharp-tailed Grouse *Tympanuchus phasianellus* **STGR** L 17in

Fairly common but declining. Found in agricultural and prairie grassland bordered by trees. Usually on ground but often feeds on buds in trees, particularly when ground is snow-covered. Leks on top of knolls, sometimes with GRPC; occasionally hybridizes. Offspring of the latter show Intermediate characters. **ID**: Easy to confuse with prairie-chicken but has colder spotted upperparts, longer pointed tail, and paler underparts with distinct streaks/chevrons. In flight, look for long, white-edged tail. ♂ makes mellow cooing sounds at leks.

ad. ♂

ad.

Greater Prairie-Chicken *Tympanuchus cupido* **GRPC** L 17in

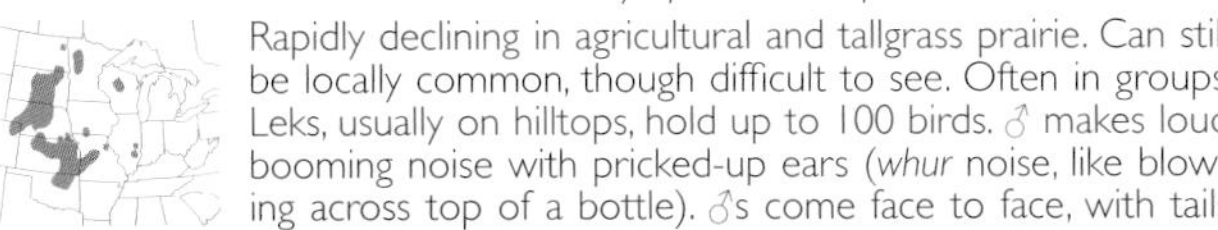

Rapidly declining in agricultural and tallgrass prairie. Can still be locally common, though difficult to see. Often in groups. Leks, usually on hilltops, hold up to 100 birds. ♂ makes loud booming noise with pricked-up ears (*whur* noise, like blowing across top of a bottle). ♂s come face to face, with tails cocked, before following each other like toy trains on a track. Multicolored neck sac (yellow in LEPC). ♂ and ♀ similar. Short, all-dark tail noticeable in flight. 'Attwater's' (TX race) 10% smaller and has shorter neck plumes (pinnae); highly endangered.

ad. ♂

ad.

Lesser Prairie-Chicken *Tympanuchus pallidicinctus* **LEPC** L 16in

Scarce and very local. Declining rapidly. Found in arid prairie grass, often in areas with shrubs. Very similar to GRPC but slightly smaller, shorter bill, paler and grayer with narrower barring, shorter neck plumes with darker borders. Realistically, most are identified by range. Occurs in flocks in winter. Gives gurgling *wump wump wump*, quieter and shorter than GRPC.

ad. ♂ drumming

ad.

Ruffed Grouse *Bonasa umbellus* **RUGR** L 17in

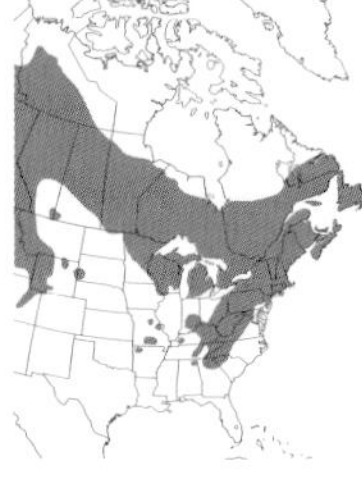

The common grouse in forested areas. Walks quietly on forest floor blending into surroundings. Usually flushed from underfoot—startling in a tranquil forest. Hunted extensively. ♂ drums mostly during early spring mornings, but sometimes all day and night from ground or flat-topped log. Sites are well worn with feathers and droppings. Difficult to hear, but drumming vibration may be felt. Can be imitated by thumping your chest progressively faster. Best located roosting in trees, feeding on tree buds, or crossing forest roads. In winter, often comes to road edge to eat gravel. Chicks leave the nest after a few days and are tended by parents. This is probably the easiest time to see them, their presence also given away by the parents' clucking. **ID:** Sexes similar. Medium-sized grouse with medium-length tail and small head. 2 color forms, rufous and gray, but many intermediate showing a mixture of both colors. The most striking features are the overall pale underparts with broad flank bars, crest, and broad band on barred tail.

Spruce Grouse *Falcipennis canadensis* **SPGR** L 16in

Fairly common in the N. Scarce in many areas in s. part of range. Found in dark, damp spruce and other coniferous forest with mossy areas and dense vegetation. Less common in other types of forest. Usually seen alone, but sometimes forms groups in winter. Famously tame allowing close approach—a real experience! Dark plumage matches habitat and makes it very easy to overlook. Found on ground or in trees (particularly when heavy snow). **ID**: Right habitat is often strongly suggestive of presence. Much darker than RUGR with boldly spotted underparts. The tail is black with rufous tips. ♀ is much less striking than ♂ and varies in color from gray to brown.

Northern Bobwhite *Colinus virginianus* **NOBO** L 9.75in

Uncommon in open, scrubby areas with cover. Sometimes in more heavily wooded areas. The most likely gamebird in many areas, but numbers are declining rapidly. Usually found in coveys of up to 15 birds. Often flushed from close quarters before being seen due to small size and preference for sitting tight in cover. Frequently flushes one-at-a-time, so check nearby vegetation. In flight, uniform upperparts, grayest on wing and tail. Small, with short, rounded wings; flies with fast flaps and short glides. More often heard than seen: emphatic *whit* preceded by quieter *bob*. Silent for much of the year and often doesn't start calling until late in spring. **ID**: Small and chunky with short tail and neck. Always with rufous tones though there is regional variation. ♂: striking head pattern with white throat and supercilium and peaked chestnut crown and dark eyestripe. Uniform rufous breastband. ♀: similar face pattern to ♂ but buffy yellow and brown. Underparts paler and more streaked.

Scaled Quail *Callipepla squamata* **SCQU** L 10in

Scarce and localized in hot, arid grassland or semidesert scrub with bushy vegetation used for shelter. Sometimes in backyards and at feeding stations. Always appears to be running for cover, either for protection or to get out of the midday sun. Often occurs in coveys. **ID**: Subtle shades of whites, browns, and gray, with distinctive black-fringed 'fish-scales' on underparts. Punky crest with a white highlight; ♀'s crest averages smaller and buffier. ♂ in s.TX has chestnut on belly. Variety of calls including squawky *weer* and rhythmical *chip-choo*.

Montezuma Quail *Cyrtonyx montezumae* **MONQ** L 8.75in

Very rare at edge of region in grassy undergrowth of wooded hillsides, mostly the Edwards Plateau, TX. Known as 'Harlequin Quail' due to striking head pattern. Usually feeds quietly, scratching at ground for food. Found in pairs or small groups, making it difficult to locate but fairly approachable, and will freeze rather than fly if disturbed. Best found by sitting quietly and listening for scratching noises. Sometimes seen at roadside. Flushes with loud wing pop. ♀ is a subdued version of ♂. ♂'s call is a loud *Wiiirrrr*.

RAPTORS

RAPTORS

Raptors are one of the most popular groups of birds in North America—and throughout the world. Many species are large and powerful, and spend a lot of time in flight, either on migration or in search of food. Many regions have hawkwatch migration sites, which are often great vantage points to see good numbers of birds and share insights and experiences with fellow birders. In recent times, the ban on certain dangerous chemicals, such as DDT, has aided in the recovery of raptor populations, including species such as Bald Eagle and Peregrine Falcon. Other species, such as American Kestrel and Ferriginous Hawk, have declined in certain areas due in part to loss of suitable habitat.

In-flight views of raptors are often distant, making it harder to judge size with accuracy. Smaller birds generally have faster wingbeats, larger birds slower. Shape and proportions are essential factors for ID of most, if not all, birds. Wing and tail shape are particularly important.

Female raptors are often substantially larger than males. In some buteos, juvs have subtly, but noticeably, slimmer wings than adults. The inner primaries of juvs are paler than those of adults in most species, creating a window that makes many buteos ageable, even at some distance. Some buteos vary in color from light to dark. Often categorized as morphs, these are usually broken down into light, intermediate (rufous), and dark. They tend to be variable, and many birds are hard to categorize specifically. In many cases, raptors show a continuum of coloration—from light to dark , and it seems that many birds are just a different shade of the same basic plumage pattern. Features such as tail bands and the presence of a dark trailing edge to the wing are identical within the color forms. Using such consistent features for ID and ageing is always better than focusing on variables such as color.

It takes several years, and multiple plumages, for larger raptors to reach adulthood. Most buteos molt into adult-like plumage after a year, though many retain a few juv feathers, making them ageable with careful observation. Accipiters take 1–2 years (2–3 plumages), eagles 4–5 years (5 plumages), and falcons 1 or 2 years, to reach adulthood. Most raptors keep juv plumage for a year. Understanding this molt sequence is key to ageing and ID.

To simplify, these birds can be broken into 5 groups: vultures, eagles, and Osprey; buteos; accipiters; falcons; and miscellaneous species.

Vultures, eagles, and Osprey. Vultures are scavengers eating carrion (dead animals) exclusively. They are often seen in groups following the pack in the search for food. Eagles and Osprey are hunters. Bald Eagle also eats carrion, scavenges, and steals food from other birds.

Buteos. This group comprises medium to large raptors with broad and typically long wings but shortish broad tails. They spend a considerable amount of time soaring. Shape, and tail and wing pattern are often keys to ID.

Accipiters. Smaller than buteos. They have rounded wings and proportionately longer tails. They also soar but are incredibly agile and are often seen dashing between trees. Some will even crawl through bushes in pursuit of prey.

Falcons. Small to medium-sized. The slimmest raptors, falcons have narrower and more pointed wings, and are designed to travel at high speeds for hunting. Peregrine Falcon has been clocked at over 200mph.

Miscellaneous species. A number of other raptors such as kites don't fit neatly into any of the above groups.

OWLS

A mesmerizing and often mysterious group of birds, particularly as they are mostly nocturnal. Their habits, and cryptic plumage, make it easy for us to walk straight past them. A collection of regurgitated pellets, whitewash (feces) on or under trees, or a group of agitated songbirds—usually led by chickadees—can often give away their presence. Even if you know where they are, they can be tricky to see. With a second and third look into foliage they often miraculously appear. Patience and perseverance is the name of the game. Some species frequently roost communally out of breeding season and have favored locations. Common sense should prevail as they are sensitive and easily flushed.

Some owls have 'ears,' actually just tufts of feathers, and facial discs that help direct sound into the hidden ear openings. Although owls have good eyesight, their incredibly sensitive hearing can determine both direction and distance, allowing some species to catch food even when they can't see it. Serrated edges to the outer primaries allow air to pass through, helping the owl to fly silently. At night, the best way to locate owls is by listening. They give a variety of hoots, whistles, barks, and other weird noises. Another way to find them is to make high-pitched mouse-like squeaks, which they will often come to investigate. Bad weather, particularly wind, makes it difficult for owls to hunt. If you want to see one, you will have most joy looking on the first calm evening after a storm.

Several species breed in the far North and are highly dependent on vole populations. With food shortages, northern owls become distressed and often invade to the south of their natural range and into the US.

Black Vulture *Coragyps atratus* **BLVU** L 25in

Very common in the SE, numbers fewer to the W and N, though range is expanding. The smaller vulture, almost tail-less with square wings. A number of hurried flaps followed by short wobbly glides on flattish wings is a giveaway. Extended glides often cause ID confusion with eagles. BLVU does not soar so well or as often as TUVU due to its short wings. Watches for other vultures dropping out of sky after it has located dinner. Sits around in packs, like hyenas, waiting to take advantage of a carcass. Happily eats our trash and roadkills. Only rarely takes live prey. Often mixes with TUVU, standing on ground, sitting on rooftops or in trees. Size, shape, and hideous gray wrinkled head apparent. **ID**: The compact vulture with a gray head. In flight, striking 'square' wing patches obvious even at long distances. Very stocky with short tail and broad wings. Adult: extensive wrinkled gray head (similar to closely related storks). Ivory-tipped bill. Juv: black unfeathered head and black bill. 2nd–3rd-yr: wrinkled skin becomes more extensive and pale tip lighter and more extensive. Flight feathers a mixture of juv and adult-type feathers.

Turkey Vulture *Cathartes aura* **TUVU** L 27in

Very familiar and often numerous bird, in many habitats particularly in the S. Most frequently seen soaring with prominent dihedral, often in groups. Appears all-black (actual color dark brown) with long pale tail and trailing edge to underwing, making the 'Turkey Buzzard' an easily recognizable bird. Often seen sunning in morning with wings open to warm up or dry out. Finds food by smell as well as sight. When flushed from roadside carrion, waits shiftily for the all-clear to go back. Vultures often fight for food 'at table.' Has a featherless head to avoid soiling after forays into body cavities—it's hard to describe vultures as anything other than ugly. **ID**: Large (only a cursory glance would confuse it with BLVU), long wings and tail prominent in flight, wings always held in 'V,' except for occasional labored flaps. Adult: featherless reddish-purple head and ivory bill. Juv: grayish, lightly feathered head with uniform juv flight feathers. 2nd–3rd-yr: red bare skin becomes more pronounced, bill ivory with progressively less black on tip. Mixture of juv and adult-type feathers.

Bald Eagle *Haliaeetus leucocephalus* **BAEA** L 31in

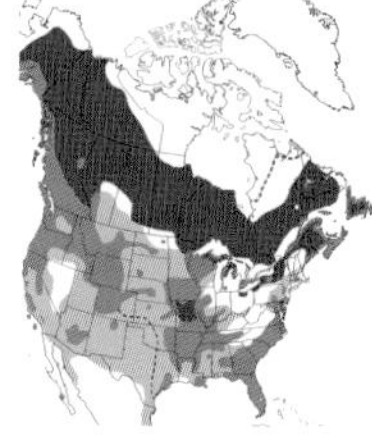

Fairly common in the SE, uncommon elsewhere. Increasing in numbers. National symbol. Builds huge nest of sticks in forested areas with lakes and rivers. In winter, more widely distributed—a common sight at garbage dumps, dams, and sites with fish or carrion. Concentrations build where a glut of food. Adults are well-known and easily identified. Imm birds are not so well known and often overlooked as hawks. Powerful flyers, they move fast on shallow, stiff wingbeats but spend much time soaring, covering large areas. Glides and soars on flat wings—the best way to pick it out versus slightly smaller TUVU or GOEA (dihedrals). Sometimes chases OSPR for its catch. **ID**: Long neck and tail, long parallel-sided wings. Adult (5th-yr): all-white head and tail. Brown body often looks black. 4th-yr: as adult with some brown in tail/head. Birds 3 years old and younger are mostly brown and easily overlooked. Juv: pied with blackish underparts contrasting with white underwing coverts and base of tail. Uniform trailing edge to wing. 2nd-yr: as juv with patchy white belly and serrated trailing edge to wing. 3rd-yr: white adult-like feathers molted in underparts and head with dark line through eye.

Golden Eagle *Aquila chrysaetos* **GOEA** L 30in

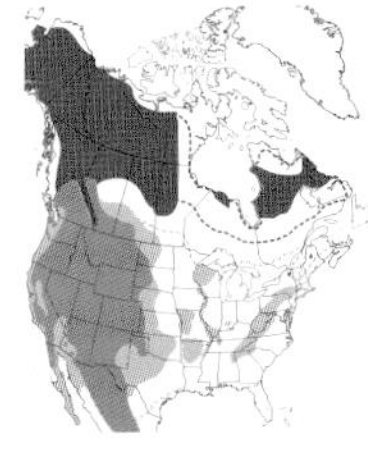

A summer breeder in n. mountains and forest, nesting mostly on cliff faces. In winter, sometimes moves to open countryside and lower elevations, particularly in the W, migrating along ridge lines. Hunts primarily in forests in the E. Much scarcer than BAEA. Stalls and stoops, sometimes from great heights, to catch large mammals, birds, and other prey. Likes sitting on rocks and utility poles on the lookout. Has a distinctive dihedral, even at a distance, but without the rocking of TUVU, and appears very dark overall. **ID**: Long and slim but with short neck. Wings are narrow-based. Imm: striking white wing patches and tail base, older birds more uniform. Golden nape obvious when seen well. Adult (5th-yr): all dark with narrow gray bands in tail. 4th-yr: as adult with white in tail feathers. 2nd/3rd-yr: combination of retained ratty white-based juv and newer dark adult-like feathers. Juv: white wingstripe on underwing and tail base, though some are mostly dark, leading to confusion. Uniform trailing edge to wing. Compared to BAEA is slimmer and shorter-necked (enhancing long-tailed appearance). BAEA soars on flat wings.

ad. ♀

ad. ♂

juv.

Osprey *Pandion haliaetus* **OSPR** L 23in

Common and familiar summer visitor to coastal areas (and resident in much of FL); scarcer inland. Winters in the S. Builds large stick nests on conspicuous man-made nest platforms, buoys, channel markers and in trees. The 'Fish Hawk' is frequently seen hunting, diving, or carrying fish from lakes, rivers, or the ocean. Patrols waterways, often in circuits, hovering intermittently. Spectacular plunge-dives, led by huge feet and talons, envious fishermen often watching from nearby. Frequently cuts out before hitting the water or comes out empty-handed. Fishes over the ocean, but moves to calmer inland water in rough weather. Sits on prominent perch to rip prey apart, but also on the ground in marshes. Head on, profile is flat-winged, or a shallow 'W.' Familiar call when flying or sitting, a high-pitched *choop*. **ID**: White-headed with a dark eyeline. Slim with wings typically angled back but straighter when soaring. White body and underwing contrast with dark carpal and barred flight feathers and tail; a unique combination. Ad ♀: averages darker breast-band. Juv: superficially as adult but with pale-fringed upperparts and buff on breast and underwing coverts.

1st-yr.
ad.

Swallow-tailed Kite *Elanoides forficatus* **STKI** L 22in

Uncommon early migrant (Feb) to forested areas in the SE, often near water. Strays n., mostly in spring. Simply beautiful; the epitome of grace (it would have been at home in the movie 'Avatar'). Flies just above treeline, never in a hurry, but deceptively fast. Loves warm weather, swooping after insects, which it eats effortlessly on the wing. Glides on bowed wings. Rarely seen perched. **ID**: Elongated, small-headed, and slim with long forked tail in simple black and white. Juv/1st-yr: buff on head and breast quickly fades by fall. Duller black, and slightly shorter-tailed than adult.

White-tailed Kite *Elanus leucurus* **WTKI** L 15in

Scarce in any kind of open country. Strays n. Often seen hovering, gradually dropping on prey with wings pointing skyward. Flies with deep graceful wingbeats gliding intermittently, wings held in a strong dihedral. Light but powerful. Perches on treetop snags or utility poles. Clean-cut appearance but black around eye gives it a sinister look. **ID**: Pallid with large black shoulders in all plumages. Adult: uniform upperparts lacks pale tips. Juv: pale tips and edges to upperparts through first year, rufous wash on breast and head fades during fall. 2nd-yr: as adult, may still have a few worn juv feathers.

Snail Kite *Rostrhamus sociabilis* **SNKI** L 17in

Uncommon in cen. FL, eating mostly apple snails. Lethal-looking hooked bill, designed for snail dinners. Hunts leisurely over reedbeds, harrier-like but with broad rounded wings held downward and protruding fingers; rounded tail. Sits on posts or bushes, often eating snails plucked from water. **ID**: Habitat, shape, and bold tail pattern distinctive. Orange legs. Ad ♂: dark gray body. Orange facial skin. Ad ♀: pale-tipped brown upperparts. Heavily streaked underparts, bold head pattern. Juv/1st-yr: as adult ♀ with buffier face, broader fringes to upperparts, finer-streaked underparts, duller facial skin and legs.

Hook-billed Kite *Chondrohierax uncinatus* **HBKI** L 18in

Very rare in Lower Rio Grande Valley forest. Usually stays hidden perched in the canopy. Mid-morning in fall and winter, is a good time to start looking for soaring birds on nice days. Striking profile: rounded wings with bulging secondaries, nipped in at base, and long tail, rounded at tip. Stonking 'hooked bill'—if you are lucky enough to get close. Ad ♂: broad gray body bars. Ad ♀: broad brown bars and rufous inner secondaries. Juv: like ♀ but narrower barring on underparts and tail, also lacks rufous in wings. Dark morph all dark.

Mississippi Kite *Ictinia mississippiensis* **MIKI** L 14in

S. breeder, increasing in numbers, expanding range (nested in NY, NH, and CT recently). Breeds in wooded areas, preferably near water. In the Great Plains found in more open areas, but always roosts in trees. Happy in towns, seeming to find them productive hunting grounds. Regularly strays n., particularly in late spring. Very agile and light aerialist, gracefully maneuvers to catch insects in its talons, ripping them apart as it floats on the wing. Appears kite-like with wings and tail spread, but when in a glide, easy to confuse with a slim PEFA. Often gathers communally where common and forms fast-moving, large flocks on migration. **ID**: Long, pointed wings with long, broad, square-ended tail that, when fanned, becomes rounded. Adult: mostly gray with dark around eye, wings, and tail. Rufous in primaries difficult to see. White secondary patches on upperwing contrast with dark tail and outerwing. ♀ has paler tail with dark terminal band. Juv: white-tipped gray/brown upperparts, indistinct pale supercilium with heavy brown blotches on underparts. 1st-s: the common spring overshoot. Mostly gray body and speckled underwing with retained barred juv tail feathers. Lacks white secondary patch.

Northern Harrier *Circus cyaneus* **NOHA** L ♂ 17in, ♀ 19in

A fairly common bird of marshes, wetlands, farmland, priarie, and other open areas. Nests on the ground. Long, slim, and elegant; everything about the 'Marsh Hawk' is lightweight and buoyant. Gracefully courses low over open areas for rodents, suddenly twisting, legs dangling, as it drops down on its prey. In flight, wings are held in a dihedral, as it rocks from side to side, like TUVU. Occasionally soars higher, slim build and dihedral always striking. Sits on the ground or fenceposts (rarely high) to eat its prey. **ID**: White rump obvious in all plumages. Owl-like facial disc, slim build and long legs obvious when perched. Adult ♂: 'gray ghost;' variable shades of gray bordered by black wing tips and secondaries. 2nd-yr ♂: variable, often similar to adult with more brown in plumage. Adult ♀: barred tail and wings with brown, streaked breast on buff background. Juv: very similar to adult ♀ but with finer streaking on rufous underbody and has rufous upperwing coverts (buffier in adults). Differences become more subtle when juv's rufous feathering fades to buff through winter.

Northern Goshawk *Accipiter gentilis* **NOGO** L ♂ 19in, ♀ L 23in

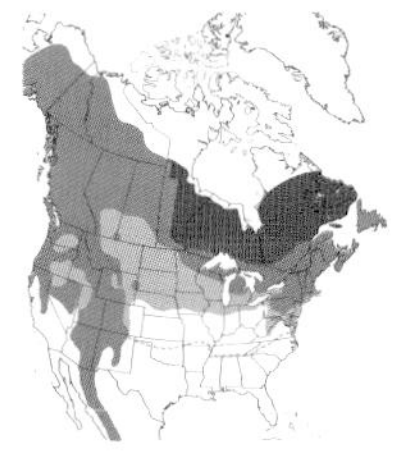

Widespread but very scarce breeder in n. forest, particularly in highlands. Some migrate s. Large and nasty, tends to stay well hidden, a stealth hunter through forest only occasionally getting high into the sky. Hunts prey nearly as large as itself, such as rabbits and grouse (and even dive-bombs humans if you get too close to the nest!). Often hunts from perches. Spends much of the day sitting quietly. **ID**: Large and heavy-chested with broad long tail and wings, sometimes recalls a buteo or GYFA. Compared with COHA, other than larger size, ihas broader, bulging secondaries, which enhances the wings' pointed look. Adult: slate-gray with finely barred underparts (coarser in ♀) and whopping white supercilium contrasting with black ear coverts. Cruel beauty. 2nd-yr: adult-like with some retained juv feathers. Often orange rather than red iris. Juv/1st-yr: similar to COHA but with stronger upperwing bar formed by pale tips to greater coverts and 'wavy' tail bands. Underparts heavily streaked, extending down to undertail coverts. White supercilium. Size and shape always key! In accipiters, larger size and bulk in ♀ is particularly noticeable, and adult ♂ has more rounded wings than ♀.

Cooper's Hawk *Accipiter cooperii* **COHA** ♂ L 15in, ♀ L 18in

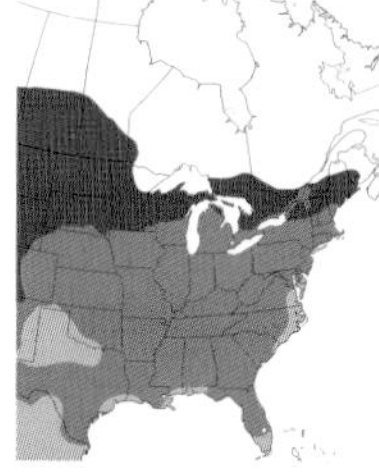

Fairly common in forested areas. Now adapting to city life, your feeders, and everywhere in between. Cleverly sneaks up on familiar feeding places from behind cover with a resulting swirl of feathers and twitters. Takes birds its own size, often struggling to drag them off into the undergrowth. Once shy, many are quite happy to sit on utility poles as you walk by. When hunting or displaying, sometimes has slow exaggerated upstrokes as if to fool potential prey. Flies with stiff, shallow wingbeats that come from the shoulder. **ID**: Similar to smaller SSHA, though no overlap in size but still a source of confusion for many birders. COHA is longer- and rounder-tailed with broader white tail tip. Wings are narrower-based, and longer with straighter leading edge. Longer neck projection. Adult: darker cap and paler hindneck. Orange-and-white barred underparts brighter in ♂ than ♀. Juv: golden buff head conspicuous in flight. Underparts white with well-defined dark fine streaks (coarse bars on SSHA). 2nd-yr: adult-like with some retained juv feathers, paler iris.

Sharp-shinned Hawk *Accipiter striatus* **SSHA** L ♂ 10in, ♀ 12in

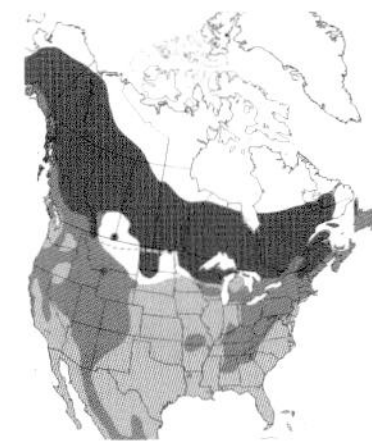

The commonest accipiter, but declining. Found in forested areas and sometimes around humans. Occasionally visits feeders, but generally less urbanized than COHA. ♂ very small, ♀ nearly as big as some COHA. Compact: small head, short broad wings, squarish tail. Fast, fluttery flight followed by short glides—seems to take a lot of effort. When soaring, more compact structure, less rounded tail without broad white tip, and short neck projection. **ID**: Compared with COHA has smaller, rounded head, shorter square-ended tail with narrow pale tip. Adult: lacks capped appearance and pale nape. As COHA, orange-and-white barred underparts similarly brighter in ♂ than ♀. Upperwing uniform in ♀, contrastingly darker primaries in ♂, these fading brownish by spring, appearing more juv-like. Juv: underparts less white with broad warm brown markings lacking the well-defined streaks of COHA. Head darker lacking paler golden buff and its back has limited white blotches compared with COHA. 2nd-yr: adult-like with some retained juv feathers, paler iris, but birds are hard to age in the field.

Harris's Hawk *Parabuteo unicinctus* **HASH** L 20in

Uncommon in s. TX. Unusually proportioned, a buteo with the feel of an accipiter. Found in arid areas. Sits upright on telegraph poles, cactus, or trees scoping out prey. Long legs suitable for chasing down prey. Usually in pairs, often approachable. **ID**: Strikingly long legs and tail. In flight, round-winged, long tail contrasts with short neck. Ad: dark chocolate-brown sets off bright chestnut wing coverts and thighs with striking white tip and base to tail. Juv: similar colors, but has streaked underparts and chestnut fringes to coverts. 2nd-yr: adult-like with a few retained juv feathers.

Common Black-Hawk *Buteogallus anthracinus* **COBH** L 21in

Rare in canyons and hilly areas in w. TX, mostly summer. Spends most of day perched, particularly near water. Extremely broad-winged, exaggerated by short tail. Adult: strikingly patterned black and white, which, in combination with shape, make ID straightforward. Juv: streaked breast with heavily barred wing and tail feathers. Pale bases to inner primaries form pale panel..

Zone-tailed Hawk *Buteo albonotatus* **ZTHA** L 20in

Scarce in the Rio Grande Valley and w. TX. TUVU look-alike: black, long-tailed with long wings held in a dihedral as it wobbles from side to side. A loner that sometimes mixes with TUVU, so look carefully! Smaller with heavily barred flight feathers. Adult: tail has broad bars, narrow in juv. Yellow bare parts.

Short-tailed Hawk *Buteo brachyurus* **STHA** L 16in

Uncommon in s. FL, very rare in s. TX. Flushes easily. Soars high in sky over open country or forest. Wings held in slight dihedral. Similar to a broader-winged BWHA: wings round-ended, and bulging secondaries nipped in at base. When gliding, carpal protrudes with wings pointed. **ID**: 2 morphs. Dark: black with heavily barred flight feathers and tail. Dark juv: underparts brown with white spots. Tail has more bands. Light ad: dark cheeks striking against white underparts. Juv: similar to adult with pale supercilium, light flecking on cheek, creamy underparts, and dark spots on sides of throat.

Gray Hawk *Buteo nitidus* **GRHA** L 17in

Scarce but increasing in the Rio Grande Valley, TX. Often seen perched on snags or circling low overhead. Soars with flat wings or slight dihedral. Accipiter-like with broad wings and fairly long tail. Vocal, call often giving away presence; a loud *ah-waah* or screamed *keyaah*. **ID**: Adult: unusually marked, gray, uniform above and finely barred below. Boldly barred tail contrasts with white vent. Juv: bold head pattern with dark eyeline and malar stripe. Underparts white with bold brown streaks and spots and with heavily barred tail. Juv BWHA lacks the longer tail and boldly striped face.

ad.

juv.

juv. s. FL

ad. eastern

juv.

ad. s. FL

Red-shouldered Hawk *Buteo lineatus* **RSHA** L 17in

Uncommon in n. of range, common and tamer in the S. In the N prefers deciduous forest, particularly near water and swamps. In summer, stays in forest, feeding on small mammals and reptiles. Outside breeding season, is more visible, sitting on fenceposts and snags, often next to the roadside. When not sitting quietly, will often circle territory, calling noisily. Far-carrying voice with BLJA-like squeal—*kee-yeer*—often the only giveaway. **ID**: Slightly larger than BWHA, with longer tail and wings. In flight, look for pale crescent bordered by dark-tipped primaries, an important feature. Adult: striking burnt-orange-barred underparts. Red 'shoulder' sometimes visible, upperparts dark mixed with white and orange. Juv: upperparts similar to adult. but broner and less striking. Very similar to BWHA but RSHA often has evenly spaced heart-shaped streaks, barred secondaries, more heavily spotted upperparts, longer legs and tail. Pale areas above and below eye give studious look. 3 races in the E with overlap. Migratory n. birds are darkest. 2 races in the S, birds in s. FL are palest.

Broad-winged Hawk *Buteo platypterus* **BWHA** L 15in

Fairly common breeder in deciduous forest, often near wetlands, though hard to find. High-pitched whistle often the best clue to its presence. On migration, forms flocks, some in the thousands, that travel along the Great Lakes, down Appalachians and exit through TX. At favored hawkwatch sites they may swirl around in the sky, gaining altitude before arching wings forward and gliding off at deceptive speed. The smallest N American buteo, compact and well proportioned. Wings are somewhat pointed, and at distance appears white with dark border, broadest at wingtip. **iD**: Adult: broad tail bands and trailing edge to wing. Rufous-barred underparts—some pale birds can appear confusingly similar to dark juv. Upperparts dark-centered with warm fringes—little or no white. Juv: underparts variable from near-white to fairly dark. On lighter birds, streaking heaviest at sides. Finely barred tail. In spring birds, molting primaries often appear to have pale crescents and are therefore easy to misidentify as RSHA. Adult-like by 2nd fall. Dark morph is very rare in the E—all dark except for flight feathers and tail, which have the same pattern as light-morph birds.

Red-tailed Hawk *Buteo jamaicensis* **RTHA** L 19in

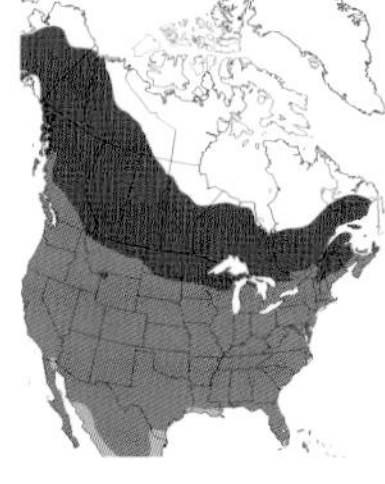

The common buteo. Large and familiar, it occurs anywhere from city centers to forest, mountains, and open countryside, mostly in ones and twos. A familiar sight next to highways, soaring above, sitting on posts, or standing over lunch. Occasionally hovers. Gives spine-chilling down-slurred scream. **ID**: Complex regional and individual variation with large overlap, but always picture it as the large buteo with broad rounded wings. In the range, nearly all birds are e. *borealis*. Other forms rare, mostly in the W, and are incredibly variable. Large, robust, with broad rounded tail and wide rounded wings nipped in at base. E. birds typically have white underparts including throat, streaked belly band, and dark leading edge (patagial) to wing. Adult: red tail with black-and-white tip. Rufous wash to underparts. Juv: most easily aged by multiple thin tail bands, usually brown but sometimes red. White underparts. Upperparts brown with white scapular patches. W. and Harlan's vary from light to dark. Harlan's, formerly a separate species, is usually dark with wavy tail bands. Great Plains birds average paler underparts and (whitish) tail with extremes known as 'Krider's'—an anemic version of e. Red-tail.

Rough-legged Hawk *Buteo lagopus* **RLHA** L 21in

High-Arctic breeder. Winters on farmland, marshes, and other open countryside in the harshest environments. S. winter movement varies in numbers from year to year. Usually seen circling or hovering over snowbound fields in search of rodents, or perched on isolated treetops with a good vantage. Feathered legs are an adaptation to n. living. Flight light and buoyant with pointed wings held in dihedral. **ID**: Large buteo with small head and long pointed wings that reach tip of tail. 2 distinct color forms. Light: (slightly commoner) ♀ and juv: dark carpal patch, belly, wing tips and tail band. Ad ♀: tail band is well defined, broader trailing edge to underwing and more heavily marked underwing coverts. Ad ♂: usually shows multiple tail bands and mottled underparts, often with bibbed effect. All light morphs have dark tail contrasting with striking white base. Dark morph: mixture of rufous and dark brown. Tail brown with narrow pale bars. Some adult ♂s are black. Best aged by checking pattern of undertail and trailing edge to underwing—as light morph.

Swainson's Hawk *Buteo swainsoni* **SWHA** L 19in

Fairly common summer visitor to MW prairies and farmland. Regular out of range. Large slim buteo with long tail and very long, pointed wings (for flying to S America) giving harrier-like feel; wings held in dihedral adds to this impression. Light and buoyant flight. Migrates in large flocks, moving effortlessly and deceptively fast. Feeds mostly on rodents and ground squirrels but also runs on ground picking up insects with bill. **ID**: Only raptor with contrasting dark flight feathers on upperwing (visible on underwing)—striking even at long range. Extremely variable. Light forms—about 80% of population. Adult: bold white underparts, contrasting black flight feathers and chestnut breast. ♂ typically shows grayer head and ♀ more streaking on breast. Juv: barred primaries and variably streaked belly, rufous-tinged background. 1st-s: mixture of juv and adult feathers (fewer)—most buteos don't have distinct 2nd-yr plumage. Dark/intermediate forms are on a sliding scale of darkness. Adult: rufous underwing coverts contrast with darker body and flight feathers. Body varies from darkest marking on breast to uniformly dark. Juv: heavy blotchy streaks, heavier in darker birds.

White-tailed Hawk *Buteo albicaudatus* **WTHA** L 20in

Scarce in open brushy areas and farm fields in s. TX. Loves sugarcane and other burns. Sits on fenceposts. **ID**: Heavy, with long pointed wings that extend past short tail when perched. Unique shape, striking in flight: wings in steep dihedral, tail almost meets trailing edge of wing. 4-yr hawk. Adult: beautiful with red shoulders, gray hood, white underparts, and bold white tail with black subterminal bar. 3rd-yr: similar with fine barring on belly and dark throat. Juv: dark brown upperparts. Underparts: body and coverts dark brown with pale tips and white central breast patch. Tail and flight feathers thinly barred and paler. Longer-tailed than adult. 2nd-yr: similar to 1st-yr with rufous in scapulars and broader tail band.

Ferruginous Hawk *Buteo regalis* **FEHA** L 23in

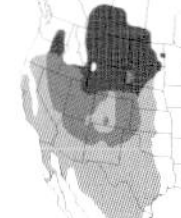

Uncommon in open and dry areas. Perches on ground. Waits for prey such as prairie dogs, snakes, and larger insects. Hunts mostly from perches and while hovering. Slow, deliberate flight. Territorial and vocal. **ID**: Large, with big head and bill, long tail, and long, pointed wings. 2 color forms (dark is rare). Light: strikingly pale underparts with bright rufous leg feathering and wing coverts. Light tail and bases to flight feathers illuminated in flight. Juv: underparts lack rufous and upperparts colder brown with narrower tail band. Dark: dark rufous with pale primaries and base of tail. Juv: narrow tail bands.

Crested Caracara *Caracara cheriway* **CRCA** L 23in

Localized and uncommon in open areas. Unusual looking, often running around or perched characteristically upright. Feeds on carrion and small prey. Everything is long: head, neck, tail, wings, and legs. In flight, appears as a 'flying cross,' with BLVU-like white patch, parallel-sided wings, white tail and black tail band. Flies low, occasionally soars with vultures. **ID**: White neck, dark cap, breast, and upperparts. Striking facial skin. 3-yr raptor. 2nd-yr: as adult, averages duller facial skin, browner upperparts, narrower neck barring. Juv: 'ghost' pattern of older birds but paler brown with dull bare parts.

Aplomado Falcon *Falco femoralis* **APFA** L 16in

Rare. Reintroduced in s. TX in grassland. Larger and more regal version of AMKE. Disproportionately long tail. Often in pairs. Adult: Orange vent, dark belly, and white breast. Bold malar stripe. ♂: has more limited streaking on breast than ♀. Juv: buff underparts.

Gyrfalcon *Falco rusticolus* **GYRF** L 22in

Scarce high-Arctic breeder; a loner with a huge territory. Rare farther s. in winter. Perches on ground, in trees, on buildings (sometimes downtown) or telegraph poles before flying low and at great speed. Like a huge PEFA but more NOGO-like in some ways. **ID**: Large, heavy-chested with broad neck but small head and long, broad tail. Wings broad-based and blunt-tipped. Wing point well short of tail tip (about equal in PEFA). 3 highly variable morphs: gray (most common), white, and dark. Note size, shape, wing point, less pronounced maler stripe, and flight feathers paler than underwing coverts.

Prairie Falcon *Falco mexicanus* **PRFA** L 16in

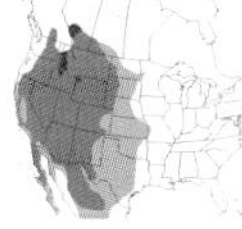

Scarce in the far W, rare to the E in prairies, desert, and farmland. Nests on cliffs. A pallid slim PEFA! Hunts mostly by flying close to the ground but also stoops. **ID**: Like more common PEFA but slightly slimmer and longer-tailed. At rest, wing point is short of tail tip. Most striking feature is dark underwing line formed by greater coverts and axillaries—prominent even at distance. Plumage browner and paler than PEFA with undertail noticeably paler. Slim mustachial stripe. Adult: spotted underparts. Slightly heavier in ♀. Juv: streaked rather than spotted underparts and blue cere.

Peregrine Falcon *Falco peregrinus* **PEFA** L 16in

Widespread, uncommon but increasing. Found anywhere, though reintroduction plan clouds picture. Traditionally nests on remote cliff faces but now found in many cities, using buildings and large bridges. Frequently use man-made hacking towers. Sits on or near nest sites. Hunts (mostly birds) in incredible stoops. Makes repeated passes at crouched prey, causing chaos among nearby birds. Stiff, shallow flaps with wings held flat and straight. Coastal migrant. **ID**: Fairly large. Broad-based wings and tail enhance compact look. Dark upperparts, broad mustache, and barred/streaked underparts. 3 races: *tundrius*, *anatum*, and *pealei*. *Tundrius*, an Arctic breeder, is widespread in winter and on migration. Adult: broad black mustache bordered by white cheek. Slate upperparts. Underparts white, sometimes buff-washed, with variable black barring. Juv: brown (sometimes gray) upperparts and streaked underparts with pale crown. 2nd-yr: mixture of adult and juv feathers. *Pealei* and *anatum* are w. breeders and rare in the E. Reintroduced birds are intergrades often showing features of *pealei*—extensive black on cheek, heavier barring in adults, largely chocolate-brown in juv.

Prairie Merlin

♀

ad. ♂

juv.

Merlin *Falco columbarius* **MERL** L 10in

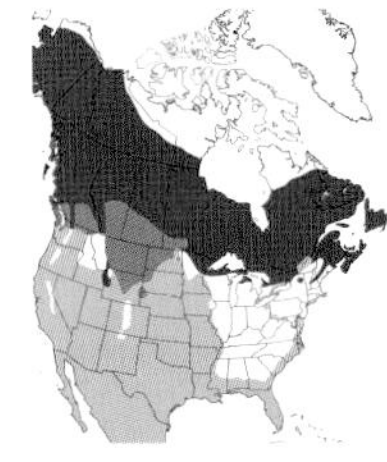

Uncommon in open areas but can be seen anywhere on migration, particularly on the coast. A feisty dasher. Often sits on prominent posts and snags (but not utility lines like AMKE), then sets off like a 'bat out of hell,' chasing its prey with fast but stiff wingbeats, following every twist and turn. Feeds on birds and insects. Like other successful raptors, it has started to adapt to humans and is increasingly moving to urban areas. **ID**: Similar to a very small PEFA with a narrow tail. It always looks more compact than AMKE. A dark-looking bird with slate or brown upperparts. Barred tail with pronounced white tip and broad black subterminal tail band. Wings heavily spotted, giving underparts a dark look. Underparts buff with extensive blurry streaks and with contrasting pale throat. Weak mustachial. 2 races: 'Taiga' (*columbarius*) and 'Prairie' (*richardsoni*). Ageing and sexing very difficult. Taiga: Ad ♂: blue-backed with dark shaft streaks and orange vent. ♀ and juv: vary from slate to brown upperparts and white to dark buff underparts. Prairie: scarce in the W. Much paler than Taiga. Head is contrastingly pale and nondescript. White tail bands are broader.

1st fall ♂

♀

♀

American Kestrel *Falco sparverius* **AMKE** L 9in

Fairly common and widespread, but declining in open countryside, particularly farmland. A beautiful falcon. Small and slim, with long tail and angled wings, has a light flight and graceful appearance. Wingbeats are deeper and more laid back than in other falcons. Lacks the straighter edges and more compact feel of MERL. Unlike other falcons, it frequently hovers before dropping on mice, insects, or other small prey. A familiar sight, perched on fenceposts, pumping its tail, in part to maintain balance, as it scopes out dinner. **ID**: All plumages show double face stripe. Ad ♂: the most colorful and boldly patterned falcon. Salmon-and-blue upperparts and orange wash to breast with beautiful head and tail patterns. In flight has checkered underwing with a trailing edge of translucent white spots (less obvious in ♀). Juv/1st-w ♂: like adult ♂ but with more heavily spotted white underparts lacking orange-streaked hindcrown and more heavily barred back. Molts into adult-like plumage through fall. ♀: upperparts uniformly brown with black bars. Underparts streaked brown.

ad. ♀

ad. ♂

Barn Owl *Tyto alba* **BANO** L16in

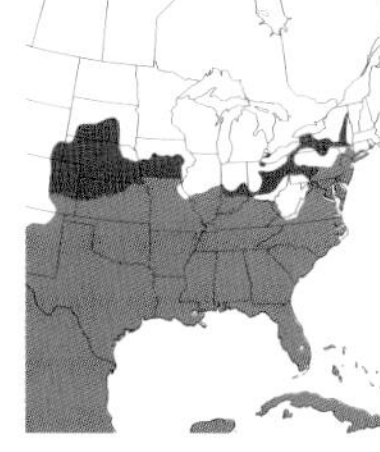

A cosmopolitan species. Uncommon and declining in agricultural and urban areas, found wherever rodents are plentiful. Roosts and nests in old buildings, barns, trees, and nest boxes. Coming out to hunt at dusk, courses open areas low to ground or sits on perch watching or listening for dinner, often seen in car head-lights. Tends to look at you when disturbed, with striking white heart-shaped face, before silently disappearing into the night. Caught in a flashlight, this is the 'white' owl, somewhat ghostly in overall appearance. Rarely sits out in the open like much larger SNOW. Call: given by adults and young at nest site or in flight, a shrieking hush-like *sssshhhhh*—like scolding naughty kids. It can be heard at long range. Begging young will call almost constantly. **ID**: Medium-sized with tan-and-gray upperparts. Underparts usually white in ♂ and with orange/buff in ♀. In silhouette, wings appear slightly longer and more pointed than in 'eared' owls.

Short-Eared Owl *Asio flammeus* **SEOW** L 15in

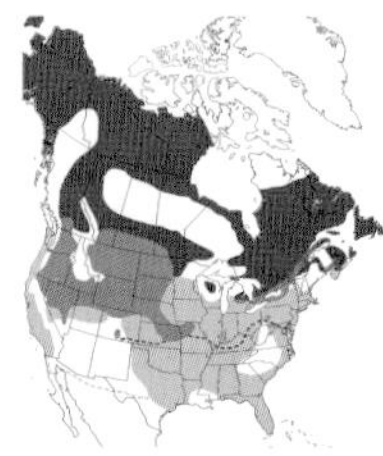

Uncommon in marshes, grassy areas, and tundra. Roosts and nests on the ground. Other than SNOW, the only owl that can often be seen hunting in the daytime, though far more likely to be seen at twilight. Gracefully courses over open areas, flying low, but often changes direction quickly or stops suddenly, dropping on an unsuspecting rodent. Sometimes seen sitting on ground, recalling NOHA, but much chunkier with different facial pattern. Often uses communal roosts in winter at favored locations. **ID**: A medium-sized brown owl, with finely streaked underparts and marbled upperparts. Short ear tufts never as long as LEOW's. Dark eye surround gives sinister look. ♀ averages darker breasted than ♂. Caribbean race occurs rarely in s. FL: slightly smaller and darker, with restricted fine streaking on underparts.

active posture

camouflage posture

Long-Eared Owl *Asio otus* **LEOW** L 15in

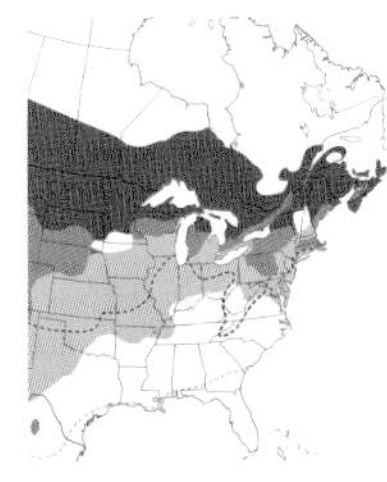

Uncommon deep inside forest. Melts into the trees—you know they are there, watching you try to find them. Nocturnal, coming out when pitch-black, rarely before. Away from breeding grounds, roosts are usually communal, sometimes with several birds in one tree. Sits upright and slim (camouflage posture). Whitewash or pellets are usually easier to spot than the bird. Easily flushed, the wings are heard hitting branches as it flies to another tree. A much closer look will often reveal more birds, still sitting. Gives a wide array of hoots, barks, and squeals that are easy to confuse with GHOW. **ID**: Superficially similar to SEOW. LEOW has longer ear tufts and orange face with black lines through eye and more extensive white in between. Breast barred as well as streaked and much darker—easiest difference to see in flight (daytime). Other differences in flight are more rounded wings, shorter tail, and finely barred outer primaries that lack contrast and dark tip of SEOW. Both species give wing-clapping displays. GHOW is much larger but very similar when flying away in dense woods.

Barred Owl *Strix varia* **BDOW** L 21in

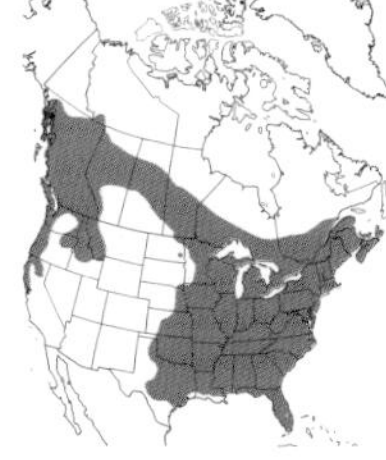

Common in the S, scarcer farther n. Found in swamps and wooded areas near water. Usually in pairs and doesn't form communal roosts. Call is a memorable, deep, throaty *who-cooks for you,* often repeated and building to a crescendo (in chorus)—the night sound of swampy forest. Sometimes very vocal, at others frustratingly quiet for long periods. Although nocturnal, is sometimes seen and heard in the day. Occasionally tame and approachable, particularly on well-traveled paths where there are lots of people. Hunts on the wing and from perches, taking mostly rodents and other small mammals. Often comes to forest edge at twilight to feed, sometimes sitting on utility poles like GHOW. **ID**: Thick body, short tail, and broad wings. Has large black eyes on brown face, and yellow bill "wrapped in a shawl," with brown-streaked breast. Has a gentle look—it is the owl of choice for cuddly toys. Brown marbled upperparts. Sexes similar. Easily flushed, brief views can make ID from GHOW difficult but GHOW is larger, lacks brown back, and the more contrasting wing pattern with white tips.

♀ eastern

ad. southwest

fledgling

Great Horned Owl *Bubo virginianus* **GHOW** L 22in

Fairly common and widespread in all habitats, perhaps most widely distributed bird in N America. A large, lean, mean killing machine that puts fear into birds and mammals alike. Quite happy to eat neighborhood birds (tough to find other species of owl in its territory), possums, rabbits, and other prey as large as itself. Hunts from perches, often high and exposed, usually coming out at dusk but sometimes in the daytime. Starts nesting in Jan/Feb (often when trees leafless), typically using same large stick nest. Offspring (brown face) frequently misidentified as LEOW (black face). Flight direct and lacks choppy up-and-down nature of other 'eared' owls. **ID**: A large, tall, barred owl with long ear tufts, yellow eyes, and black bill. Stereotypical owl call, a mellow, hooted *hoo-hoo-hoo hoo-hoo*, usually given at twilight. Large regional variation in color. E. populations are brown, becoming grayer to the w. in more open habitats. ♀ averages larger and darker than ♂.

Great Gray Owl *Strix nebulosa* **GGOW** L 27in

Resident of boreal forest and meadows. Rarely ventures s. of Canadian border. Hunts rodents at dusk, occasionally during daytime, either from perch or in flight. In winter, incredible hearing allows it to catch rodents using plunge-dives through snow and powerful talons. Tame and approachable, the circular rings on facial disc give soft expression. Small eyes (face has cross-eyed look) stare you down. **ID**: Massive, large-headed, and chunky, with finely patterned gray plumage. Black-and-white 'bow-tie' under bill highlights face.

Northern Hawk Owl *Surnia ulula* **NHOW** L 16in

Inhabitant of n. open spruce forest or taiga. In winter in more open country. Rare s. of border. Sits at top of trees, with condescending glare. Long tail and narrow, pointed wings add to hawk-like appearance (hence name). Hunts from perch, though frequently hovers while focusing on prey. Fast flight as it swoops up to perch. **ID**: Distinctive shape and habits. Pale head with two bold black lines contrast with brown-barred underparts and distinctly spotted upperparts.

ad. ♂

1st-w. ♀

ad. ♀/1st-w. ♂

Snowy Owl *Bubo scandiacus* **SNOW** L 23in

High-Arctic beauty that regularly wanders s. of border. Found on farm fields, barrier beaches, marshes, and airport fields. Nocturnal/crepuscular but often seen roosting in daytime on ground, fence posts, blinds, or buildings. Prefers perches with commanding views but hunkers down in bad weather. Feeds on rodents, ducks, gulls, and small animals. **ID**: Large and white with yellow eyes and dark markings, a combination that always attracts admirers—and perfect for a snowbound lifestyle. Legs heavily feathered for warmth. Ad ♂: from almost pure white to limited dark markings. Ad ♀/1st-yr ♂: intermediate number and extent of dark markings. 1st-yr ♀: striking, with broad, dark bars, white face and underwing.

Rufous

Gray

Western Screech-Owl

Eastern Screech-Owl *Megascops asio* **EASO** L 8.5in

Common resident, though often overlooked. Found in wooded areas, from the deepest forest to suburban gardens. Strictly nocturnal but sometimes seen in daytime at nest or roosting in tree hole or nest box. Sunny weather on cold days is the best time. Feeds on small birds and rodents. Other birds will scold it if found roosting. Call: frequently imitated by birders to attract other species. Song: a descending whinny or even-pitched trill, given by both sexes: whistle through saliva in the back of your throat to imitate. **ID**: Small, with ear tufts and green bill. Intricately patterned plumage. 2 color morphs, rufous and gray, though many intermediates exist. Also regional variation, with n. birds averaging larger and paler. Almost identical to Western Screech-Owl (*Megascops kennicottii*) (WESO)—the latter occurs in w. TX, w. OK, and sw. KS. WESO has dark bill, broader streaks on breast, and different call—like an accelerating bouncing ball *pwep pwep pwep pweppwep....*

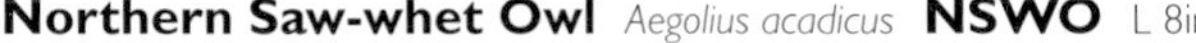

Northern Saw-whet Owl *Aegolius acadicus* **NSWO** L 8in

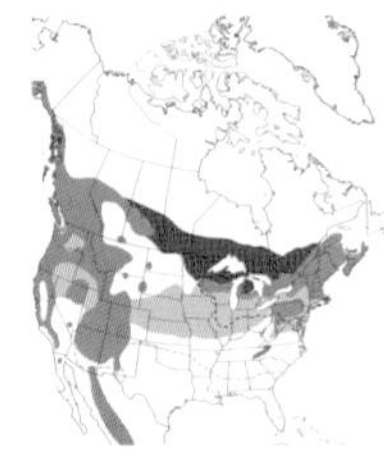

Fairly common in evergreens, mixed forest, and white pines near streams. Pocket-sized and cuddly, you would love to take it home. Banding stations on migration routes often catch hundreds. Finding them in the daytime seems impossible—often in dense, tangled areas or close to tree trunk, but always with access from below. Remarkably approachable—if you can find them. Nocturnal, feeds on rodents. Song easily imitated: a steady monotone whistle repeated over and over again, twice per second. Also cat-like *shreek*. **ID**: Small and very broad with short tail. Warm brown plumage. Finely streaked head, pale brown face, with white 'V' between eyes, and yellow-orange iris. This bird has an inquisitive look. Underparts have broad brown streaks. From behind, lacks pale back spots of larger BOOW.

Boreal Owl *Aegolius funereus* BOOW L 10in

Scarce n. owl found in evergreen and mixed forest. Some move s. in winter, particularly when there is a shortage of food. Hungry birds such as these can often turn up in unexpected places, for example, birdfeeders that attract rodents. Rare s. of border. Song: a series of about 10 deep, rapidly whistled *pu* notes. **ID**: Similar in habits and appearance to NSWO, but is larger and colder brown. Focus on the face pattern—whiter with contrasting darker border and dark line through the eye giving it an angry expression. Also paler bill, yellow iris, and spotted upper breast, head, and back.

Ferruginous Pygmy-Owl *Glaucidium brasilianum* FEPO L 6.75in

Rare in open habitat with trees (s.TX). Often heard and seen in daytime. Listen for repeated, whistled *pip* (2–3 per sec) or scolding birds. Small, with long, barred tail, bold brown-streaked underparts, and spotted brown upperparts. Note the intriguing cryptic pattern on the back of its head.

Elf Owl *Micrathene whitneyi* ELOW L 5.75in

Rare summer visitor to the Lower Rio Grande Valley, TX. Breeds in riparian woods. Nocturnal, is more often heard than seen—a sharp single *pew*, or given in series, a steady *pe pe pe pe pe pe pe*. Tiny, the smallest owl in N America. **ID**: Intricately marked plumage, small rounded head lacks ear tufts;, has yellow iris, and white eyebrow on a relatively bland face.

ad. eastern

ad. western

Burrowing Owl *Athene cunicularia* **BUOW** L 9.5in

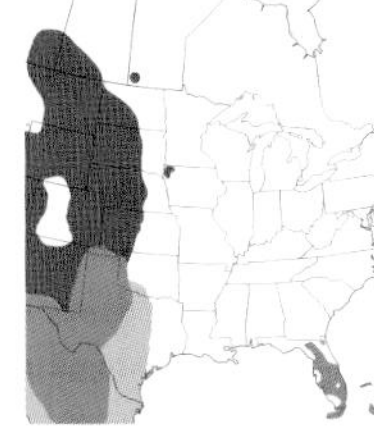

Scarce in arid areas. Declining. Identifiable by behavior alone. Lives in burrows, sometimes colonially, in dry open fields (the type that prairie dogs like), also airports, golf courses, etc. Head can often be seen poking out as it scopes its surroundings, often with its partner for company, and sometimes appears more interested in you than you are in it! When safe, but particularly at twilight, will stand out in the open or venture to a nearby post. Rangy but direct flight with long, slim wings, sometimes stopping to hover for mice. ♂: gives a variety of single- and double-note calls. **ID**: Fairly small but with long legs. Brown plumage with pale spots or bars, white throat, yellow eyes. FL race is smaller, darker brown, has less buff in underparts, with extensive barring on belly, and heavily streaked forehead.

MISCELLANEOUS LARGER LANDBIRDS

Nighthawks and nightjars (collectively called goatsuckers) are mostly nocturnal. Goatsuckers sing primarily at dawn and dusk. All are a similar cryptic brown. Nighthawks have wingtips that reach or go past the tail (pointed in flight). Nightjars have wingtips that are shorter than the tail (rounded in flight). Nighthawks have unbarred primaries; in some species the primaries are inticately barred and marked and serve as superb camouflage. Species in this group are not only difficult to find, due to their cryptic plumage and nocturnal habits, but are also tough to identify. As with other cryptic groups, such as owls and grouse, the dominant colors are usually gray and rufous, but species don't fit neatly into either group, and birds frequently show a variable amount of both colors. They are most easy to find at night by eye reflection, using a flashlight.

Pigeons and doves usually build shallow stick nests. They call frequently and are often more easily heard than seen. Many species' calls are soft and rhythmic. Their wings often make noise on takeoff.

The cuckoo family includes anis and roadrunners as well as the better known Yellow and Black-billed Cuckoos. Anis are cooperative breeders. North American cuckoos use nests built by other species but are not brood parasites.

Parrots, parakeets, and macaws comprise a large and fascinating group. Species are brought to this country as captive birds, but many escape and live in the wild. Several of these species have created sizable breeding populations. South Florida, and to a lesser degree south Texas, have been a magnet for this 'colonization,' opening up a new world of birding—regularly occurring species are covered in this book but many more are seen only occasionally. Difficult to see in the daytime, they gather at favored roost sites, some of which involve hundreds of birds. They are typically heard before seen; calls vary very little between species. Some species hybridize, and a few parakeets are best left unidentified based on current lack of ID knowledge. 'Psittacides' can be broken into groups that have distinct shapes. In flight, it is easy to work out to which group birds belong, based on these specific shapes. *Brotogeris* are very small with slim wings and long, pointed tails. Wingbeats are very rapid, and birds twist from side to side in flight. *Aratinga* parakeets are medium-sized with long tails. There are several very similar-looking species that probably hybridize. *Amazona* parrots are large and heavy with rounded wings and short, wide, slightly rounded tails. In flight, they have yellow-tipped tails, black trailing edges to wings, and bold red patches at the base of the secondaries, but all have different head patterns.

Kingfishers are fish-eating birds that plunge-dive for their prey. They dig burrows in river banks. The larger the bird the higher the river bank needed for nesting. Typically seen alone, they often chase other individuals of the same species.

Woodpeckers are a popular group of birds, with a number of species frequently found in backyards. Their toes are long and strong for clinging to tree trunks, while their bills act as chisels to get to grubs and insects. They excavate holes for nesting—a new one each year. In spring, they repeatedly tap trees at great speed (drumming). The speed, length, loudness, and intonation make each drum identifiable to species, just like songs. They have a strikingly undulating flight pattern shared by few other species—a key ID feature. For the smaller species, calls are typically short single *pik* notes. Range alone can ID many species. However, while most species are considered sedentary, individuals are often found out of range and some move south to winter.

Jays are fairly large with rounded wings and long tails, and are usually brightly marked. They typically travel in flocks in wooded areas, often feeding in trees but sometimes dropping to the ground. All quite happily visit feeders. Loud and vocal, they can also at times be frustratingly quiet and hard to find. They are mostly resident but will irrupt when food shortages occur. Jays often store seeds for hard times and GRJA even uses its saliva to help 'clump' its food for easier storage. Calls are usually loud and raucous—learn one and the intonation and quality are somewhat similar for all species.

Crows are not everybody's favorite birds. They are essentially black, very similar, and often described as "evil-looking." They are intelligent. Research has shown they can count and do addition/subtraction. Some pose tough ID challenges, compounded by their monotone color. For the beginner, other black birds, such as grackles, may also cause confusion. The only way to id these birds is by size, shape, call, and often range/habitat.

ad. ♂

ad.

Chuck-will's-widow *Caprimulgus carolinensis* **CWWI** L 12in

Fairly common in a variety of open dry coniferous and deciduous forest. Often seen at s. migration watch points in spring. Roosts on the ground or on branches. Flushes from underfoot with a loud annoyed croak, usually disappearing just out of sight. Named after song, a repeated *chip wil WHID ow*—a loud and rushed *whid* at a distance. Heard at twilight, sometimes during night. Best seen when appearing at dusk, gliding over the treeline 'wing-clapping' in display or hunting insects. **ID**: Easily the largest nightjar, with long and broad rounded tail. Wings also long and rounded, wing point short of tail tip at rest. Perched birds are large-bodied, neckless, with huge flat-crowned broad head—the best features to focus on. Plumage usually brown-toned (EWPW usually grayer). ♂: similar to ♀ other than white inner web to outer 3 tail feathers, ♀ has buff tips. ♂ EWPW has more extensive white in tail.

♀

ad. ♂

Eastern Whip-poor-will *Caprimulgus vociferus* **EWPW** L 9.75in

Recently split into 2 species, Eastern and Mexican (*C. arizonae*), the latter found in the SW. Common but declining in most areas. Found in most types of forest, though being driven out by CWWI in some areas. CWWI usually prefers more open habitats. Named after its song; we usually hear a fast repeated *poorWILL*. Best heard at twilight, but can often be heard everywhere at night, though at other times bird is frustratingly quiet. Sits on the ground or on branches and is very difficult to find. Can perch vertically and look like a small owl. In flight, long tail and wings rounded at tip but not so long as in CWWI. **ID**: 20% smaller and slighter version of CWWI with more rounded head and usually grayer tones and darker throat. Plumage patterns similar though averages paler and more contrasting scapulars (braces). Often confused with easier to see CONI, the latter has much smaller head, slimmer build, and wings extend past tail, also angled pointed wings in flight. ♂: white across tips of outer 3 tail feathers (only inner webs in CWWI) and dark throat. ♀: buff throat and buff tips to outer 3 tail feathers.

Common Poorwill *Phalaenoptilus nuttallii* COPO L 7.75in

Fairly common in rocky open areas, usually with some vegetation. Roosts on ground in daytime. Song an echoing sonar-like *poor-will*—note on end difficult to hear. At night, sings or flies up to catch insects from rock perch and sits in dirt tracks. True status in winter unknown, some resident, others migratory. **ID**: The only nightjar with a really short tail, and note small size—not much more than a ball of fluff! In flight, wings are short and rounded. More uniform spotted gray upperparts than other nightjars. White throat line with black border. ♂: white corners to tail, buff in ♀.

Common Pauraque *Nyctidromus albicollis* COPA L 11in

Common locally (s.TX) in thorny woods. Roosting birds melt into leafy ground habitat. Approachable and doesn't usually fly far when flushed. At night, frequently sits on road, red eye shine and white bar across wing visible when cars flush birds. Song: a whistled *pwiYYerr*. **ID**: Very long-tailed, and has long wings with white bar across rounded tip. Heavy-chested. Prominent white outer tail feathers in ♂, restricted to tips in ♀. White-spotted coverts and pale-fringed coverts.

Common Nighthawk *Chordeiles minor* **CONI** L 9.5in

Fairly common and the most easily seen goatsucker in many areas. Declining. Found mainly over cities, towns, prairies, and wooded waterways. Note bat-like flaps followed by swooping glides as it hunts insects, mostly at twilight. Nests on the ground, and in well-lit areas such as flat-roofed shopping malls. Check floodlit ballparks or anywhere insects are attracted to lights. Usually roosts on tree branches—check protruding lumps that look like bark. Sometimes flies around in the daytime and sits on fenceposts and other open perches. Frequently in groups, while round-winged nightjars always alone. Traditional migration routes are along rivers. Call: a woodcock-like *pint*. **ID**: Great variation in color within populations and regionally. Most have gray tones, but some are browner and warmer. Sitting birds small-headed and slender with wing point extending past forked tail. White patch on primaries well short of longest tertial tip. In flight, slim with forked tail and thin pointed wings angled at elbow. Conspicuous white bar halfway down outerwing. ♂: white throat and subterminal tail bar. ♀: buff throat, lacks tail bar and has smaller primary patch.

Antillean Nighthawk *Chordeiles gundlachii* **ANNI** L 8.5in

Scarce summer visitor to the FL Keys in open areas. Vagrant as far n. as NC. Distinctive *pity-pit-pit* call only sure way of telling from CONI. Slightly smaller than CONI. In flight, appears slightly more compact, head larger, wings straighter and less pointed, giving a MERL-like impression. Plumage brown rather than gray, with buffy undertail coverts—as some CONI.

Lesser Nighthawk *Chordeiles acutipennis* **LENI** L 9in

Uncommon in s. TX, but increasingly regular winter visitor to s. FL. Vagrant as far n. as NJ. Roosts on ground or in low vegetation, feeding over ponds, open fields, and arid areas at twilight. Song: a quiet trill. **ID**: Very similar to CONI but averages smaller, larger-headed, and shorter-winged. On sitting birds, wing patch equal to or past tip of longest tertial (also squarer in shape). On CONI the wing patch falls short of longest tertial. Vent/undertail coverts buff. In flight, patch is closer to wingtip than carpal. Flight is more fluttery, body rocking from side to side. With great views, P10 (outer primary) shorter than P9, giving a blunter-winged look than in CONI. Feathers also broader and less pointed with small buff spots at base (diagnostic). Prominent white spots on coverts and scapulars. Adult is only nighthawk to molt in N America in fall.

Rock Pigeon (Feral Pigeon) *Columba livia* **ROPI** L 12.5in

Common and familiar just about anywhere from city centers to arable areas, particularly where food is plentiful. Originates from cliff-nesting wild European stock. Domesticated worldwide—aptly named Feral Pigeon. Nests and roosts on buildings, ledges, and bridges. Often in large flocks making trademark *cooing* noise, before loud whirring of wings when flushed. In summer, large flocks disperse to nesting sites. One of the most common birds seen flying overhead, usually in tight groups. In most areas is the only pigeon. Where overlaps with other pigeon species, it is the only one with wing bars and pale underwing. **ID**: Wild stock are gray, darkest on head with 2 black wing bars and white rump. Centuries of domestication have resulted in many color variations ranging from all white to nearly all black. All are the same size and shape, and with dark-bordered pale underwing and white rump. Juv: duller, lacks iridescence, and has dull orbital ring.

1st-w.

ad.

White-crowned Pigeon *Patagioenas leucocephala* **WCPI** L 13.5in

Scarce in s. FL, particularly the Keys. Commoner in spring and summer. A treetop dweller, more often seen flying than sitting. Inhabits mangroves and fruiting trees, but sometimes sits on wires in towns. **ID**: Like a large dark but slim ROPI with white crown, dark rump, long tail, and large bill. Ad ♂: white crown, yellow iris, iridescent nape. ♀: averages duller crown and nape, browner tones to plumage. Juv/1st-yr: browner plumage with fine pale fringes, partial dull cap, and brown iris.

Red-billed Pigeon *Patagioenas flavirostris* **RBPI** L 14.5in

Scarce and declining along the Lower Rio Grande, s. TX. Migratory, fewer in winter. Most often seen flying to fruiting trees early and late in the day, where it sits up high but often out of sight, or crossing river to Mexico. Falcon-like in flight, powerful and direct. **ID**: Bill more pale than red! At distance, bird often appears all dark and is sometimes mistaken for WWDO. Seen closer, it is 2-toned: dark gray with maroon coverts, head, and breast. ♀: averages duller than ♂. Juv: duller with little or no maroon, fine pale fringes, brown iris.

Band-tailed Pigeon *Patagioenas fasciata* **BTPI** L 14.5in

Very rare visitor from the W, usually in late fall/winter. Prefers higher-altitude forest but also found in lowlands, where also at feeders. **ID**: A large, lanky, long-tailed bird with dark-tipped bill. Gray with maroon breast. Broad pale tail bar is obvious in flight or at rest. White-and-green nape patch equally prominent. ♀: averages duller. Juv: gray-naped.

White-tipped Dove *Leptotila verreauxi* **WTDO** L 11.5in

Common in s. TX on forest floor in thorny woods. Tough to see, listen for leaf-litter moving. Walks quite fast with head jerking back and forth. When disturbed lifts head and tail. Visits feeders with PLCH. Song: like noise produced when blowing into a bottle—deep and resonant. **ID**: Medium-sized, robust, and well-proportioned. Bland pale plumage blends in with habitat. Cream below, darker on breast, pale head with iridescent nape, and brown gray upperparts. In flight, brown underwing and black tail with white corners. Juv: similar, but even blander.

Eurasian Collared-Dove *Streptopelia decaocto* **EUCD** L 13in

Increasingly common and successful invasive species enjoying rapid range expansion, as it did in Europe. Found around towns, villages and farms and in the countryside. Perches on fences, wires, and trees before dropping to the ground. Often glides on flat wings, in seemingly casual and unhurried manner. Display flight—gentle arcs side to side and swooping. On the ground, struts purposefully as it feeds. Often mixes with MODO and other doves. Song: a rythmical, repeated *coo coo cup*. **ID**: Noticeably larger, heavier, and paler than MODO. Pale brown with a black half-collar. From below, broad square-ended white tail with black base is striking both sitting or in flight (formerly colonizing Ringed Turtle-Dove has brown base to tail, is smaller and paler). MODO, with growing tail feathers, can have similar-shaped tail. Upperparts have gray covert panel; white corners to tail. Color variants, both lighter and darker than typical, are regular. Juv: has reduced or lacks nape collar; well-defined pale fringes to upperparts.

White-winged Dove *Zenaida asiatica* **WWDO** L 11.5in

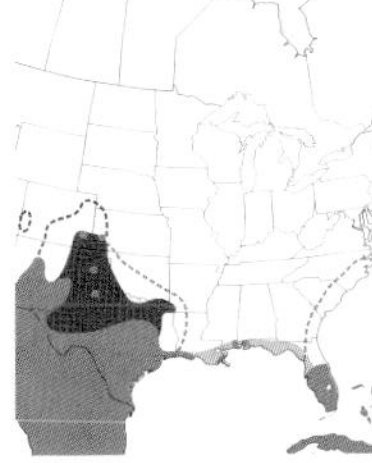

Fairly common in the far S and expanding range. Regular vagrant anywhere n., particularly May and late fall. Often found with other doves in towns, farms, and the countryside. Although often perched on wires in the open, can also be secretive, sitting quietly in mesquite and other trees. Find one and you will probably find a few. Feeds on grain and seeds on the ground but will also hang from corn stems. to feed Often seen flying powerfully overhead in small groups, when white wing panel, set off by dark outerwing, is particularly striking. Looks as much like a pigeon as it does MODO. Song a repeated *who cooks for you.* **ID**: Heavy-set with a short square tail giving compact feel. Similar to EUCD from below but shorter tail with narrower white tail band. Sitting birds show white outline on forewing. Long bill gives 'snouted' look. Red iris and isolated black cheek whisker. In flight, bird comes alive with contrasting wing and tail colors, hopefully removing any ID doubts. Juv: duller iris and well-defined pale fringes to upperparts. Doves keep juv plumage briefly.

1st-w.

juv.

ad. ♀

ad. ♂

Mourning Dove *Zenaida macroura* **MODO** L 12in

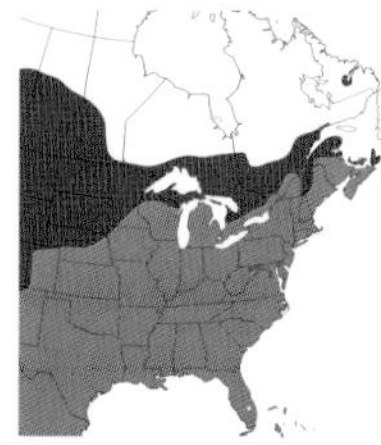

The common and familiar dove in most areas. A regular garden and feeder bird but found almost anywhere except densest forest and mountain tops. Can breed year-round. Sometimes in flocks, but more often in ones and twos, particularly summer when in pairs. Frequently sits on telegraph wires, posts, and other exposed perches; wings make whirring whistle when flying off. Feeds quietly on the ground, soon melting into the background. Jerky yet purposeful walk, head rocking backward and forward. Frequently seen in flight. Distinctive pointed tail, small head, fairly uniform coloration, and narrow-based pointed wings. Direct steady flight. Display flight is a short descending glides on more rounded wings. Song: a soft *wahoo hoo hoo hoo*. **ID**: A small-headed, slim dove with long pointed tail (some square-tailed birds due to molting or broken tail feathers). Tail has black subterminal and white terminal tips. Black-spotted wing, gray rear neck on otherwise brown bird. Ad ♂: iridescent pink neck patch and warmer underparts. ♀: duller and lacks iridescent neck patch. Juv: stronger face pattern and well-defined pale fringes to upperparts. Molts quickly into adult-like plumage. Extended breeding season.

juv.

ad.

Inca Dove *Columbina inca* **INDO** L 8.25in

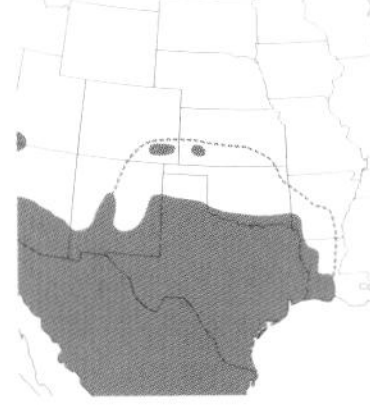

Uncommon in dry brushy areas, particularly around villages and farms. Often tame and confiding, not going far when approached or disturbed. When resting, perches head height in bushes, usually in groups cuddled up together. Spends a lot of time on the ground feeding in backyards, at feeders, in scrubby patches, and sometimes more open fields. Tends to run away with legs and head moving at full tilt. When flushed, wings make a hollow rattlesnake-like rattle. Song: a rhythmical *there's hope*. **ID**: A subtly patterned tiny dove with very long, thin tail (CGDO has a very short all-dark tail). Pallid and strikingly scaly. Bright rufous outerwing is pronounced in flight; just be careful you see the long tail as ground-doves have similar wing pattern. Outer tail feathers white. Sexes similar. Juv: reduced scaling with white covert spots. Plumage held briefly.

Common Ground-Dove *Columbina passerina* **COGD** L 6.5in

Uncommon in towns and countryside, preferring dry open areas near scrubby cover. Tiny size and scurrying ground-dwelling habits make it easy to overlook. Bright rufous wings striking in flight. Views often brief: a few flicked wingbeats and glides before it drops into cover. Song: a repeated *wer up*.

ID: Small and fat with short broad tail; distinctive. Spotted coverts, scaled crown and breast, bicolored bill. Black tail with white corners. Ad ♂: gray nape contrasts with salmon-pink face and underparts. Ad ♀: much duller and more uniform. Juv: similar to ♀ with white-fringed upperparts and dark bill.

Ruddy Ground-Dove *Columbina talpacoti* **RUGD** L 6.75in

Very rare near Mexican border, often with INDO. Favors damp areas. Bright male COGD regularly misidentified as RUGD. From COGD by slightly larger size (fairly hard to determine in the field), lack of head and breast scales, brighter colors, longer tail, black underwing coverts, dull bill, and black wing spots that extend onto scapulars. ♂ brighter than ♀.

Smooth-billed Ani *Crotophaga ani* **SBAN** L 14.5in

Nearly extirpated in s. FL from scrubbiest, most marginal habitats. A black crow-like cuckoo with long, graduated tail, and a weird honker of a bill. Travels in family groups, huddled together like the best of friends. Hops around on ground with tail cocked and wings drooped. In flight, a few wingbeats followed by short glides, always in a straight line. **ID**: Paler fringes to black upperparts and head peculiar to anis. Very similar to GBAN but ranges don't overlap. Large 'smooth' bill with subtly different shape (see GBAN). Juv: browner and lacks pale feather fringes. Call: long, whining, single, rising note.

Groove-billed Ani *Crotophaga sulcirostris* **GBAN** L 13.5in

Fairly common in s. TX in summer. Local fall and winter. The ani that wanders n. of range most regularly. Behavior as SBAN. Feeds on insects in bushes or on the ground in scrubby countryside. Similar to SBAN but is slightly smaller; ranges don't overlap. Bill has distinct grooves, less curved dark ridge (top edge to bill) that extends straight onto forehead. Lower edge of bill is straight or slightly upturned (slightly downturned or convex in SBAN)—the overall impression is of a narrower and more pointed bill. Bill size is variable, and juv lacks grooves initially. Multiple short GTGR-like calls, often disyllabic.

Yellow-billed Cuckoo *Coccyzus americanus* **YBCU** L 12in

Fairly common in deciduous woodland in the E, scarce in riparian corridors in the W. Secretive and easily overlooked, often sitting motionless for periods of time when is hard to detect unless it calls. Puts its head down as if to hide and peek at you through the leaves. At other times it will slowly rotate its head scanning the area. Moves with purpose, a few hops or strides. Agile in flight—as if its slim profile were made for squeezing between branches. Loves caterpillars but also other insects such as praying mantises. Numbers in a given area fluctuate from year to year in response to prey abundance. Often seen flying across clearings—look for rufous outerwing on angled wings, and long tail. More often heard than seen, night and day, variable, rapid *cuck* or *coo-cuck* notes, frequently starting slowly and accelerating at end. Sometimes harsher single *koowp* or in sequence. **ID**: Large and slender with long, graduated tail, slim downcurved bill, bold white underparts, and brown upperparts. Always look for yellow lower mandible, rufous in wings, and large white spots on undertail. Juv: similar to adult but often briefly with all-dark bill; dull eyering, and subdued tail pattern.

juv.

ad.

Black-billed Cuckoo *Coccyzus erythropthalmus* **BBCU** L 11in

Uncommon in open deciduous woodland and fields, often near wet scrubby areas. Territories are large and birds can easily disappear on breeding territory. Behavior is very similar to YBCU. Find lots of caterpillars and you will find this bird, for as long as there is food—then they will disappear. Tends to be more vocal at night or twilight than in the daytime, with a LEBI-like, fast monotone *cu-cu-cu-cu*, often given on the wing. In flight, slightly smaller and more compact than YBCU with brown wings lacking rufous. **ID**: Slightly smaller and slimmer than YBCU with more rounded crown, finer all-dark bill, uniform upperparts (lacking rufous in wing). From below, has small white spots on undertail and seems to lack strong contrast of YBCU. Adult: red orbital ring. Juv: most similar to YBCU with dull yellow orbital ring and hint of rufous in primaries. Look for pale-fringed wing feathers, buff throat, blue-gray lower mandible, and dull undertail.

Mangrove Cuckoo *Coccyzus minor* **MACU** L 12in

Scarce in mangroves and hammocks, mostly in summer. Rare in winter. Easiest to see in rain, hence its nickname, 'Rainbird.' Mannerisms as other cuckoos. Makes a deep growled *agh agh agh agh agh*. **ID**: Size, shape, and bill are similar to YBCU. Most obvious differences are dark mask, buffy underparts, and lack of rufous in wings. Yellow restricted to lower mandible. Juv: black mask faint or absent. If in doubt, look for completely black base to undertail.

Greater Roadrunner *Geococcyx californianus* **GRRO** L 23in

Uncommon in arid habitats with some cover and s. pine clearouts. In the hottest hours often seen panting in the shade. In comical cartoon style, runs around catching lizards, snakes, and just about anything that moves. In flight, usually to perches, has striking black underwing, though generally prefers to run for cover. Perches in the open on walls, posts, and other elevated objects, often to sunbathe. Frequently raises crest, lowering it in slow motion as if to say "Look at me." **ID**: Large and long-tailed with athletic legs. Streaky brown and white with bushy crest and blue facial skin.

Budgerigar

Melopsittacus undulatus **BUDG** L 7in

Small and long-tailed with striking wing pattern. Common escape. Wild form is green and yellow, but many color variants possible.

Cockatiel

Nymphicus hollandicus **COCK** L 12in

Occasionally occurs as an escape. Crested with long tail. Wild stock typically yellow-faced with orange cheek patch, gray body, and white wing patch.

Peach-faced Lovebird

Agapornis roseicollis **PFLO** L 6.5in

Tiny green parrot with blue tail and peach face. NB: Other lovebirds do occur occasionally.

Yellow-chevroned Parakeet *Brotogeris chiriri* **YCPA** L 8.75in

Scarce but more numerous than WWPA around Miami, FL. Behavior as WWPA. Green-feathered lores accentuate pale eyering on green head. Small pink bill and yellow wing coverts, lacking white, easiest to see in flight.

White-winged Parakeet *Brotogeris versicolurus* **WWPA** L 8.75in

Increasingly scarce in date palms around Miami, FL. Usually in small flocks. Small size enables it to fly into trees and disappear. Formerly conspecific with YCPA. Told from latter by unfeathered gray lores and grayish eyering, bigger bill, and white wing bar.

Monk Parakeet *Myiopsitta monachus* **MOPA** L 11.5in

The most widespread parakeet throughout the US, even found in dense urban environments, as far n. as the border. Populations are localized. Tough and hardy, adapts well to the coldest climates—other parrot species don't. Builds massive dense nest of sticks with windows for watching the world pass by. Often visits feeders. **ID**: Bright green upperparts with blue flight feathers. Gray forehead bisects eye and extends down breast. Pale green belly and vent. Never shows red in wings in flight.

Red-masked Parakeet *Aratinga erythrogenys* **RMPA** L 13in

Uncommon, though true status unknown. **ID**: Very similar to MIPA but slightly smaller, legs average pinker. Adult: red mask varies in extent. Importantly, fairly clean border with rest of head and with few red patches elsewhere. Extensive red on wing bend. Imm: less extensive red on wings. Head initially all green, lacking red forehead.

White-eyed Parakeet *Aratinga leucophthalmus* **WEPA** L 13in

Probably scarce in FL, though true status is unknown. Very similar to GRPA. Green with red on forehead, off-white to pale gray eye with a few orange-red spots on head in adult (juv all green). The main difference from GRPA is orange feathers with pale yellow tips on underwing—all green in GRPA. Hybridization could no doubt create birds with this appearance.

Green Parakeet *Aratinga holochlora* **GREP** L 13in

Fairly common in towns along the Lower Rio Grande river, s. TX. Status in FL not clear; seems scarce but similarity to juvs of other *Aratinga* clouds issue. In TX, feeds quietly in palm trees in daytime and is more easily seen in flight. At dusk, they appear from nowhere in large numbers, flying to a noisy communal roost site where they line up on wires before disappearing into a few favored palm trees at nightfall. Everything then goes quiet! Completely green underwing separates from other species. Otherwise like WEPA with all-green plumage, orbital ring that varies from off-white to gray, and a few scattered red spots. Iris red in adults, brown in imms.

Mitred Parakeet *Aratinga mitrata* **MIPA** L 15in

The commonest *Aratinga* in FL. Very similar to RMPA, and many birds probably best left unidentified. **ID**: Large with a big pale bill, broad white eyering, and orange iris. Legs gray, pinker in RMPA, variable and tough to judge. Adult: extensive ragged red area on face and many isolated patches of red on head and neck and even orange/red blotches on breast and wings. 1st-yr: red limited to forehead, brown iris, and green head. Many intermediate with adults as they grow older. Orange-red on wings limited to narrow strip on leading edge.

Black-hooded Parakeet *Nandayus nenday* **BHPA** L 14in

Uncommon in FL with well-established populations, mostly on the cen. Gulf Coast. Usually found in urban and suburban areas in palm trees and perches conspicuously on telegragh wires. Visits feeders. A striking bird, it causes few ID problems, with bold dark head, sky-blue breast, and red 'socks' on otherwise green underparts. Green upperparts with hard-to-see red bordering the black hood. In flight, the wing feathers are black, which, combined with the black head, make it stand out.

Sulphur-crested Cockatoo
Blue-crowned Parakeet
African Gray Parrot
Rose-ringed Parakeet
Chestnut-fronted Macaw
Blue-crowned Parakeet

Red-crowned Parrot *Amazona viridigenalis* **RCPA** L 12in

Fairly common in Lower Rio Grande Valley towns, scarcer in FL but still the commonest *Amazona*. Usually in noisy suburban flocks. Like other parrots, often quiet in the daytime, but birds fly from everywhere to gather at noisy roost sites. A great spectacle. **ID**: Adult: a green parrot with yellow tip to tail and red crown. Red is restricted to forehead in imm. Rear crown and neck lilac-blue, very similar to Lilac-crowned Parrot. Lilac-crowned has red restricted to forehead and, more importantly, has a longer, slimmer tail. These 2 species are often found roosting together, and individuals of both species are easy to overlook.

Lilac-crowned Parrot

Yellow-headed Parrot

Mealy Parrot

♀ Yellow-lored Parrot

Red-lored Parrot

Yellow-naped parrot

Blue-fronted Parrot

♂ Yellow-lored Parrot

Orange-winged Parrot

Yellow-crowned Parrot

Green Kingfisher *Chloroceryle americana* **GKIN** L 8.75in

Uncommon along the Lower Rio Grande river, streams, and nearby ponds. Very small, making it easy to overlook as it perches low in dense vegetation near water. Dives at angles, picking prey from near surface rather than submerging. Most easily found by call—a single or double *tik* note (given in 2s and 3s like Morse code). Punky and cute. **ID**: Striking green-and-white pattern with bold white outer tail feathers. Long-billed for such a small bird. Ad ♂: rufous breastband. Ad ♀: 2 weak green breastbands. Much smaller than other 2 kingfishers found in our region.

Ringed Kingfisher *Megaceryle torquata* **RIKI** L 16in

Uncommon along the Lower Rio Grande river and on nearby ponds. A large bruiser of a kingfisher that commonly sits high on branches and telegraph poles as it watches for fish. Wary and easily spooked. Flies high, moving between perches. Call: a loud machine-gun rattle, like BEKI, but deeper in tone and notes more isolated. **ID**: Huge size and boldly colored underparts make it unmistakable. Ad ♂: rufous underparts extend to throat. Underwing coverts white (rufous in ♀). Ad ♀: slate-blue breastband bordered by white. Young birds similar to adults but show rufous mixed with blue in breastband.

juv. ♀

ad. ♂

ad. ♀

juv. ♀

Belted Kingfisher *Megaceryle alcyon* **BEKI** L 13in

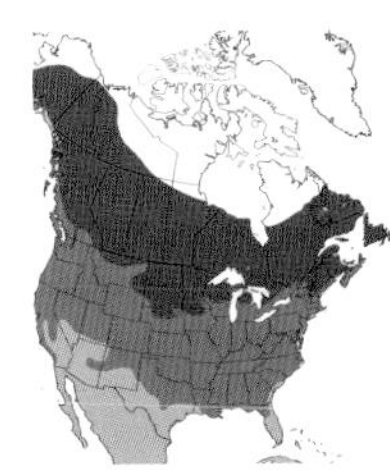

The widespread common kingfisher, but seen only in ones or twos. Found wherever fish present. Nests in hole in sandbank, often away from water. Plunge-dives for fish from exposed branches, signs, wires—anywhere providing a good vantage point. Also hovers. Easily spooked, noisy rattled call given as it flies off. Often seen flying overhead as it moves considerable distances between feeding areas—white from below with a dark collar. Flight is powerful but not smooth with choppy wingbeats. Distinctive profile with large head and bill but disproportionately short square tail. Prominent white patches on upperwing. Its call carries a long way, and bird is often heard and not seen. Looks studious as it examines the water, head tilted at 45 degrees. Occasionally raises crest, usually when agitated. **ID**: Thick-set bird with large bill and a spiky crest. Slate-blue upperparts, head, and breastband. White collar, underparts, loral patch, and tiny spots on upperparts. Ad ♂: blue-gray breast sometimes tinged with rufous (more rufous in juv ♂). Ad ♀: rufous lower breastband (narrower in juv ♀) extends down flanks. Rufous is more extensive in imm birds, but judging this is often difficult.

juv.

ad.

Red-headed Woodpecker *Melanerpes erythrocephalus* **RHWO** L 9.25in

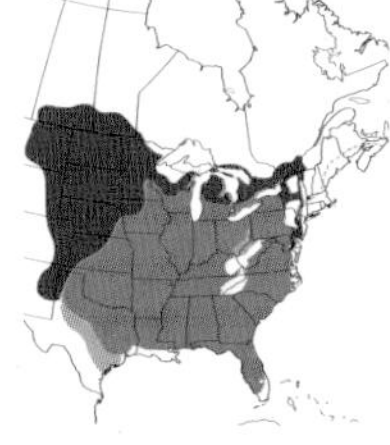

Uncommon in wide-open deciduous woodland, and suffering some local declines. Often in town parks and campsites, particularly in oak trees. In the MW found in more isolated trees and happily feeds in surrounding fields or catches insects on the wing. Also glides to the ground to get food. In flight, large white wing patches and rump are unmistakable at any distance. Sits quietly for periods of time, always looking around so it knows what's going on. Sometimes in small colonies (clusters). In winter can be in small groups, also with other woodpeckers, particularly on dead trees. Nest hole neat and tidy, often taken over by EUST. The commoner RBWO is often mistaken for this species. Call: similar to RBWO but harsher. **ID**: HAWO-size but looks bigger with particularly broad wings in flight. Strikingly simple but beautifully bold red, white, and black. Juv: dingier brown version of adult with dark secondary bars. Molts through winter so by spring it is similar to adult, but often slightly duller with a few retained dark bars in secondaries.

Red-bellied Woodpecker *Melanerpes carolinus* **RBWO** L 9.25in

The common larger garden woodpecker in most areas, often at feeders. Conspicuous and vocal, it is the woodpecker many nonbirders notice. Prefers open deciduous woodland such as parks, gardens, golf courses as well as larger forest. Poorly named after barely noticeable red on belly. Often mistakenly called 'Red-headed Woodpecker.' A better name would be 'Bar-backed' or 'Red-naped Woodpecker.' In flight, underwing finely patterned black and white, adding to barred appearance. White patch at base of primaries obvious even at distance. Call: a piercing *qwe-eer*, rising at end, a common sound in e. woodland. **ID**: Medium-sized. Slim head, long bill, and spiked tail give it a slender appearance though it is quite large-bodied. Barred back can look surprisingly uniform but always contrasts with red head. Buff underparts, and white rump with faint black bars. Orange-red on head variable in extent. Birds in S average smaller and darker, with white in tail restricted to rump, but much variation. Ad ♂: red from nape to bill, often paler on forehead. Diffuse red belly patch. Ad ♀: gray crown, pale belly. Juv: limited orange nape and nasal tufts (forehead); quickly molts into adult-like plumage through fall.

Golden-fronted Woodpecker *Melanerpes aurifrons* **GFWO** L 9.5in

Common RBWO look-alike of cen. and s. TX. Ranges barely overlap but some hybridization. Found wherever trees. Active, always shuffling up and down branches, also at feeders. Vocal, calls similar to RBWO but louder and more raucous. Differs from RBWO by all-black central tail feathers, white rump lacking black, grayer underparts with yellow wash on belly, and different head pattern. Nasal tufts always yellow. Ad ♂: red cap isolated from orange nape. Ad ♀: orange nape. Juv: duller version of ♀.

Red-naped Sapsucker *Sphyrapicus nuchalis* **RNSA** L 8.5in

Occurs in Black Hills, SD. Very rare elsewhere in our region. Differs from similar YBSA by red nape (occasionally absent), narrower black malar (sometimes broken with red touching white), more extensive red throat (rarely white), and more extensive black in center of back. YBSA occasionally show a touch of red on nape. Hybrids occur regularly. Calls: as YBSA.

Lewis's Woodpecker *Melanerpes lewis* **LEWO** L 10.75in

W. woodpecker, rare in the E, mostly winter. Found in open woodland, particularly pine and oak. Often in a big dead tree in someone's garden. Sallies after insects on broad dark wings, returning to top of favorite dead tree or telegraph pole. Stockpiles acorns. Large and superficially all dark—the only woodpecker to lack white. Subtle greens, grays, and pinks can be seen with good views.

Yellow-bellied Sapsucker *Sphyrapicus varius* **YBSA** L 8.5in

Fairly common breeder in deciduous or mixed forest, often at higher elevations, though its retiring nature means it is easily overlooked. In winter, possible anywhere with trees, often in gardens. Loves trees with sap wells, eats sap and insects attracted to them. Often returns to favored tree. Look for Chinese elm, aspens, and other trees with lines of holes dug by hungry birds. Often sits quietly in trees for periods of time and is easy to overlook. Highly migratory. Very bouncy flight with conspicuous long, white wing patch. Also finely checkered pale underwing, and spiky tail. Call: a squeeky nasal *wheer*. **ID**: Fairly small, intricately marked woodpecker with a bold head pattern and mottled underparts lacking the cleaner look of HAWO and others. Bold white line across wing is diagnostic for sapsuckers. All plumages have barred back and underparts. Ad ♂: red throat, sometimes mixed with white. In winter, browner and duller than summer. ♀: throat white. Juv: brown ghost of adult with heavy mottling and bars. By mid-winter looks adult-like.

Downy Woodpecker *Picoides pubescens* **DOWO** L 6.75in

The smallest and commonest woodpecker in most areas. Found anywhere with trees, and a regular garden bird, often at feeders where happily mixes with other species. Small enough to feed on larger plant stems and reeds and from the smallest branches, hanging acrobatically. Occasionally in feeding flocks. Reasonably tame and approachable. In flight, look for small size, tiny bill, and narrow-striped head. Compared with HAWO is substantially smaller and less powerful, more compact and darker-headed and -tailed. Call: a *pik*. **ID**: Small and cute with tiny pointed bill, indeed bill size is the best way to eliminate HAWO. Clean white underparts, black upperparts with white back, spotted coverts, and barred flight feathers—a very fresh look. Striped head pattern. Ad ♂: red nape patch. Ad ♀: white nape. Juv: very similar to adult, but duller, more flecking on underparts, and sometimes red flecking (more on ♂) on crown. W. populations have less white in wing.

Hairy Woodpecker *Picoides villosus* **HAWO** L 9.25in

Fairly common in mature woodland with large trees, often in damp areas. Less likely in sparsely wooded areas. Typically feeds on its own fairly high on larger limbs, rarely joining mixed flocks. Visits feeders; it is partial to suet, particularly in colder regions. Quite secretive and shy. Often flies considerable distances between feeding areas with powerful bounding flight, when its larger size than DOWO is apparent. Widespread but nearly always outnumbered by DOWO. Call: a *peek*, stronger than DOWO—tends to go very quiet for parts of the year.

ID: Very similar to DOWO in color pattern. Noticeably larger and more muscular with much larger bill, giving meaner facial expression. White cheek stripe broader at rear, giving overall impression of whiter head. This and its extensive white outer tail stand out in flight. DOWO usually has black spots in tail. Ad ♂: red nape patch. Ad ♀: white nape. W. populations have less white in wing. Briefly held juv plumage has orangey red on crown like juv DOWO.

Red-cockaded Woodpecker

Picoides borealis **RCWO** L 8.5in

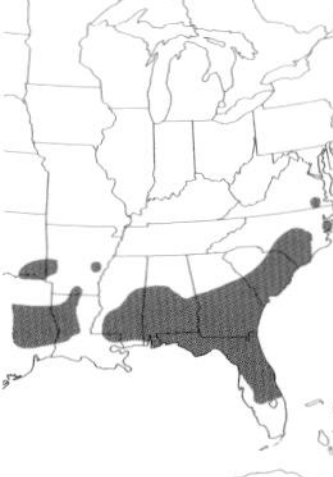

Rare and local in old-growth long-leaf pine. Lives in family colonies, or clans, usually in managed areas (look for painted circles on trees). Nest/roost holes are protected by drill holes that ooze sticky resin. Forages mostly in surrounding forest during day. Chips off bark flakes—a giveaway when you're below a feeding bird! Wings sometimes held distinctively in air. Call: an unusual squeeky *sherp*. **ID**: The only smaller woodpecker with white cheek, bordered by black cap and thick malar. Streaked flanks and heavily barred upperparts. Red 'cockade' at upper rear edge of ear coverts on ♂ barely visible.

Ladder-backed Woodpecker

Picoides scalaris **LBWO** L 7.25in

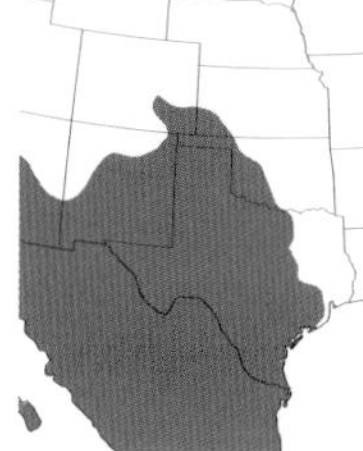

Fairly common in arid wooded areas, particularly mesquite and cactus. Habitat often seems marginal at best with few trees. Behavior a lot like DOWO, very active on smallest branches and plants, joins feeding flocks and quite approachable. Range barely overlaps with DOWO and is the small woodpecker in most of its range. Call: a deeper version of DOWO. **ID**: Very small. 2 black lines across face often join around back of cheek. Heavily barred upperparts. Buff underparts with barred flanks give 'busy' look. Ad ♂: extensive red crown. Ad ♀: black crown. Juv: red crown.

Black-backed Woodpecker

Picoides arcticus **BBWO** L 9.5in

Scarce in dense evergreen and mixed forest with plenty of dead trees (due to insect infestations and fires). Chips off bark looking for insect larvae—flakes and bare patches on tree sometimes a give-away. Call: a flat *kuk* like HAWO. **ID**: Chunky bird with thick neck. Black head and back lacking white spots make it dark overall. Heavily barred flanks. Distinctive white lines from forehead through ear coverts give an unusually shaped capped appearance. Ad ♂: yellow crown. Ad ♀: black crown. Juv: duller than adults; yellow crown reduced in ♀.

American Three-toed Woodpecker

Picoides dorsalis **ATTW** L 8.75in

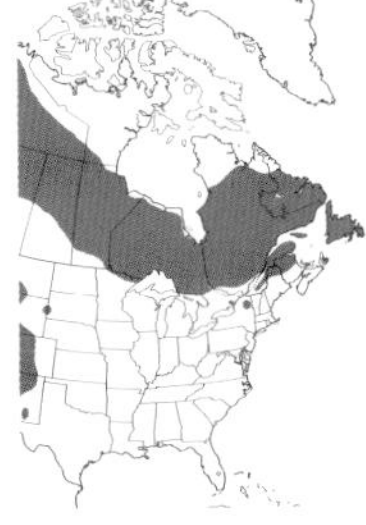

Scarce. Habits and behavior as BBWO; both have 3 toes. Both also occasionally irrupt s. following food. Subtly smaller and less muscular than BBWO. Although superficially similar, ATTW has a different face pattern with 2 white lines and white barring on back (however, at the wrong angle it can look black-backed). Like other woodpeckers, often best found by listening for tapping as it feeds. Call: a *pik*, like a dull HAWO. Ad ♂: yellow cap. Ad ♀: white-flecked crown. Juv: duller with some yellow on crown, less in ♀.

Pileated Woodpecker *Dryocopus pileatus* **PIWO** L 16.5in

Fairly common in mature forest, but most readily found in swamps, though range extends to more marginal habitats, even suburban areas. The largest woodpecker and a majestic bird. Feeds on fallen trees, digging large holes for carpenter ants and beetles, often in habitat more suited to canoes! Slow steady thumps chisel out lumps of wood leaving large cavities. It will also sit high in the largest tree. Commonly seen in deep undulating flight, with exaggerated wingbeats, as it crosses roads and rivers. Its territory is large and mostly inaccessible. Drums loud enough to knock a tree down, and its rapidly repeated *kek* notes can be heard at distance. Best found by flicker-like emphatic *kek kek kek kek* call. Often misidentified as presumed extinct Ivory-billed Woodpecker. **ID**: Strikingly large and black with white neck line. An odd shape with large body but skinny neck and pointed tail. The red crest and large bill are always striking. In flight, morphs into a black-and-white woodpecker with striking white underwing and wing patches above. Ad ♂: red malar, and red runs from crest to bill. ♀: red is restricted to crest. Juv: as adult, some duller.

♀ Yellow-shafted

♀ Red-shafted

♂ Yellow-shafted

♂ Red-shafted

♀ Red-shafted

Northern Flicker *Colaptes auratus* **NOFL** L 12.5in

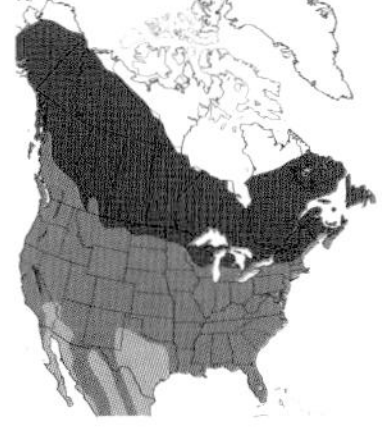

Fairly common in open forest, parks, and gardens. Typically feeds on the ground looking for ants, and this, combined with its atypical color pattern, often confuses beginners. Conspicuous on migration, often traveling in groups. Vocal, frequently heard shrieking in fear as raptors fly past. 2 forms: Red-shafted in the W; Yellow-shafted in the N and E. Intergrades occur where ranges overlap. The latter show a mix of head markings and underwing/undertail coloration that can match either parent or be intermediate between the two. **ID**: 'Yellow-shafted': large, brown, with eye-catching pattern. Spotted underparts with black crescent collar and barred upperparts with red-crescent nape. In flight, striking yellow underwing and white rump enhance bold appearance. Ad ♂: black malar stripe. Ad ♀: brown face, lacking malar stripe. 'Red-shafted Flicker' differs by reddish underwing (beware of early morning light!) and gray cheeks. ♂ has red malar stripe and lacks red nape crescent.

Blue Jay *Cyanocitta cristata* **BLJA** L 11in

Common and spectacular, one of our most recognizable birds. Some migratory. Occurs in the deepest forest but just as happy in parks and gardens. Feeds on acorns, robs nests, and is regular at garden feeders, quietly sneaking in, and quick to leave. Raucous array of calls often announces presence but can be quiet and frustratingly difficult to find. Often feeds in treetops but just as happy hopping along on the ground. Some populations migrate. Usually in flocks, sometimes in 100s, as they migrate high overhead, contrasting black-and-white colors standing out. Rounded wings with bold white secondaries; tail tip and body contrast strongly with black underwing, collar, and base to tail. Bold white secondaries also prominent on upperwing. Does a great RSHA imitation among its many calls. **ID**: Superficially blue and white, but actually a remarkably complex variety of colors and patterns. Crested and thick-set with fairly long, graduated tail. Unique black throat collar. Juv: briefly held plumage, similar to adult but duller with blue or gray lores. Adult-like by fall but retains juv greater primary coverts, and alula lacks black bars.

Steller's Jay *Cyanocitta stelleri* **STJA** L 11.5in

Very rare irruptive fall/winter visitor from the W to the w. Great Plains. Found in evergreen or mixed forest. Often around campgrounds. BLJA-size but slimmer with longer crest and not so boldly patterned. 2-toned blue and dark gray with white lines above eye and on forehead. Sneaks through trees, sometimes sitting in the open. An array of raucous calls, some imitating both birds and animals.

Green Jay *Cyanocorax yncas* **GREJ** L 10.5in

Common in woods in s. TX. Usually in small groups and very vocal. Exotic and unmistakable appearance. Common at feeders. Beautiful combination of lime green and yellow with a blue-and-black face mask. Yellow outer tail feathers hard to miss in flight.

Brown Jay *Cyanocorax morio* **BRJA** L 16.5in

Numbers declining and only occasionally seen in the Lower Rio Grande, TX, woods. Visits feeders. Typical jay, easy to miss but often vocal and obvious. Large and grungy dark brown with buff belly. Adults are dark-billed. Juv: yellow bill becomes progressivly darker through first 2–3 years.

Western Scrub-Jay *Aphelocoma californica* **WESJ** L 11.5in

Scarce and local in open oak, pinyon, and juniper woods in TX Hill Country and irruptive to the w. Great Plains. Visits garden feeders. Usually in small flocks, often feeding on acorns. Sits on treetops before dropping. Often moves quickly with bounding leaps when on the ground. Slimmer and smaller-headed than BLJA. A crestless jay with subtle patterns of gray and blue. Ear coverts always darkest, emphasized by thin, pale supercilium. Juv: briefly held plumage, paler and grayer, particularly around the head.

Florida Scrub-Jay *Aphelocoma coerulescens* **FLSJ** L 11in

Uncommon and localized endemic in transitional 10ft-high scrubby palmetto and oak in cen. FL. Drastic loss of habitat and birds—both now endangered. Communal, one usually seen perched up, the rest of the group feeding nearby but out of sight. Can be very tame, sometimes visiting feeders. The only crestless jay in the SE. Slim with pale forehead, has blue necklace and pale gray belly. Juv: briefly paler and grayer, particularly around the head. Sometimes found with BLJA. Short rasping calls.

Pinyon Jay *Gymnorhinus cyanocephalus* **PIJA** L 10.5in

Very rare visitor from the W. Somewhere between a crow and jay in overall appearance. Scarce in the W in pinyon pines and pinyon–juniper forest. Call is often the best way to locate it—nasal, almost nuthatch-like. Travels in fast-moving flocks, sometimes in hundreds. Feeds on seeds from trees or ground. Slim, long-winged, short-tailed with straight, pointed bill. Mauve-blue all over with darker blue head and paler chin. Juv: paler and grayer, particularly around the head.

Clark's Nutcracker *Nucifraga columbiana* **CLNU** L 12in

Rare irruptive fall visitor from the W. Prefers coniferous forest. Moves between trees feeding on pine seeds but regularly sits in open on treetops. **ID**: A pudgy bird with short tail. Black eye and long, pointed black bill are emphasized by white face. Black wings and tail on gray body come alive in flight with bold white secondaries and outer tail feathers.

Gray Jay *Perisoreus canadensis* **GRAJ** L 11.5in

Uncommon in n. coniferous forest, often near bogs. Usually travels in groups of 2 to 4, feeding on seeds or fruit. Tends to be quiet, making it harder to find than other jays. Reacts to human noises and responds readily to pishing, suddenly appearing from out of nowhere. Although usually within trees, it will sit on treetops. In flight, makes a few flaps followed by long glides. Reasonably quiet for a jay. Call: a soft whistled *wheeo*, sometimes in sequence. **ID**: A friendly looking, fluffy jay with small bill, round head, and gentle facial expression. Cream forehead, pale underparts, darkest on belly—pale tail tips and dark gray upperparts. Juv: dark-headed with white malar stripe. Much regional variation, w. birds being paler-headed.

juv.

ad.

Black-billed Magpie *Pica hudsonia* **BBMA** L 19in

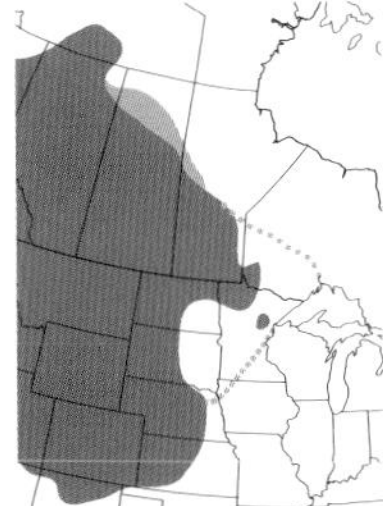

Common in open areas in the W. Uncommon in much of range in the E. Very rare vagrant farther e., where status clouded by occurence of escapes. Noisy and raucous, usually seen flying around alone or in small groups. Opportunistic, scavenges, eats carrion and just about anything it can find. Often seen crowded around food or sitting on fenceposts. Walks a lot with confident gait, often hopping when it needs to go faster. Builds dome-shaped nests of spiky twigs that sit precariously in the tops of large bushes. **ID**: Pied plumage and very long, graduated tail make it easy to id. With good light and views, black wings and tail turn iridescent blue and green. In flight, white outerwing, shoulders, and belly make it unmistakable. Juv: duller with pale gape and shorter tail.

American Crow *Corvus brachyrhynchos* **AMCR** L 17.5in

The common crow through most of lowland America. Commonest in agricultural areas. Scavenges on garbage dumps but found just about anywhere. Out of breeding season, often forms huge night roosts in urban areas. In daytime, found alone or in loose groups. Feeds mostly on the ground, walking with a purposeful strut. When necessary, speeds up to a hopping jog. Around food, such as a carcass, numbers will build, birds often having to wait their turn behind vultures. Eats eggs and chicks, so it is always being chased 'out of town.' May show pale patches at base of flight feathers when it is in wing molt. **ID**: Much individual and regional variation, making ID from similar FICR tricky. Call, a familiar *caw*, is the best way to separate them. AMCR is larger, with longer legs and bill and rangier gait but differences subtle. With good light and views feathers look scaly; freshly molted birds have glossy upperparts and broad, black fringes contrasting with dull nape (FICR unicolored). In flight, longer neck, broader-based and more rounded wings—FICR is smaller, more compact. Juv: uniformly dull brown with pink gape (briefly). 1st-yr: worn juv feathers contrast with newer adult feathers creating a patchwork.

Fish Crow *Corvus ossifragus* **FICR** L 15in

In summer, the common crow in coastal marshes and surroundings, scarcer inland. In winter, moves inland, gathering at landfills and other suitable feeding places, sometimes with AMCR. Behavior much as AMCR—very aggressive and opportunistic. Frequently mobbed by 'worried' smaller birds, in turn chases passing raptors or harasses owls. Roosts are often large. Forms small groups, feeding on ground or perched on posts or wires. Often found near much slimmer and longer-tailed BTGR. Common call: a nasal double note *uh-uh*—a kid with a deep voice saying "No No." Trickier is a single note *ock*, higher-pitched and more emphatic than AMCR. **ID**: From AMCR with difficulty. FICR smaller and more compact with shorter neck, bill, and legs. In flight, subtly shorter neck, pointed wings, narrow at base, make tail proportionally long. Some individuals easier to separate than others both in flight and standing—some are best left unidentified. Juv/1st-yr: uniformly dull brown-black with pink gape. Like AMCR, many feathers molted in fall. Retained juv feathers have increased contrast through year as they become more worn and faded. Molts Jul–Oct (AMCR earlier, May–Aug).

juv.

ad.

Common Raven *Corvus corax* **CORA** L 24in

Uncommon but widespread in a variety of habitats, particularly at higher altitudes: forest, desert, tundra, rocky slopes. Often in ones and twos, rarely in larger groups. Patrols its large territory by wing, its deep, croaking *bronk* call often the only sound in harsh environments. Has adapted to scavenging trash around humans, also carrion. Often nests or perches on prominent radio towers, power lines, cliffs, or trees. Seems to love flying and often performs aerial displays—rolling dives with wings tucked in before swooping up again. **ID**: 25% larger than AMCR but can cause confusion (rarely seen together except at landfills). CORA is a fiercer looking beast with a big bill, a flatter head, and long wings and tail. In flight, the long wedge-shaped tail becomes more obvious, though sometimes looks rounded and even square-ended. Wings are narrower, angular, and more pointed. Often soars with feathers spread, becoming much more crow-like. Long, shaggy, throat feathers and gray base to neck feathers. Juv/1st-w: uniformly dull black with brown cast and pale gape. Mixture of new adult-like and retained juv feathers through first year.

Chihuahuan Raven *Corvus cryptoleucus* CHRA L 19.5in

Uncommon in lowland arid areas. Seen in ones and twos but more often in flocks where food is abundant. Birds can frequently be seen searching low over flat terrain. Range occasionally overlaps with CORA. Very similar to CORA but 20% smaller with shorter bill, longer nasal bristles (cover 60% of bill, CORA 50%). Slightly shorter and more rounded tail. Bases of neck feathers white but almost impossible to see, unless ruffled in a breeze. Call: more crow-like.

Eurasian Jackdaw *Corvus monedula* EUJA L 14.5in

Very rare visitor to the NE. Small and quite cute for a corvid. Size, small bill, white iris, black face with pale gray neck and dark gray body make this a striking and easily recognizable bird for anyone who encounters it. Calls frequently—a distinctive *kayah*, often given in flight.

Tamaulipas Crow *Corvus imparatus* TACR L 14.5in

Formerly regular near Brownsville, TX, particularly at the landfill. Now very rare (there are no other crows in this area). Small and slender with long wings and tail. Bulbous throat and glossy plumage. Call croaky but still crow-like. CHRA is much larger and is a different shape.

AERIAL LANDBIRDS

Most of us are familiar with the special beauty of 'hummers.' Their tiny size, tenacity, and incredibly rapid wingbeats (up to 80 times a second) fascinate and entertain. Although, for most of our region, Ruby-throated Hummingbird is the species that occurs in summer, in recent times the number and variety of hummingbirds wintering in the Southeast has increased. A number of species seem to be expanding north with birds showing up as far as Canada in late fall and winter. The list of potential species is almost limitless. From late October until Christmas, plants, such as blooming pineapple sage, and active hummingbird feeders have a reasonable shot at attracting a hummingbird, and there's definitely a very good chance that it will be something other than a Ruby-throat. ID is very difficult, so assume nothing.

One of the most challenging groups, hummers are not all safely identified, and mistakes and confusion abound. Focus on features such as shape, size, wing length relative to tail tip, bill length and shape, tail pattern, and general coloration. A number of species (the *Archilochus* genus) have inner primaries noticeably narrower than outer primaries, e.g., Ruby-throated, Black-chinned. Others, such as Anna's, Broad-tailed, Rufous, Allen's, and Calliope, all have primaries of similar width. Working out to which group a bird belongs makes ID a whole lot easier! Camera and video images are helpful, and often essential, for positive ID of rarer species.

Adult males are often stunning. Brilliant iridescent gorget (throat) colors can switch on and off in a second, depending on the angle that light hits the feathers. Immature males molt iridescent feathers through fall and winter. Determining the color of these feathers is a big help in ID, but be careful that you have seen the color correctly—iridescence can be misleading.

Attracting hummers to your yard for a closer look is rewarding, and easily done using feeders with sugar water, and planting favored flowers. Birds are often pale-faced or pale-crowned with pollen picked up while feeding. Hummers also eat insects and can often be seen flycatching from exposed twigs. They can obtain lift on the up- and downstroke. Because they fly incredibly fast and in straight lines, they are far more likely to be confused with insects than other birds. Hummers give a variety of *tic* notes, some of which are distinctive to the trained ear.

Swifts, martins, and swallows are most often seen zooming around, hawking insects. As harbingers of spring in the North, their graceful flights and swoops can mesmerize. They often form mixed flocks, particularly out of breeding season. These can be large to the point where they seem like clouds of gnats in the distance. Nesting on buildings, bridges, and other man-made structures, many use dried mud to build their homes. ID can be tricky, but with practice most species have a noticeably distinct individual structure. Patterns of color are also different but often difficult to see on flying birds.

Green Violet-ear *Colibri thalassinus* **GREV** L 4.75in

Very rare, mostly TX, but can turn up anywhere. Prefers higher elevation. Fairly large with slightly downcurved bill, which is always black. All plumages iridescent green mixed with blue—darkest around ear coverts. Ad ♂ brightest, imm ♀ dullest.

Broad-billed Hummingbird *Cynanthus latirostris* **BBIH** L 4in

Rare in TX and LA but recorded as far n. as ON. Fairly small and stocky with a forked tail when perched. Its 2-toned red-and-black bill stands out, though less obvious in ♀. Ad ♂: iridescent greens and blues, darkest on throat and tail with white dot behind eye. ♀/imm: dingy gray below. Diffuse pale supercilium borders gray ear coverts. Call: a rough chatter.

Buff-bellied Hummingbird *Amazilia yucatanensis* **BBEH** L 4.25in

Fairly common in s. TX, rare elsewhere in the S. A distinctive, medium-sized hummingbird with buff or cinnamon belly, downcurved bicolored bill, and striking, long rufous tail. Tail can look forked or rounded. All plumages are similar but with ♂ averaging darker-throated and more richly colored. Imm ♀ palest and dullest.

Green-breasted Mango *Anthracothorax prevostii* **GNBM** L4.75in

Very rare lowland hummer, most records from TX. Fairly large with striking downcurved bill. All except adult ♂ have unique dark line through throat and breast. Ad ♂: dark green, bluer on throat and breast (reverse of GREV) with purple tail. Imm: like ♀ but with brown border to throat and flanks.

White-eared Hummingbird *Hylocharis leucotis* **WEHU** L 3.75in

Very rare. A much bolder version of BBIH but slightly smaller, even chunkier, and has a shorter bill. White supercilium longer, broader, and strikingly brighter in all plumages. Ad ♂: dark throat, browner back, and shows very little blue. ♀/imm: green spots on whiter underparts.

Blue-throated Hummingbird *Lampornis clemenciae* **BTTH** L 5in

Very rare in the S. Large and long-billed with striking white corners to broad, dark tail. Ad ♂: blue throat and dark gray ear coverts bordered by white; underparts otherwise dull gray. ♀/imm: pattern similar to adult ♂ with bolder double white face stripes, but lacks blue throat.

Magnificent Hummingbird *Eugenes fulgens* **MAHU** L 5.25in

Very rare. Similar to BTTH but longer-billed, slightly forked tail with much smaller white tips. Ad ♂: appears green and black but with a twist of light can also appear purple-crowned and blue-throated. Ad ♀: dingy gray below with green on flanks and indistinct white line behind eye. Imm: strong face pattern like BTTH but with green spots on underparts and less white in tail.

imm. ♀

ad. ♂

ad. ♀

Black-chinned Hummingbird *Archilochus alexandri* **BCHU** L 3.75in

Fairly common summer visitor to dry woods such as those on the Edwards Plateau, TX. Scarce but regular winter visitor through the S and E. Often pumps its tail (more so than RTHU). **ID**: Very similar to RTHU. Slightly duller upperparts including the crown, dingier underparts, and longer bill. ♂'s iridescent throat colors are distinctive. Positive ID of other birds must be based on good views of wing shape: primaries are broader, and the outer edge of the wing is distinctively curved toward the tip (RTHU has gradual, even curve along outer edge of wing). Ad ♂: black throat, lower part iridescent violet (other hummingbirds with violet in throat have differently shaped gorgets). Tail less forked than RTHU and wing point just short of tail tip. Ageing and sexing of imm and ♀ as RTHU. Violet feathers, molted in through winter in imm ♂, are diagnostic. Wing point averages closer to tail tip than in RTHU.

imm. ♂
imm. ♀
ad. ♀
ad. ♂

Ruby-throated Hummingbird *Archilochus colubris* **RTHU** L 3.75in

Common summer visitor to wooded areas, parks, and gardens. Uncommon winter visitor in the SE, rare elsewhere. With global warming, plentiful feeders, and late-blooming sage, more are showing up in the N and a few attempt to winter. Easy to attract with flowers and feeders and also eats insects. **ID**: Small and compact with medium-length bill. Ad ♂: brilliant ruby throat typically looks black. Dark green breast sides and flanks. White spot behind eye and black forked tail. Ad ♀: tail shorter than ♂ and square-ended. Lacks ruby throat, underparts white with less extensive green on breast sides and only a few green flecks on throat. Juv: similar to adult ♀ with buff fringes to upperpart feathers; these wear off through winter. Imm ♂: heavily spotted throat with iridescent 'ruby' feathers increasing through winter. Imm ♀: throat mostly unmarked. Some birds are orange-buff on breast sides, recalling *Selasphorus*.

Rufous Hummingbird *Selasphorus rufus* **RUHU** L 3.75in

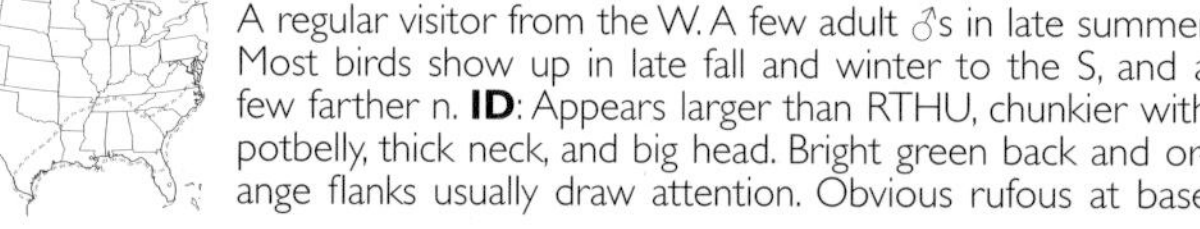

A regular visitor from the W. A few adult ♂s in late summer. Most birds show up in late fall and winter to the S, and a few farther n. **ID**: Appears larger than RTHU, chunkier with potbelly, thick neck, and big head. Bright green back and orange flanks usually draw attention. Obvious rufous at base of tail in flight. The most likely *Selasphorus* but compare and check others. Ad ♂: iridescent red throat, extensive rufous on breast and upperparts. Ad ♀: green-spotted throat densest in center. Imm ♂: evenly spread green spots on throat, some iridescent in winter. Imm ♀: a few green spots on throat.

Allen's Hummingbird *Selasphorus sasin* **ALHU** L 3.75in

Very rare in late fall and winter but previously overlooked. In most plumages, almost identical to RUHU. Only safely separated from RUHU by narrower tail feathers with more extensive white tips. **ID**: Ad ♂: narrower tail feathers, broadest in imm ♀—it is always necessary to age and sex birds for comparative purposes. This ID is extremely difficult in the field but is possible. ALHU molts earlier, so often looks faded by late fall and averages more extensive red on rump. Ad ♂: greener back and crown than RUHU.

Anna's Hummingbird *Calypte anna* **ANHU** L 4in

Very rare. A dull, medium-sized hummingbird with generally average proportions. Bill fairly short and straight. Always look for light gray underparts mixed with green. Holds body in a straight line with tail motionless. Noisy with distinctive *szit*. **ID**: Ad ♂: mind-blowing gorget that extends onto crown. ♀/imm: focus on underpart pattern.

Broad-tailed Hummingbird *Selasphorus platycercus* **BTAH** L 4in

Very rare winter visitor. A *Selasphorus* with even-width primaries and long (less tapered) tail. Subtly longer, more tapered, and paler than RUHU with smaller head. Ad ♂: white line behind eye separates rose gorget from crown. ♀/imm: color patterns similar to CAHU. Subtly distinctive face pattern: gray-faced with speckled throat.

Calliope Hummingbird *Stellula calliope* **CAHU** L 3.25in

Rare but underrecorded. Easy to confuse with *Selasphorus* due to similar color patterns. **ID**: Smaller and squat with deep belly emphasized by short square tail. Short bill. Flight feathers broad and of even width—importantly, the wing point is at least up to and typically past the tail tip. Limited rufous at base of tail. Often shows narrow white line above gape. Ad ♂: streaked purple gorget, pointed at corners. ♀/imm: pale peach flanks. Throat often finely streaked.

Chimney Swift *Chaetura pelagica* **CHSW** L 5.25in

Common over forest, towns, and just about anywhere with food. A high-flyer, in rainy or cold windy weather flies lower, often over water hawking for insects. Hurls around the sky, short glides before flapping like a bat out of hell. Noisy, with sharp twittering and chattering calls. Nests in chimneys or similar structures. Usually in small loose groups. After breeding, gathers at dusk in large numbers before descending down a chosen roost site. Never perches other than to cling to walls, having to drop before flying off. **ID**: Bullet-shaped with wings angled back. At times, appears almost headless compared with long broad rear end. Frequently soars with wings and tail spread. Spikes at tip of tail tough to see. Usually appears black. Better views show it palest on throat, darkest on wings, between rump and belly, with a dark line through the eye. Juv: white tips to secondaries, just about impossible to see in the field.

White-throated Swift *Aeronautes saxatalis* **WTSW** L 6.5in

A bird of w. cliffs and canyons, but can be found anywhere. Casual slightly e. of normal range. Medium-sized and slim with deeply forked tail that usually looks pointed. White throat (extending down belly), flank patches, and secondaries make ID straightforward with good views. At a distance it can look all dark.

Vaux's Swift *Chaetura vauxi* **VASW** L 4.75in

Very rare. A number of late fall/winter records in the SE probably this species. Very similar to CHSW but smaller with faster even more frantic flight. Subtly slimmer with proportionally longer tail. More extensively pale underneath. Pale rump contrasts more strongly with back. Different call (higher and buzzier).

White-collared Swift *Streptoprocne zonaris* **WCSW** L 8in

Very rare stray from Central America.. Very large size should be apparent. Powerful, long, and slim with slight fork on closed tail. White collar noticeable with good views but indistinct in juvs.

juv.

ad. ♀

ad. ♂

1st-s. ♂

Purple Martin *Progne subis* **PUMA** L 8in

Fairly common and widespread in suburban and rural areas with open grassy areas for feeding. Breeds almost exclusively in man-made boxes—multichambered houses or gourds where colonies create squabbling choruses. Travels farther afield to feed but always calls on the way home. Early to arrive in spring, they depart en masse in late summer. Oct martins in the N are rare and should be checked carefully. **ID**: Large and powerful in flight but stiff, recalling a stocky falcon. Broad wings and tail with shallow fork. Glides more than other smaller swallows. Ad ♂: uniformly dark blue, often appearing black. Ad ♀: variable amounts of blue and brown in upperparts with paler collar contrasting with dark head. Underparts also variable: dingy gray-brown, darkest around throat and palest in central belly and with smudgy streaks. Dark centers to undertail coverts. Juv: grayer and paler with fine streaks on underparts and pale undertail coverts. 1st-s: fairly common in breeding colonies, retains juv fine streaks on belly and vent with many adult-type feathers elsewhere; more blue in ♂.

Barn Swallow *Hirundo rustica* **BARS** L 6.75in

Common and widespread in open areas but often near humans. Builds a conspicuous nest of mud with an open top underneath the eaves of homes and sheds. Noisy, a talking 'twitter'—to many farmers, the sound of summer. A graceful flyer, it swoops down low to the ground over fields and ponds, hawking insects before rising in the air again. **ID**: A larger swallow with noticeably swept-back, long wings and very long slim body. Long pointed tail has white band across it, noticeable when spread. Adult: upperparts dark, steely blue with white patch to bases of long, forked tail. Underparts variably dark-tinged in spring, becoming much paler through summer. Throat orange with dark lower border. Juv: a very washed-out version of adult and with a much shorter tail.

ad. br.

nonbr./1st-w.

Cave Swallow *Petrochelidon fulva* **CASW** L 5.5in

S. species, expanding range n. Nests in culverts, under bridges, and uses other man-made structures, often with CLSW and BARS. Regularly wanders n. Seen in the NE from late Oct onward. In fact, the common red-rumped swallow in the NE, with numbers often in the dozens and occasionally hundreds. Roosts on buildings, hotels, or anywhere they can keep warm. **ID**: Very similar to CLSW, size and structure almost identical but subtly slimmer and lighter. Upperparts not so dark as CLSW and are paler and brown-winged. In fall, rump becomes washed out and is significantly paler than the rest of the upperparts (unlike CLSW). On closer views, dark cap with brown forehead bordered by pale buff collar and throat. The pale collar and throat highlighting the cap is the most striking difference from CLSW. Although throat is much paler than CLSW, at a distance it can look deceptively dark and is easy to misidentify the bird. Caribbean ssp, *fulva*, breeds in FL, but most wintering birds are of the Mexican race. *Fulva* averages slightly smaller and darker but very difficult to separate from Mexican race Molts on breeding grounds.

juv.

southwest

ad.

Cliff Swallow *Petrochelidon pyrrhonota* **CLSW** L 5.5in

Uncommon but widespread, mostly in rural agricultural areas and near nesting sites, often bridges and old buildings, where they build a mud house with small, rounded entrance at the top. **ID**: It's the summer red-rumped swallow away from the far S. Can be identified at a distance from other swallows (except CASW) by compact structure with square-ended tail. Orange/red rump becomes paler through summer. Mantle and wings dark steely blue with white lines down the back. Crown is the same color. Dark orange-brown on throat wraps around back of neck. In flight striking buff collar. Forehead from white to buff. Birds from the SW and Mexico have rufous forehead. These are occasionally seen in the N. Juv: also has orange-buff rump, but upperparts much browner and more muted. Pale forehead often lacking, throat also brown. Much less contrasty than adult. Molts on wintering grounds.

juv.

ad.

Northern Rough-winged Swallow *Stelgidopteryx serripennis* **NRWS** L 5.5in

Fairly common and widespread but rarely in large numbers. Often near water, nesting in holes in quarries, river banks, cliffs, and man-made structures. Long glides between pronounced deep wing flaps (slow on down-stroke) make it look jerky but buoyant. Dark underwing adds to distinctive flight appearance. Calls frequently—a buzzy insect-like *prrtt*. **ID**: Fairly large and compact swallow with very broad-based wings. Uniformly dark brown above, lacking paler rump of BANS. Brown throat and upper breast, brown extending onto under-wing. Never shows clear-cut breastband like BANS. Juv: warmer brown upperparts with broad rufous fringes to wing feathers.

juv.

ad.

Bank Swallow *Riparia riparia* **BANS** L 5.25in

Fairly common but local near nest sites. Always excavates hollows in river banks, quarries, or other sandy areas. Feeds over nearby water. A light and buoyant flyer, it mixes with other larger swallows on migration. **ID**: Easily the smallest swallow, emphasized by very slim build from head to tail. Wing bases particularly narrow with wings swept back. In all plumages has narrow but well-defined breastband, emphasized by white throat. Upperparts medium pale brown, noticeably grayer and paler on the rump. Other swallows do not show this contrast. Juv: similar to adult with pale-fringed wing feathers.

ad. nonbr./1st-w.

ad. br.

juv.

Tree Swallow *Tachycineta bicolor* **TRES** L 5.75in

Common and widespread breeder in open areas, nesting in holes in dead trees (often near water) and nest boxes. A tough swallow, many winter in the S, with a few hardy ones much farther n. On migration, often lines telegraph wires or forms tight flocks that look like a swarm of gnats in the distance. The common swallow in late fall. **ID**: A neat-looking medium-sized swallow with balanced proportions. Tail square-ended or slightly forked. 2-toned with clean divide between dark upperparts and pale underparts. A few have dark breastband but are not so dark or well-defined as on slimmer BANS. Adult: iridescent blue upperparts, darker on wings, tail, and through eye. Brown replaces blue in some ♀s. Molts on fall migration, creating a patchwork of brown and blue feathers. Tertials with bold white tips. Juv: upperparts uniformly brown, darker through eye, and with narrow white tips to tertials. Many show a buffier collar and breast sides, and these molt adult-type feathers through fall and appear as adults.

ad. ♂

ad. ♀

Violet-green Swallow *Tachycineta thalassina* **VGSW** L 5.25in

Rare visitor from the W. Small with disproportionately long, pointed wings. White underparts extend onto rump and above eye, both features conspicuous at a distance. Adult: clean white underparts, violet-green upperparts, and duller on head and wings. ♀: often duller. Juv: brown rather than green.

SONGBIRDS

Small flycatchers, mostly *Empidonax* (often known as empids), have always been the scourge of bird ID, for beginners and experts alike. Just when you think you have it worked out, you encounter birds that completely stump you. However, with practice and good views, most birds can be identified. Part of the problem is that views tend to be very brief. Birds are often in bright light one minute and shade the next. All empids have wing bars, eyerings, and variable coloring. For these reasons, always scrutinize size, proportions, and bill length. Color, however, is particularly important when dealing with Yellow-bellied and, to a lesser degree, Acadian Flycatchers. Least Flycatcher is the most likely late fall empid, but this is also the season when 'western' birds show up and nothing can be taken for granted.

In spring Willow and Alder Flycatchers don't arrive on their breeding grounds in the North until at least mid-May. Least and Acadian Flycatchers are a week or two earlier. However, these same birds turn up approximately 2–4 weeks earlier along the Gulf Coast and other places in the extreme South. The difference in time is approximately how long it takes them to migrate from the South to more northerly breeding grounds. This is also the case for most medium- and long-distance migrants.

Kingbirds tend to be found in open areas. Immature birds are similar to adults. Adult males have very narrow outer primaries, slightly broader on females, Immature birds have broader outer primaries—this can be seen in photos and sometimes in the field.

Catharus thrush identification is usually made more difficult by frustratingly brief views in dark shady conditions, where determining true upperpart color is difficult. Perseverance usually pays off and determining color is important to identify some of the species. It is always best, however, to first focus on the head pattern.

All species show similar bold underwing patterns. Wing bars created by pale tips to the greater coverts are a feature of juveniles; however, many adults in fall also show this feature, so most birds are best left unaged. On juveniles the pale tips extend down the feather shaft. In adults, these tend to be smaller and only across the tip of the feather. The juvenile wingbar is often still faintly present in first spring. You will find patient observation will often turn up many more individuals than you expected. Check holly bushes and damp areas, as these are popular on migration. *Catharus* are very vocal in flight at night.

Wrens are secretive, generally fidgeting around in the low undergrowth, where they stay well hidden. They are generally brown with downcurved bills and rounded tails that they often hold cocked. Few birds have more 'attitude.' Although mostly very small, they are very vocal and pound for pound are world-champion songsters.

Vireos vaguely resemble warblers but have fairly uniform plumages, bluish legs, thicker bills with a hard-to-see hook at the tip, and a more muscular build, and they move more slowly and deliberately. Some have spectacles and wing bars. Their songs are often burry and repeated for extended periods of time. They have deep chatter and scolding alarm calls that draw the attention of other nearby birds. Plumages remain the same year-round; often only scruffy-looking fall birds can safely be aged as adults. The softer and 'punky'-looking feathers of recently fledged juveniles can sometimes differentiate them from adults. In general, the softer and looser-textured feathers of juveniles of many species can be used for ageing birds for a short while. Gray Catbird is a good species to practice on.

Warblers are for me, and many other birders, a favorite group. Small and beautiful, they always generate interest. Although there are nearly 50 species, their ID is generally straightforward, and the term 'confusing fall warblers' is in itself misleading. Fall birds are the same size and shape, and most retain the same colors. The rest have generally similar but subdued patterns. Part of the problem is their small size, variation in plumage, and the fleeting or poor views one gets. The key is to learn their size, shape, behavior, and just one or two plumage features for each species. This should id nearly all birds. For example, while Nashville Warbler is often described as a bright yellow bird, I see it as a small, fat thing with a bold white eyering, though no doubt I'm subliminally taking in the yellow underparts. From below and behind, it is the same size and shape but has white between the legs, bordered by yellow on either side. We all see things differently but try to narrow it down to simple elements. When looking at the plates, focus on the points that you see clearly, and preferably those that remain constant between individuals.

In describing size and shape, it is necessary to have a point of reference. For me it is Yellow Warbler. I see lots of them so I know them well, and I consider them average in size and proportions. Remember, long-tailed, short-winged, small or large are subjective terms that can only be truly measured by you. If, like me, you love the challenge of flight ID, the dynamics remain the same, based on size, shape, and basic patterns of color. In some respects, flight ID can be easier since there is no vegetation in the way!

There is much overlap in plumage between ages and sexes. While the age and sex of many of the birds captioned are certain, others are not so clear. Some captions are used to give a general idea of appearance. In most species, adult males are brightest and have the highest contrast in color patterns. Features such as wing bars and superciliums tend to be the boldest. Immature females are at the opposite end of the spectrum. Adult females and immature males are intermediate and tend to be very similar in fall in general color patterns and tones. By spring, males (1st-summer) show a number of adult-male-type characters while retaining some old immature feathers—these are contrastingly worn and usually faded brown. For example, 1st-summer American Redstart will be similar to a female in its fall plumage, but will have a number of black adult male-type feathers, particularly around the head and breast. This makes birds quite easy to age and sex, which tends to impress beginners even though it is not difficult to work out. They also sing. These general patterns are shown in many other families.

Juvenile and adult feathers have many different characteristics. These features are used by banders for ageing and sexing birds in the hand. However, with careful field observation (or photographs), we can often see these. Tail and flight feathers are typically narrower and more pointed than adult feathers. They also tend to be browner rather than black. Juvenile greater primary coverts and alula (small feathers at the bend of the wing) are usually more pointed and paler.

As with other families, southern and western species often show up out of range. Once birds set out the wrong way, they can easily end up thousands of miles out of range, so anything is possible just about anywhere. Most vagrants show up in late fall. It should be pointed out that many more regular warblers seen in late fall are not in fact late-lingering birds but ones that have gone south and been carried back north in southerly airflows. At this time of year never rely on assumptions!

Sparrows pose the same problems as warblers and for the same reasons, though many of them tend to be in cover on or near the ground. Size and proportions will always be key ID factors. Typically, the face and breast patterns are especially important. Habitat, including during migration, is a major help in ID. While many of the calls are similar, they can be helpful in locating the bird in the first place. Learning the local birds intimately, in my case Song and Savannah Sparrows, should be your foundation.

Finches eat seeds and have pointed conical bills to help them do this. They are often in tight flocks that can be seen overhead in their characteristic deeply undulating flight and showing their forked tails. Many stay in the cold North in winter so are known as 'winter finches.' They feed mostly in coniferous trees but also on other fruit and seeds. Dependent on food supplies, they tend to be nomadic and occur a long way south of their usual range in some years. Knowing the success of cone crops can lead to fairly good predictions of future bird movements.

The word 'blackbird' when mentioned in conversation seems to conjure up a negative reaction. Perhaps it is their dark sinister look, or that they all seemingly look the same based on color. On the other side of the coin, it is easy to think of them as spectacular. They form large flocks, particularly when going to roost, that can number tens of thousands. Such congregations are one of the great spectacles in birding. I have heard the term 'ugly' used to describe them, but how can the stunning red on the wing of a Red-winged Blackbird as it sticks its wing out, throws its head back, and sings be anything short of beautiful. And if you take the time to look at them, most have very distinctive sizes and shapes. They are one group of birds that has adapted well to humans. They form mixed flocks, particularly in winter.

Over the years, as knowledge of ID has increased, so has the realization that hybridization occurs more regularly than previously thought. It is most common where ranges of two similar species overlap, for example, Black-capped and Carolina Chickadees, or Baltimore and Bullock's Orioles, or Indigo and Lazuli Buntings. Most unusual-looking birds are generally just that, birds that fall outside the realm of the typical. However, take careful field notes or photographs, as hybrids do occur in just about every family.

If you travel to other parts of the country, it is interesting to see that a number of species differ subtly from the ones back home. Birds such as Song and Savannah Sparrow have many subspecies throughout North America that not only look slightly different but may have a different dialect—just like people. We can sometimes see this variation in the migrants that pass through where we live.

Ornithology is still in its infancy, and one of the most exciting aspects of it is that there is so much that we can all learn. So look and study closely. What you might be learning for yourself may also be valuable to others. As this book goes to press, Winter Wren in the western part of North America has been give full-species status, separate from birds in the East. More 'splits' are sure to come.

Olive-sided Flycatcher *Contopus cooperi* **OSFL** L 7.5in

Widespread but uncommon flycatcher of n. evergreen and mixed forest, particularly near bogs and burns. Declining. Sits upright on the tallest snag on the tallest tree to scan the surroundings. After flycatching, typically returns to same snag. Only pewees show similar behavior. It stays out in the open, not feeding from inside woods as EAWP often does. Voice: loud for a flycatcher, the song a ringing *quick-3-beers*. Call: a loud *pip*, often repeated. **ID**: A large flycatcher with a noticeably bulky head, thick neck, and broad-based bill. Long-winged with relatively short, wide tail. Dark breast sides contrast strongly with pale throat and central breast. Occasionally shows white tufts near tertials. EAWP and WEWP are slimmer with narrower base to bill, smaller head, and longer tail that is strongly forked at the tip. Structure is distinctive, but underpart contrast is also significantly stronger on OSFL. Juv: very similar to adult but feathers show less wear and has buffier fringes to coverts.

Eastern Wood-Pewee *Contopus virens* **EAWP** L 6.25in

Common flycatcher in deciduous and mixed forest. Often confused with *Empidonax*, but unlike them it will sit on exposed snags for long periods, scanning the area. After spotting its prey, it makes long sorties to snap it up, often returning to the same perch. It gives a little 'wing shiver' motion upon alighting (it never flicks its tail like empids). Vocal, it often gives its presence away by a drawn-out, 2-note *peee-weee*. Several calls: *chip, peeyer, peer* and much variation. **ID**: Larger than empids with peaked crown and long, usually forked, tail, which enhances slim appearance. Wings are long and pointed with primary projection equal to tertial length. If in doubt, make sure you check this feature. Looks dark-faced, compared to empids. Has indistinct or no eyering with mainly dark lower mandible, though this is variable. Undertail coverts show dark centers, and bird never flicks its tail. Underparts darkest on breast sides, creating a subdued version of OSFL. Juv: similar to adult but fresher plumage and broader, buff wing bars.

Western Wood-Pewee *Contopus sordidulus* **WEWP** L 6.25in

Barely enters w. part of region. Rare migrant on the Gulf Coast. True status in the E unknown due to almost identical appearance and behavior to EAWP. Probably only safely identified by song: *bree* or a *bree-yurr*. Calls: very similar to EAWP.

ID: Averages darker, expansive brown across breast and less orange on lower mandible. Wing bars, and particularly greater coverts, narrower than in EAWP. These features are all subtle at best, and birds not safely identified by sight alone.

Northern Beardless-Tyrannulet *Camptostoma imberbe* **NOBT** L 4.5in

Uncommon in the Lower Rio Grande Valley, usually in mesquite woods. A small flycatcher that appears more like WAVI. Sits upright and moves deliberately. Bland with short, thick, blunt-tipped bill, bushy crest, indistinct supercilium on otherwise plain grayish face. Most often heard, a loud ringing whistled *peer* often repeated in a descending series. Frequently in mixed flocks.

Greater Pewee *Contopus pertinax* **GRPE** L 8in

Very rare in TX; often perches on treetops. A long super-sized pewee with a spiky crest. Long bill with orange lower mandible. Otherwise, tends to be dull with gray breast and indistinct wing bars. Juv typically has brighter buff underparts and wing bars. Calls year-round, *pip pip*. Song: a whistled *José Maria*.

Hammond's Flycatcher *Empidonax hammondii* **HAFL** L 5.5in

Very rare in late fall or winter. Small but big-headed and stocky. From LEFL by narrower bill with more extensive dark on lower mandible, gray head and vest contrasting with yellow-washed belly. Always focus on long primary projection. Bold eyering, overall structure, and tendency to flick wings and tail give kinglet-like feel. Call: a *pip*—different from other small empids.

Dusky Flycatcher *Empidonax oberholseri* **DUFL** L 5.75in

Very rare in late fall and winter. Structurally closest to GRFL with slim bill and long tail that never dips. Plumage more like HAFL, with olive-gray breast contrasting with yellowish bill, with bill slightly longer and broader-based. Also longer tail and shorter primary projection. Only occasionally flicks tail and wings. Molts on wintering grounds. Call: a *whit*.

Gray Flycatcher *Empidonax wrightii* **GRFL** L 6in

Very rare in late fall and winter. A slim-looking flycatcher with long, narrow tail emphasized by short wings. Narrow-based bill appears long. Lower mandible pale with well-defined dark tip. Deliberately dips tail like a phoebe—this is diagnostic for an empid. Plumage tends to be pallid and nondescript. Juv: buffy wing bars and averages brighter plumage. Call: a *whit*.

Cordilleran Flycatcher *Empidonax occidentalis* **COFL** L 5.5in

Very rare in late fall and winter. Formerly Western Flycatcher, now split with Pacific-slope Flycatcher, and this latter species probably accounts for many e. records. Essentially identical except for ♂'s disyllabic *sew-wis* call, and best treated as one species unless heard well. Like YBFL but slightly slimmer and longer-tailed. Browner-winged, less distinct wing bars and tertials. Eyering tends to be rounder and back browner. Frequently flicks tail.

Least Flycatcher *Empidonax minimus* **LEFL** L 5.25

Common breeder in deciduous woods, often near water. Scarcer to the W. The first empid to arrive in spring. On migration, found in field and wood edges, where can be frustratingly difficult to see as it moves in and out of cover. Often fidgets, flicking wings and tail upward. Song: a sharp *chi-bek*. Call: a *whit*. Typically the commonest empid, the one to learn well. **ID**: The smallest e. empid. It appears large-headed and chunky but with a short, noticeably narrow tail. Other than size and shape, short bill is the most important ID feature. The small size, chunky proportions, and short bill give a cute impression overall. Primary projection is short. Eyering is usually bold. Plumage variable, often gray-headed and contrastingly green-backed, others more uniformly brown/gray. Underparts off-white but sometimes with a yellow wash. Imm: buff wing bars. ALFL has very similar color patterns but is larger, broader, and longer-tailed, with a similar-shaped bill that is also larger. Birds in reflected light can look yellow like YBFL—always make sure the colors are true. Variable, with many seeming straightforward to ID, but others confusing. Brief views never help!

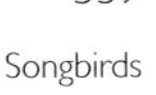

Yellow-bellied Flycatcher *Empidonax flaviventris* **YBFL** L 5.5in

Uncommon summer breeder in coniferous forest and bogs. Often stays fairly low on breeding grounds. On fall migration usually stays inside woods and flits around in the canopy—judging the color becomes very tricky here. Song similar to LEFL but less emphatic. Call: a loud *swoow* like ACFL. **ID**: Large-headed with solid build, a slightly larger version of LEFL, with a longer primary projection. Bold eyering noticeably pointed at rear. Bill shortish and broad-based, usually with all-bright, lollipop-orange lower mandible. Black wings with bold buff-white wing bars. These are worn and paler by fall. The brightest empid with yellow underparts, including throat, and with dark wash across breast. Upperparts olive-green—the brightest of the e. empids. Colors are diagnostic and quite striking when seen well, but birds can appear dull in poor light, and reflection from green leaves can make other species seem brighter than they are in reality. Juv: orange/buff wing bars, similar to adult in fresh plumage.

Alder Flycatcher *Empidonax alnorum* **ALFL** L 5.75in

Fairly common n. breeder in wet areas near scrub, often with scattered trees. LEFL prefers more wooded areas. Favors forest and field edges on migration. Arrives on breeding grounds mid-May onward. Most migrate later than WIFL in fall. Formerly conspecific with WIFL. Song: *fee-beo*, more monotone than similar but more emphatic *fitz-bew* of WIFL. ALFL occasionally gives a more WIFL-like song—beware! Call: a *pip*—different from other empids and often likened to HAWO. **ID**: In many respects, intermediate between LEFL and WIFL. Larger than LEFL with longer and broader tail. Bill is broader-based and longer. Longer primary projection. Plumage similar to LEFL: variable, often gray-headed and contrastingly green-backed, others more uniformly olive-brown with fairly prominent eyering. WIFL looks larger and slimmer with peaked crown and blander more uniform colors. As with all empids, differences are slight. Juv: buff wing bars.

Willow Flycatcher *Empidonax traillii* **WIFL** L 5.75in

Fairly common breeder with more s. distribution than ALFL. Often found together where ranges overlap, but favors even scrubbier and wetter areas. First arrives on breeding grounds mid-May onward and starts moving s. in late Jul, typically earlier than ALFL. Tends to stay in lower vegetation, flitting around in reeds and grasses as well as bushes. Song: an emphatic *fitz-bew*. Call: a *whit*, a little stronger than LEFL. **ID**: Large, with long broad tail; long-billed and often has slight crest—pewee-like in many respects, enhanced by indistinct or total lack of eyering. Slightly smaller than EAWP with shorter primary projection. Also lacks darker head pattern and breast. Plumage usually olive-brown, lacking the contrast between head and back that ALFL usually shows. Underparts off-white but sometimes with yellow wash. Juv: fresher plumage in fall with buffier wing bars.

late summer

ad. spring

Acadian Flycatcher *Empidonax virescens* **ACFL** L 5.75in

Fairly common in wet swampy s. woodland, hemlock glades, but also drier deciduous forest in the N. Tough to see as it flits around, usually higher in the canopy than other flycatchers, rarely visiting woodland edges. On migration found inside canopy like YBFL. Also has similar call. Usually most easily found by listening for its distinctive song, an emphatic *wake-up* or *pizza*. Also gives rapidly repeated buzzy *pip* notes. Adult molts late summer on breeding ground. **ID**: A larger empid with a long broad-based bill (mostly orange) and sloping crown peaking behind eye. Long, broad tail rounded at tip. Primary projection long. Fairly distinct white eyering with extensive pale loral area—a very good feature to look for. Plumage varies but usually with green upperparts just slightly duller than YBFL. Underparts usually have yellow wash across breast. Adult: broad buff wing bar but often wears white through summer. Juv: orange-buff wing bar. YBFL is smaller and more compact with shorter primary projection, bill, and tail. YBFL also lacks the steep forehead and buffier wing bars.

1st-w. ♂

1st-s. ♂

juv.

ad. ♂

ad. ♀

Vermilion Flycatcher *Pyrocephalus rubinus* **VEFL** L 6in

Fairly common but local in open areas, often near farmsteads, cattle, or water. Can be tame. Rare winter visitor to the SE; vagrant elsewhere. This s. stunner is unforgettable. Daintily sits on snags and fenceposts before making a darting foray for insects. Often dips tail and spreads wings phoebe-like. Call: a sharp *pseep*. **ID**: Small and flat-headed but often with a hint of a crest. Thin bill enhances gentle expression. Black tail, with narrow white border, often held flared. Ad ♂: mind-blowing red with dark mask and upperparts. Ad ♀: pale underparts with pale streaks and pinkish red vent. Juv: similar to ♀ but bolder streaks and pale fringes to upperparts. Imm ♀: vent yellowish becoming adult-like in summer. Imm ♂: similar to ad ♀ but molts new red feathers through winter and first summer. As adult by second winter.

Black Phoebe *Sayornis nigricans* **BLPH** L 7in

Uncommon in the W, usually near water. Very rare in the E. Sits low and darts out over water to catch insects. Slim and elegant with long tail and fine bill. Tail is pumped up and down, though less deliberately than in EAPH. In all plumages, 2-toned color pattern distinctive. **ID**: Adult: gray-black, slightly paler on back, but sometimes with a hint of brown. White belly and vent sharply demarcated, extends to center of breast—a very clean look. Juv: similar with rusty-fringed coverts and tertials. Molts into adult-type plumage quickly in late summer.

Say's Phoebe *Sayornis saya* **SAPH** L 7.5in

Fairly common in dry open areas, often rocky landscapes in the W. Early migrant that regularly wanders e. in fall. Sometimes hovers kestrel-like while hunting. Dips tail. **ID**: Similar to EAPH in size and shape but smaller-headed. Cinnamon-buff lower breast and belly, becoming brighter orange on vent. Rest of bird relatively bland gray-brown with black tail. Juv: similar to adult but with orange-buff wing bars.

juv.
fall
ad. spring

Eastern Phoebe *Sayornis phoebe* **EAPH** L 7in

Fairly common and widespread in a variety of habitats from woodland and farmland to suburbs, fields, and also waterways. Not shy. Like other phoebes, it builds a cupped nest under eaves or bridges, often in busy traffic areas. The first flycatcher to arrive in spring and last to go in fall. Often sits low, characteristically dipping its tail nice and slowly as if to make sure you see it. It is often approachable. Song: a gritty *phe-be*. Call: a *chip* (like SWSP). **ID**: A large flycatcher, it has soft lines: rounded crown, belly, and back. Fairly long dark tail is square-ended. Dull upperparts. Distictively dark-capped—the easiest way to id it. Lacks supercilium and bold wing bars. Underparts variably white and yellow with smudgy markings on breast. Ad br: off-white, often with yellow wash. Smudgy dark markings on breast sides. Molts in late summer. Fall birds in fresh plumage have more uniform and yellower underparts. Juv: browner with rusty fringes to wing coverts. Molts into adult-like plumage in late summer.

ad. summer

fall

Great Crested Flycatcher *Myiarchus crinitus* **GCFL** L 8.75in

Common and familiar in ones and twos anywhere with deciduous trees. Scarce winter visitor in FL. Lives in the canopy, where it catches insects, or feeds on fruit and other goodies. Tough to see but vocal, far more often heard than seen, giving a raucous and far-carrying *reep*, or rapid, repeated *reep reep reep*. Often sits quietly, or flits around and hovers as it picks fruit. Like other *Myiarchus*, tilts head to one side inquisitively. Frequently flies across open spaces to other trees, sometimes with a second bird in pursuit. **ID**: A large, sturdily built flycatcher with bright yellow belly. Long and slim, it can also appear muscular with thick neck and large broad-based dark bill. Medium-gray throat and upper breast contrast nicely with lemon-yellow lower underparts, often creating a hooded effect. Upperparts browner with obvious rufous tail and primary panel. Tertials with bold white edges. Juv: rufous fringes to coverts. These are quickly molted to adult-like plumage.

1st-w.

ad.

Ash-throated Flycatcher *Myiarchus cinerascens* **ATFL** L 8.5in

Fairly common sw. species, favoring bushy open areas. The only breeding *Myiarchus* in much of its range. Rare winter visitor to FL. Regular but rare migrant to the E. In late fall (Nov) any *Myiarchus* in the N and E is most likely to be this species. Typically feeds closer to ground than GCFL in bushes and weedy fields and is rarely seen in canopy. Call: a rolling *ka-whip* (heard on breeding grounds) or *prrrt* with soccer-whistle qualities. **ID**: Smaller and noticeably slighter than GCFL with paler colors. Bold rufous in wing and particularly bold median covert bar create strong contrast in wings. Bill is shorter and finer. In areas of overlap, these features should be enough. If in doubt, undertail has a dark border that extends around tip—limited to sides on GCFL—and white tertial fringes are narrower. Scarce BCFL has similar color patterns but is significantly larger with heavier bill, more deeply colored underparts, and the uniform rufous tail lacks a dark tip.

Brown-crested Flycatcher *Myiarchus tyrannulus* **BCFL** L 8.75in

Uncommon in s. TX riparian woods. Very rare winter visitor to s. FL and LA. Tends to stay deep and high in vegetation. Call: is a usually emphatic *whip*. Song: is a rolling *whip-will-do*. **ID**: Same size or slightly larger than GCFL, usually with noticeably larger bill. Plumage is like a bright ATFL but with duller wing bars and tertial fringes. GCFL has darker upperparts and throat, deeper yellow on belly and bolder white fringes to tertials. Slightly more extensive dark edge to undertail than GCFL. W. race (*magister*)—larger with bigger bill than TX race (*cooperi*)—has occurred in region.

La Sagra's Flycatcher *Myiarchus sagrae* **LSFL** L 7.25in

Very rare Caribbean visitor to FL. Smallish with short wings and long narrow tail. Pale underparts, capped appearance, low evenly domed head, and slow movements in lower vegetation, all give it a phoebe-like feel. Head disproportionately large and bill narrow but long. Extensive dark outer edges to tail feathers. A distinctive *Myiarchus*.

Dusky-capped Flycatcher *Myiarchus tuberculifer* **DCFL** L 7.25in

Very rare in wooded areas in s TX. Smallest *Myiarchus* with peaked crown and slim build. Bill is strikingly long but narrow on small brownish head. Tail appears long and, importantly, is mostly uniform brown with little rufous. Dull wing bars and tertials show only slight contrast with upperparts. Call: a somber *wheer*. Also gives a sharp *whit* and *I'm over here*.

Great Kiskadee *Pitangus sulphuratus* **GKIS** L 9.75in

Locally common in woodland in s.TX, particularly near water. A flycatcher that also eats fish. Often sits on open perches. The life of the party, it gives its loud and noisy *kis-ka-dee* at regular intervals. **ID**: Large and blocky with a big head and short tail. Bold head pattern, bright yellow underparts with bright rufous wings, are particularly striking in flight. Juv: similar to adult with dull crown and more extensive rufous in upperparts.

Rose-throated Becard *Pachyramphus aglaiae* **RTBE** L 7.25in

Very rare, mostly in s.TX. Canopy feeder in woodland. Moves around slowly, sitting upright. Fairly small with black cap that is sometimes punked up. Ad ♂ gray, ♀ brown, paler below than above. Rose throat on ♂ not always obvious.

Sulphur-bellied Flycatcher *Myiodynastes luteiventris* **SBFL** L 8.5in

Very rare visitor possible anywhere. Squat with broad head and short tail. A streaky-looking bird with bright rufous tail and striped head. Distinct breast streaking. Similar Piratic, Variegated, and Streaked Flycatchers from farther s. have been seen in N America—focus on size, shape, bill length.

Tropical Kingbird *Tyrannus melancholicus* **TRKI** L9.25in

Uncommon but increasing resident in s. TX. Fall vagrant as far n. as Canada, spring as far as Fl. Sits on treetops and telegraph wires to flycatch for insects. Will eat fruit. Almost identical to COKI. Call: high-pitched twittered trill, never a single note. Large 'yellow' kingbird with green back that contrasts with gray head. Focus on long dark bill on large flat head, but also note long black tail that usually appears notched. WEKI is smaller with stubbier bill, grayer upperparts, and white outer tail feathers on shorter, square-ended tail.

Couch's Kingbird *Tyrannus couchii* **COKI** L 9.25in

Fairly common in s. TX, particularly in summer. Strays to the N less than TRKI. Almost identical to TRKI. Arguably only safely identified by call: an upslurred *qwe-uurr*, or single/repeated *pip* notes. Bill often looks noticeably broader-based and shorter. Some brighter and green-backed with stronger contrast to gray head, but differences subtle at best.

Cassin's Kingbird *Tyrannus vociferans* **CAKI** L 9in

Scarce in the W where woods meet fields. Range overlaps with more common and similar WEKI. Size and shape as WEKI, sometimes with rounder crown. Always look for darker head, throat, and upper breast, creating hooded effect and strong contrast with pale malar and throat. Has paler-edged wing coverts, sometimes noticeable at long distances. Tail all black with pale tip; WEKI has white outer tail feathers, which is diagnostic but not always easy to see.

Western Kingbird *Tyrannus verticalis* **WEKI** L 8.75in

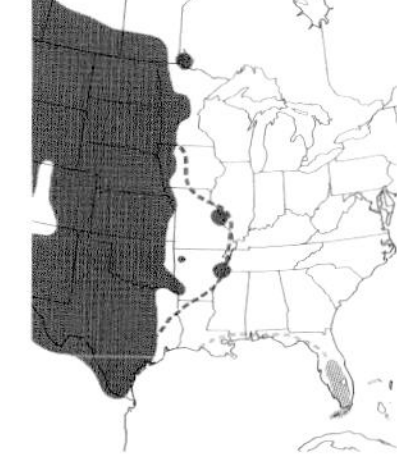

Common in the W, scarce but regular in the E, particularly in late fall. Scarce winter visitor to s. FL. Favors open fields and areas with scattered trees. The default 'yellow' kingbird. Sits on open perches—fences, bushes, trees, flower heads. Often found with or near STFL. Usually in ones and twos, but occasionally small groups; sits around waiting for unsuspecting insects, returning to perch to eat prey. Seemingly happy, will often up and fly to another area. **ID**: A medium-sized yellow kingbird with average proportions. Appears paler than other yellow kingbirds. Bill looks quite stubby, crown is peaked. Tail rarely appears long and is usually square or even rounded. Diagnostic white outer tail feathers, though most visible in flight, are often tough to see. Slightly larger and slimmer with more angular and pointed wings than EAKI. Colors intermediate between CAKI and TRKI/COKI. Mantle usually a mix of green and gray—it can appear either color. Juv: paler with broad pale fringes to coverts. Molts into adult-like plumage late summer.

Scissor-tailed Flycatcher *Tyrannus forficatus* **STFL** L 10in

Fairly common in grassland with scattered vegetation. Rare but regular out of range. A few winter in s. FL. Often in small groups, sits delicately on perches in fields or on wires waiting to flycatch. Long and elegant with soft colors to match, there are few more striking birds. In flight, crimson axillaries with peach underwing and flanks. Tail is spread as it lands, revealing lots of white. Silhouetted birds have straight tails, slightly downcurved in FTFL. Call: a *kip*. **ID**: Ad ♂: similar to ♀ but has longer tail and brighter peach on flanks and underwing. Juv: much shorter tail, paler underparts, and browner wings. 1st-s: many retained juv wing feathers, medium-length tail, and dull underparts; ♂ brighter than ♀.

Fork-tailed Flycatcher *Tyrannus savana* **FTFL** L 10in

Very rare, mostly the E Coast spring and fall. STFL-size with long tail that bends slightly downward. Color pattern similar to EAKI with paler mantle and thus more capped appearance (it curves up behind ear coverts). Tail length as STFL: longest in ♂, shortest in juv. Most records are of S American birds, (*savana*), but some (TX) are pale-mantled Cen. American birds (*monachus*).

Eastern Kingbird *Tyrannus tyrannus* **EAKI** L 8.5in

The common and widespread kingbird in most of the region. Found in open fields with some cover, but also parks, gardens, and even clearings within heavily wooded areas. Noisy, sitting and in flight, a variety of sharp *zeeet* notes, often repeated rapidly. Sits on perches high or low, often roadside wires, waiting for food to fly by. Flies back from sortie with shallow fluttery wingbeats, calling triumphantly (probably trying to impress a nearby admirer). Frequently in small groups. On migration, found in larger groups; gathers in large evening roosts. Aug 25th is peak in the NE. **ID**: Striking diagnostic white-tipped black tail. Medium-sized, solidly built with rounded crown and broad tail. All plumages 2-toned black and white. Black cap creates a clean line with white cheeks, easy to see at long distances. Red crown feathers rarely seen. Juv: similar to adult but duller, often with brown mixed into upperparts and cap, and with smudgy border between cap and white cheeks. Buff wing bars soon wear paler. Underparts, including underwing coverts, are white, often with gray or yellowish wash. Adult has pointed outer primary, more rounded in juv. All kingbirds can be aged using this feature (see plate).

Gray Kingbird *Tyrannus dominicensis* GRAK L 9in

Fairly common summer visitor to FL, particularly on the Gulf Coast. Very rare as far n. as Canada, usually May or Sep. Found in open coastal areas with trees or mangroves. Twittering notes similar to TRKI. **ID**: Like a large pale EAKI with a massive bill. All plumages show a dark mask through eye emphasizing paler gray head and upperparts. All-dark tail is fairly long and usually appears forked. Juv: as adult, but slightly duller with buffier covert fringes. Look closely at any out-of-season kingbirds! Sometimes mixes with EAKI.

ad.

1st-w.

Northern Shrike *Lanius excubitor* NSHR L 10in

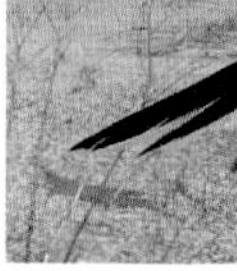

Uncommon and incredibly hardy n. species found in open areas with trees. Some move s. in winter, but generally a bird of frozen landscapes. Sits on the highest snag, unafraid of the cold, looking for rodents, insects, small birds, and other prey. Usually unapproachable. Often disappears for stretches of time, probably sheltering in bushes. **ID**: Compared with similar LOSH, has narrow face mask, longer bill, is larger and paler with an eyering. All birds show barring on underparts. 1st-w: similar to adult with some juv wing coverts, others much browner and more heavily barred, often with pale base to bill.

1st-w.

ad.

Loggerhead Shrike *Lanius ludovicianus* **LOSH** L 9in

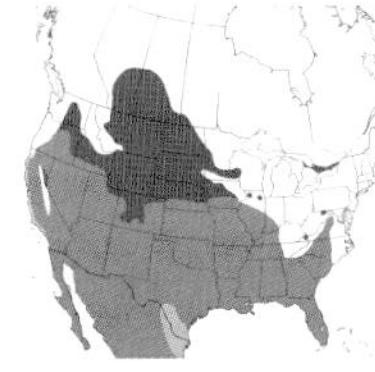

Common in the SE, rare to the N, but declining almost everywhere. Found in open areas with trees. Behaves like NSHR, perching on tops of snags looking for mice, insects, and just about anything its size or smaller. Sometimes hovers (it is kestrel-like in many ways), using hooked bill for killing prey, carrying it off to be impaled on a barbed wire fence or pointed snag—hence the name 'Butcher Bird.' Flight is fast and direct with rapid wingbeats. Superficially mockingbird-like and easy to overlook. **ID**: Compact with a big muscular head and long tail. Clean-cut gray, white, and black with white wing patch striking in flight. Stubby all-black bill and extensive dark mask that extends across forehead. Adult: Underparts gray-washed and without barring. Upperparts clean-cut black, white, and gray. Juv: plumage held briefly—buff-tipped coverts and scaly underparts. By late summer most are similar to adults, with a few retained juv coverts, larger white tips to tertials, darker scapular tips, more white above face mask, and subtly duller.

Northern Mockingbird *Mimus polyglottos* **NOMO** L 10in

Common in ones and twos just about anywhere, from desert to open woodland, but particularly around gardens. Sits on posts, mailboxes; hops on the ground and chases anything that dares come into its territory. Is just as happy in denser vegation nabbing berries and insects. Gray and relatively nondescript, in flight large white wing patches bring it alive. Flight is direct but with very deep languid wingbeats. Often feeds on the ground by jerkily opening its wings, like images from an old movie, perhaps confusing its potential prey. A singer with a big voice that has driven my wife mad (and I'm sure many other people) as it endlessly goes on at any time of night or day. A great mimic of local birds, notes are usually in batches of 2–4. Call: a *chak*. In summer 'skylarks' from high perches as it sings, returning to same perch. **ID**: AMRO-length with a long tail and smaller rounded body. Head seems disproportionately small. Pale gray upperparts with off-white underparts, it can look brownish. Black eyeline, white wing bars, and white outer tail feathers stand out on an otherwise bland bird. Juv: plumage held briefly, duller and browner with spots on breast.

juv.

ad.

Gray Catbird *Dumetella carolinensis* **GRCA** L 8.5in

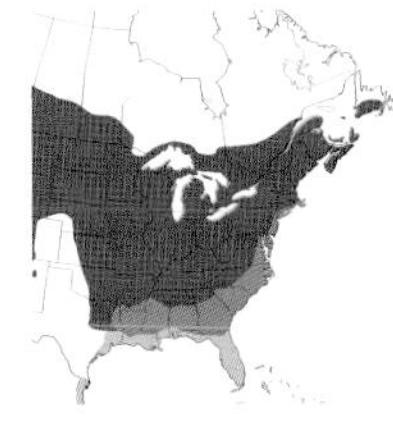

A common and familiar bird. Favors denser vegetation with berries but found anywhere with trees or bushes. Although tends to stay hidden, can be tame and approachable, particularly in gardens. A bird that is usually found around head height, it is quite happy to hop on the ground, often with tail cocked high and wafted from side to side like a feather duster as it looks for insects. Sits in the open looking friendly and inquisitive, tilting its head at an angle as if inquiring about something. Tail is often spread, sometimes tilted down as if it is trying to get it between its legs. In colder weather will sneak around in denser vegetation, particularly where there are berries. Flight direct and low, usually to the nearest cover. Very vocal, its call is a mournful cat-like *meeh*. Song: a continuous ramble of different scratchy notes, often given from treetop. Distressed birds give a rattle. **ID**: Smoothly rounded undercarriage and head. Can look potbellied, but the overall impression is of a sleek bird with a fairly long rounded tail. Uniquely gray with a black cap and tail. On close inspection has a bold rufous undertail. Birds fresh out of the nest appear bedraggled and have fluffy soft feathers.

Sage Thrasher *Oreoscoptes montanus* **SATH** L 8.5in

Breeds in sage plains in the far W. Some winter in TX. Very rare to the E, a few in spring, most late fall/early winter. Perches on fenceposts and bushes, particularly when singing, but more often on or close to the ground. Tends to run between vegetation, but when it flies shows white corners to tail. The smallest thrasher, gray above and warmer buff below but with typically obvious brown streaks. Streaking limited on worn birds (beware of NOMO).

Curve-billed Thrasher *Toxostoma curvirostre* **CBTH** L 11in

Fairly common in dry scrubby areas, often in suburbs. Occasionally wanders. Sits singing from prominent perch, but just as happy on the ground. Not so shy as some thrashers. Large size and bulk, with long thick downcurved bill, usually distinctive. Bland with blotchy/spotted underparts, grayer on flanks. Juv: much subtler spotting.

Long-billed Thrasher *Toxostoma longirostre* **LBTH** L 11.5in

Fairly common in dense vegetation in s. TX. Similar to BRTH in behavior and appearance. Differences include duller upperparts, more boldly streaked underparts extending to undertail coverts. Iris is brighter orange, and bird has an all-black bill that is narrower-based and strongly downcurved. Ranges can overlap in winter.

Brown Thrasher *Toxostoma rufum* **BRTH** L 11.5in

Fairly common in forest edges and hedgerows. Early to arrive in spring, late to leave in fall with some hardy enough to winter in the N. The only thrasher in most of the region, it's most likely to be confused with a poorly seen WOTH. Walks or hops on the ground tossing things out of the way in search for food; you can often hear it rustling! Usually holds its long tail slightly cocked as if not to let it touch the ground. Stays partially hidden in cover but will suddenly skip along to the top of a bush, belting out a sweet, warbling ramble, musical phrases given in 2s and 3s. In other seasons announces presence with a loud *chack*, or a rising, somewhat weird, *ooweeee*. Direct flight, usually for cover. **ID**: Large with a long tail, sturdy legs, and big downcurved bill. Large beady eye, that always seems to be looking at you, on a pale face. Bold black streaks (not spots) to underparts. Upperparts rich brown, dull in the shade, with prominent black-and-white wing bars. Juv: plumage held briefly; blander-faced with brighter cinnamon upperparts.

juv.

ad. ♂ br.

1st-w.

ad. ♀

ad. ♂ nonbr.

Eastern Bluebird *Sialia sialis* **EABL** L 7in

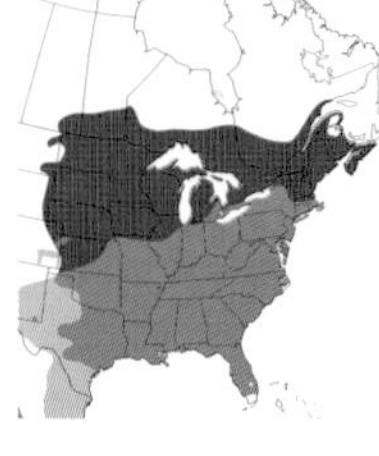

Fairly common in open countryside, golf courses, parks, orchards, and anywhere with scattered trees. Happy in larger gardens and can be attracted by nest boxes or natural cavities, though you may have to kick out HOSP, TRSW, and EUST. A familiar and very popular bird. Sociable, it sits on exposed branches, fenceposts, and telegraph wires and drops to the ground to eat insects and grubs before returning to its perch. In late fall, gathers in larger flocks to migrate s., its light, mellow shorebird-like *tewlee* regularly heard overhead. Flocks loose and evenly spaced. In winter, often in mixed flocks including PIWA and CHSP. Song: a warbled chatter. **ID**: Fairly small and plump. Often seems neckless with shortish tail. All plumages have variably blue or blue-gray upperparts. Underparts are orange with white belly and vent. Plumages are duller in winter, with buff fringes. Ad ♂: beautiful blue upperparts brightest in spring. Ad ♀: duller below, grayer above with pale throat and partial eyering. 1st-w: similar to ♀ with some retained juv coverts and flight feathers. Ageing and sexing is difficult! Juv: dark-spotted plumage held briefly.

ad. ♂

ad. ♀/imm.

Mountain Bluebird *Sialia currucoides* **MOBL** L7.25in

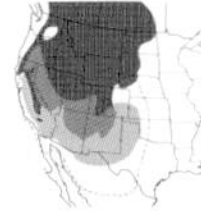

A bird of open country. Regular stray to the E in late fall. Often hovers over fields, otherwise behavior similar to EABL. Noticeably longer and slimmer than EABL with smaller head; longer, more strongly notched tail. Longer-winged with long primary projection. ♂: beautiful sky-blue, darker on upperparts. ♀/1st-w: much duller (fall birds often buff-chested), palest on belly; darker flanks with contrastingly pale blue wings and tail. Tail has dark tip. 1st-w: aged by retained juv coverts.

Townsend's Solitaire *Myadestes townsendi* **TOSO** L 8.5in

Lives in conifer-clad mountains, moving lower in winter. Regularly strays e. in winter, often to berry-bearing trees such as cedars and junipers, where mixes with CEDW and AMRO. Sometimes elusive, at other times sits upright on treetops as is if standing to attention. Eats berries and insects. Slim, plain gray bird with a long tail and gentle demeanor. Small, rounded head with a bold white eyering. Buff bands at base of flight feathers and wing bar on otherwise bland bird.

Swainson's Thrush *Catharus ustulatus* **SWTH** L 7in

Fairly common n. breeder in dense mixed forest. Widespread migrant, the commonest *Catharus* in many areas. Feeds on leafy forest floors, moving a few hops at a time. Often jumps up onto a log or branch, where the first thing to look for is its bold eyering and lores ('spectacles'). Often sits in trees, particularly when flushed. Like all *Catharus* thrushes, can be difficult to see well, and judging color in a shady forest is tough. Call: a *plip*, like the drip of water. Song: a series of spiraling fluty notes rising off the scale. Flight call is spring-peeper-like *heep*. **ID**: Always look for the spectacled appearance created by bold eyering and pale supraloral. They are easy to see. Compared to VEER and GCTH, is subtly smaller, also smaller-headed, slimmer-billed, and with a more tapered rear end. Upperparts uniform olive-brown, between VEER and GCTH in tone. Breast usually distinctly buff with fairly bold spotting. HETH is smaller and chunkier with rufous tail and less pronounced 'spectacles.'

1st-w.

ad.

Veery *Catharus fuscescens* **VEER** L 7.25in

Usually the commonest *Catharus* in deciduous forest. First to head s., usually in Aug (others Sep/Oct). A single downslurred *veer*, *vyur*, or *cheroot* often gives its presence away, sometimes followed by several birds calling in response. This can be a common nocturnal flight sound. Song: a series of soft spiraling fluty notes gradually going down the scale. **ID**: The most muscular-looking *Catharus* with a thick neck and broad tail. It has noticeably warmer brown upperparts than other *Catharus*: in bright light it can appear like WOTH. Just beware, it can also look dull in poorly lit conditions. Has a bland face pattern and hint of gray cheeks, so initially it's easy to think GCTH. The spotting on the underparts is always more diffuse than in similar species. It has the palest flanks of any *Catharus*. The buff on breast and throat extends onto the head, creating a more featureless look than other *Catharus*.

Gray-cheeked Thrush *Catharus minimus* **GCTH** L 7.25

Uncommon breeder in damp evergreen forest or willow/alder thickets. Long and thick-set: closest to VEER in size and shape. **ID**: The darkest *Catharus* with dull olive upperparts and even darker 'gray cheeks.' Boldly spotted breast. Face pattern recalls VEER, but breast pattern and upperpart color different. Flight call, often given on the ground, a thin rising *where*. Song: shorter, scratchier than other species, and seeming to run out at end slightly (downslurred!). Scarce NL race similar to BITH in size.

Bicknell's Thrush *Catharus bicknelli* **BITH** L 6.75

Split from GCTH. Breeds in stunted spruce on highest mountain tops. **ID**: Recalls HETH: small and fat with tail often cocked. Also contrastingly rufous tail and bases to primary coverts and primaries. Always has much colder tones and grayer ear coverts than HETH. From GCTH by smaller size, shape, rufous tones, subtly browner upperparts and less contrast between ear coverts and nape. NL race of GCTH is a problem! Focus on duller colors. Song: similar to GCTH, slightly more even-toned with a stronger ending. Call: similar, to GCTH, but averages higher-pitched. All differences subtle.

Hermit Thrush *Catharus guttatus* **HETH** L 6.75in

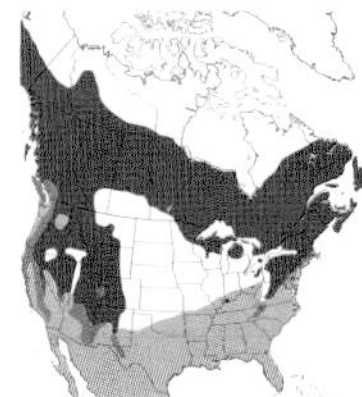

Common in a variety of forest, often where there is an open understory. The last thrush to migrate in fall, the only one that winters in N America, and the earliest n. in spring. Spends much of its time hopping on the ground, quite happy to come into the open, particularly in the early morning sun. Frequently flicks its wings and cocks its tail in a deliberate fashion. Often seems to be trying to stand as tall as it can. Its distinctive *chuck* call is a good one to learn. Also a rising whiny *weou*. 2-part song, first part trailing off, second part higher-pitched than first; and at a different pitch when repeated. **ID**: A fairly distinctive thrush. Small and chunky. Upperparts variably gray or brown but always with contrastingly rufous tail and wing panels (bases to primaries and greater primary coverts). Bold spots to throat. Some have bold eyering and lores approaching SWTH—size, shape, and upperpart color never overlap. BITH is always dark.

Wood Thrush *Hylocichla mustelina* **WOTH** L 7.75in

Uncommon in mostly deciduous woodland and even densely wooded backyards. Declining in most areas. Formerly considered a *Catharus*, it has a more southerly distribution than other brown-backed thrushes. Usually stays well hidden and is most easily seen coming onto roads to feed, though is always quick to fly off. Beautiful fluting song comprises a series of phrases and is always the pick of the dawn chorus. When disturbed, gives a series of angry machine-gunned notes. Flight call: a dull flat buzzy *neep*. **ID**: A really fat thrush with a short tail and big head. Larger than *Catharus* but not always obvious in the field. Usually identifiable by its richer-colored upperparts, brightest on the nape. If in doubt, look for black-flecked ear coverts and a bold eyering. From the front, large black spots on a white background are much bolder than in other thrushes. BRTH is similarly colored and marked but has totally different structure.

Clay-colored Thrush *Turdus grayi* **CCTH (CCRO)** L 10in

Formerly called Clay-colored Robin. Increasing winter visitor and recent breeder in s. TX, mostly the Lower Rio Grande Valley. Found in trees, understory, or hopping on the ground. Size and shape like a longer rangier AMRO. All plumages have the deep apricot-buff underparts, darker on upperparts with dark-based pale bill. Throat finely streaked. Pallid uniformity rules out other species.

White-throated Thrush *Turdus assimilis* **WTTH** L 10in

Very rare in s. TX, mostly in winter. Usually found in trees and easy to overlook as dark CCTH. Size and shape similar but has dark striped throat bordered by white collar and tan underparts, a bold combination. Striking yellow-orange eyering. From behind, appears a rich dark brown.

Fieldfare *Turdus pilaris* **FIEL** L 10in

Very rare winter visitor in the NE. Often with AMRO. Large, heavy-chested *Turdus* with a long tail; often found on the ground. Strikingly patterned with bold gray rump and head contrasting with brown back. Striking white underwing. Unusual chattering call. A beautiful and distinctive bird.

Redwing *Turdus iliacus* **REDW** L 9in

Very rare winter visitor in the NE. Often with AMRO. Small compact *Turdus* that often has a lightweight feel. Often found on ground. Striking bold head pattern and underparts with large blurry streaks. Red flanks and armpits obvious on ground or in flight. Frequently calls in flight, a drawn-out high-pitched *tsee*.

1st-w.
imm. ♀
juv.
imm. ♂
♂
♀/imm. ♂

American Robin *Turdus migratorius* **AMRO** L 10in

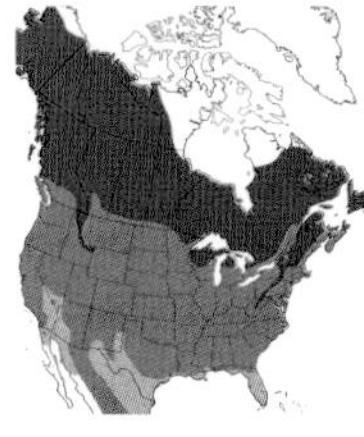

Very common and familiar in gardens, fields, and just about anywhere with trees or bushes. Often tame and approachable. Takes a few hops and stops to pull out a worm. Sometimes will stay motionless for a while with head tilted as if looking with one eye. Follow it and it will often go to its nest, usually just above head height in the fork of 2 branches. In fall, migrates s., often in waves of hundreds or thousands in evenly spaced flight, which is strong and direct, when it gives its thin *tsee* call. They will suddenly descend on trees laden with fruit or berries and glut themselves. Newly cultivated or flooded fields are also favorites. Musical song: *cheerily cher-up cheerio*. Several different calls. **ID**: Powerfully built *Turdus* with a fairly long tail and legs. Ad ♂: gray upperparts. Black head with bold eyelids, bright burnt-orange underparts. Ad ♀/imm ♂: all colors duller, underparts pale-tipped. Imm ♀: dullest with brown rather than gray back, extending onto head, and with indistinct supercilium. Pale orange underparts. Juv: dark-spotted underparts, pale-spotted upperparts with variable amounts of orange. Plumage molted out in a few weeks, though remnants can often be seen through fall—these birds can usually be aged and sexed.

Varied Thrush *Ixoreus naevius* **VATH** L 9.5in

Rare but regular winter visitor. A real skulker in the densest forest. In cold weather often found in gardens or at feeders but still plays hard to get. A loner, it usually feeds on the ground but quickly disappears to higher cover. Call: HETH-like *chup*. Although it doesn't often sing in our region, it has a distinctive long monotone whistle, periodically repeated at different pitches. **ID**: Smaller than AMRO with a shorter tail and potbelly. Boldly patterned head and wings in orange, gray, and black make it hard to miss or mistake. ♀/imm: browner upperparts and indistinct breastband.

Bahama Mockingbird *Mimus gundlachii* **BAMO** L 11in

Very rare visitor to s. FL in dense native vegetation. Unlike NOMO, a skulker, staying hidden or walking on the ground. Superficially similar but is larger with pale throat bordered by dark malar, pale supercilium, browner upperparts, streaking on flanks and undertail coverts. Lacks white wing patches. Beware of juv NOMO, which also has streaks on the underparts.

American Dipper *Cinclus mexicanus* **AMDI** L 7.5in

Scarce in Black Hills, SD. A distinctive bird, dark, round, and fat with nary a tail. With legs angled back, it bobs on rocks before standing in or under water looking for food in fast-flowing streams. Flies up- or downstream, wingbeats almost too fast to see.

juv.

1st-w.

ad. ♂

Cedar Waxwing *Bombycilla cedrorum* **CEDW** L 7.25in

Common in a wide variety of habitats where berries and other food available. Often around gardens, popping in for a bit before bombing off, seemingly in a great hurry, with direct flight and very fast wingbeats. In winter, usually in flocks from a few individuals up to 100 birds. Will descend on bushes, sometimes mixed in with AMRO and bluebirds, to scoff ripened berries. Sits upright but will hang upside down or hover to grab dinner. Often seen drinking from puddles. Late nester, coinciding with food supply. Vocal: a high-pitched, almost off-the-scale *tseeeee* (many notes to create one sound!). In flight, forms tight packs and can quite easily be dismissed as EUST. **ID**: Punky-crested, and a heavy-set bird with long wings and a shortish tail. Mostly pale brown, yellow on belly with a bold yellow—sometimes orange—tail band and a Zorro-type mask. Ad ♂: extensive black chin, red 'rackets' to secondaries, and yellow tips to primaries. Ad ♀/ 1st-w ♂: smaller, duller chin, with less red and yellow in wing. Imm ♀: dullest. Much variation, and tips to wing feathers often worn away, increasing difficulty. Juv: heavily streaked underparts lost through fall.

ad. ♂

1st-w.

ad. ♀

Bohemian Waxwing *Bombycilla garrulus* **BOWA** L 8.25in

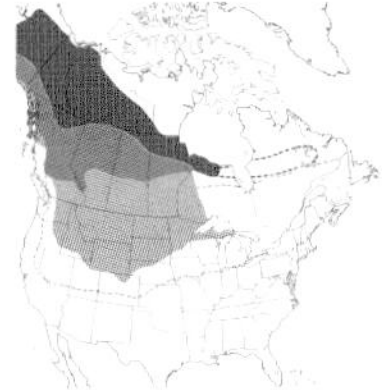

An irruptive n. bird that mostly stops at the border but occasionally comes s. in numbers. Much bigger than their cousins, they quite often party together. S. wanderers are nearly always with CEDW, but sometimes with EUST and AMRO. In winter, usually in flocks, some numbering hundreds. In a harsh environment, they can often be seen swarming around fallen fruit, seemingly gorging themselves as if they expect to go hungry for a while. Oddly, they frequently flycatch en masse from the tops of trees—a good way to find them. Mobile, large groups can appear and disappear surprisingly quickly. **ID**: From CEDW by large size, gray coloration, and rich brown vent. Also more extensive pale tips to flight feathers, more extensive dark throat, and deeper trilled call. Ad ♂: sharp edge to black throat. Broad yellow edge to primaries. Broad yellow tail band and red 'rackets' to secondaries. Ad ♀: soft edge to throat, narrower yellow inner border to primaries, reduced dark on chin. 1st-w: lacks yellow inner web to primaries. Juv: streaked underparts lost through fall.

Black-capped Chickadee *Poecile atricapillus* **BCCH** L 5.25in

The common and familiar n. chickadee, in gardens, parks, forest, and just about anywhere there are trees. Although primarily resident, occasionally wanders s. in winter, particularly when food shortages occur. A hardy and spunky bird, like CACH, it hangs upside down and flits from tree to tree, usually in small flocks, sometimes with other species. Friendly, at times tame but easily agitated, particularly by owls, and is a common feeder bird. Song: a *fee bee ey*. Named after its call, a familiar *chickadee-dee-dee*, averages coarser and slower than CACH, but there is overlap. **ID**: Like CACH but larger and chunkier with a big head and longer tail. Dark bib, black cap, and buffy flanks. Marked ssp variation but always strikingly frosty due to dark wing and tail feathers with broad white fringes. If in doubt always look for dark-centered greater coverts. By late summer, adults are very worn and can look scruffy. Tail often flared at tip. Range barely overlaps with CACH. Hybrids are reported where it does.

molting late summer

ad. fresh

Carolina Chickadee *Poecile carolinensis* **CACH** L 4.75in

Common and familiar. The s. counterpart of BCCH with ranges barely overlapping. Resident, almost never moves. Behavior and habits are the same as BCCH, though stays away from higher altitudes. Song: usually 4-note *fee bee fee bee*. *Chika-dee-dee-dee* similar to BCCH, averaging higher-pitched, sweeter, and slightly faster. Juv often slower and burrier like BCCH. **ID**: From BCCH with care, though most can safely be separated by range. Not so large, thick-necked, 'fluffed-up' or rotund. Shorter tail is usually held narrower at tip. Bill slightly shorter. Noticeably less frosty and contrasty than BCCH. Most importantly, if in doubt, centers of greater coverts are gray, not black, contrasting less with narrow pale fringes. Pattern of reduced contrast is also similar in the tertials and outer tail feathers but is not so easy to determine. Flanks average less buff, rear cheek grayer, and lower border to bib neater, but there is large variation and overlap in these features. Hybridizes with BCCH.

Boreal Chickadee *Poecile hudsonica* BOCH L 5.5in

Uncommon in n. forest, particularly spruce. Occasionally irrupts s. Often with BCCH flocks. Seemingly quieter than its commoner cousin, it can be hard to find. Also more secretive though individuals may be tame. **ID**: Slightly larger than BCCH. An 'obvious' chickadee, its brown cap, bright flanks, and dull back make ID straightforward. The white cheek patch is narrower and gray at rear, adding to its overall dull appearance. Voice more nasal than BCCH's.

Black-crested Titmouse *Baeolophus atricristatus* BCTI L 6.5in

Uncommon. Restricted to woodland and scrub. Formerly conspecific with TUTI. Almost identical but adult with black crest and white forehead: the reverse of TUTI. Juv: pale flanks; forehead and crest gray. Adult-type black feathers molted in through fall/winter. Ranges barely overlap, but hybrids reported where they do. As with most hybrids, they show intermediate characters of both species. Call and song similar to TUTI but faster.

Tufted Titmouse *Baeolophus bicolor* **TUTI** L 6.5in

Very common and familiar bird in woods, parks, and gardens and at feeders. Usually in ones and twos, often with chickadees and other species when not breeding, and can be quite tame. A heavy-set bird but very agile, hanging from all angles. Its movements are slower than in chickadees. Hardy, able to find food anywhere, and has no problem chiseling at trunks or holding nuts with its feet while it smacks them open with its bill: not everything that sounds like a woodpecker is one. It has a surprisingly weak and undulating flight for such a muscular bird. Typical song is a loud *peter, peter, peter,* but has quite an amazing range of calls. Often the first to scold owls. **ID**: Quite large and bulky, noticeably bigger than a chickadee with a long tail and distinctive crest. Gray above, pale below with markedly buff flanks and with a big, black, beady eye that stares at you. Juv: paler, lacking buff flanks and black forehead, these molted in through fall.

ad.

1st-yr.

Brown-headed Nuthatch *Sitta pusilla* **BHNU** L 4.5in

Fairly common in pine woods. Very rarely shows up out of range. Small and very active, often heard first, with its squeaky-toy-like call. Also gives a call note similar to AMGO. Shimmies up and down branches, hangs upside down, always on the move and at high speed. Usually in small groups, sometimes in mixed flocks. Visits feeders. **ID**: A miniature nuthatch with bluish upperparts, brown cap; often shows white nape. Underparts mostly off-white. Juv: similar to adult but with fluffier undertail coverts and more brown mixed in the wing.

Pygmy Nuthatch *Sitta pygmaea* **PYNU** L 4.25in

Very rare visitor from the W that rarely travels far out of range. Similar to BHNU but even smaller. Upperparts all gray-toned, with badger-like dark line through eye. In breeding range always lives in groups in pine trees, and with out-of-range birds look for same habitat. Very active, often hanging upside down, swinging from side to side as it shuffles along branches and pries into pine cones. Gives wide array of high, sharp peeping notes, quite different from 'squeaks' of BHNU.

1st-yr.

ad. ♀

ad. ♂

White-breasted Nuthatch *Sitta carolinensis* **WBNU** L 5.75in

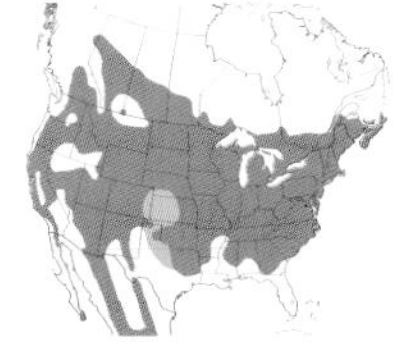

Quite common but sometimes elusive. Local migrant. Found primarily in deciduous forest, but also where evergreens, particularly in the W. Most often heard before seen. Noisy with distinctive *mee-mee-mee*, nasal but not so deep as RBNU, and slower and louder. Call: a *yenk*, also nasal. Regular at feeders. Often in mixed flocks with chickadees, warblers, and RBNU. Walks down tree trunks just as often as up, stopping to put head up to check what's going on. Movements are slower than other nuthatches, but tends to go in a straighter line, and its larger size means it covers more ground. Shape and bold pattern distinctive in flight. **ID**: A large, sturdily built nuthatch. White face, outlined by black, stands out a mile away. Clean-cut whites, grays, and blacks always give it a sharp spiffy look. Long, thin, dagger-like bill slightly upturned. Underparts appear strikingly white, but are actually gray on belly with rufous vent area. Back blue-gray. ♂ has black cap, ♀'s cap is grayer, darkest at sides so can look black. 1st-yr: as adult but subtly duller with wing feathers browner and with less contrasting fringes. Birds on e. edge of range are darker gray; ♀ has darker crown and calls a little differently.

1st-yr. ♂

♀

imm. ♀

ad. ♂

Red-breasted Nuthatch *Sitta canadensis* **RBNU** L 4.5in

Fairly common n. breeder in evergreen forest, particularly spruce and fir. Somewhat irruptive to s. in winter, movements based on food supply. Small flocks fly overhead as early as Aug. Prefers pines but can be found in all types of woodland and at feeders. On migration often forms small flocks when can be heard giving its deep nasal *neep neep neep*; deeper, shorter and more nasal than WBNU. Call: an *eenk* has same qualities as WBNU's. Also found in mixed-species flocks. Very active, often switches direction as it walks, hangs, and descends branches. Much smaller than WBNU, is frequently found at the end of twigs, hanging on cones looking for food. In flight, striking, short square-ended tail and broad wings. Bouncy flight takes it nowhere quickly. **ID**: All plumages have distinctive bold head markings, orange underparts, and blue-gray upperparts. Ad ♂: most brightly colored with deep orange underparts, blue-gray upperparts, and dark flight feathers. The crown is black. Ad ♀: paler with blue-gray crown. 1st-yr: like adult but slightly duller with browner wings, ♂ brighter than ♀.

Brown Creeper *Certhia americana* **BRCR** L 5.25in

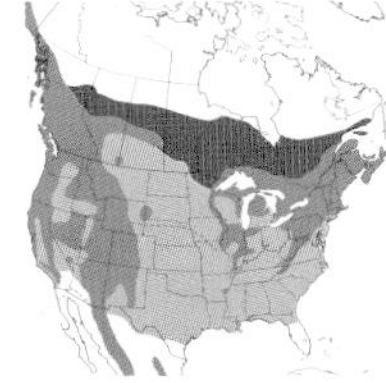

Fairly common in woodland but easily overlooked. Prefers big trees with coarse bark, perhaps easier for prying its bill under for grubs and insects. Very quiet; listen for its thin high-pitched kinglet-like *tseee* call, repeated several times in song. Creeps or shimmies quietly up tree-trunks, much like a brown-and-white mouse with a long, spiky tail. Usually starts at tree base, working up and wrapping around tree and along branches before dropping back to ground or a new tree to start all over again. Joins mixed-species flocks in winter, when is often the trailing bird. **ID**: Behavior alone is usually all you need. Small, slim, with a long spiky tail and very fine-based, downcurved bill. Underparts white, intricately marbled upperparts with lines in wings. Always strikingly 2-toned. All plumages similar.

Bushtit *Psaltriparus minimus* **BUSH** L 4.5in

Scarce in the far W. Found in a variety of woodland, but prefers thorny habitat. In nonbr season usually found in packs that stick tightly together, zipping quickly through vegetation, giving chattering calls as they go. Often hangs upside down in very chickadee-like fashion. **ID**: Small, chunky, neckless, gray bird with a long tail. Brown ear coverts contrast with gray cap. Stubby dark bill. ♂ and juv have dark iris, ♀'s is pale.

Verdin *Auriparus flaviceps* **VERD** L 4.5in

Fairly common in mesquite and scrubby vegetation in dry areas, often around head height but hard to see, staying well-hidden. Always on the move, very fidgety. Tail often flicked up. Rounded nest reasonably conspicuous. **ID**: Small. Chunky-bodied, muscular-necked, and round-headed with a slim parallel-sided tail; looks cute with its dark lores and pointed bill on spiffy yellow head. Rufous patch on bend of wing on otherwise dull body not always visible. Juv: paler headed, quickly becomes adult-like.

Rock Wren *Salpinctes obsoletus* **ROWR** L 6in

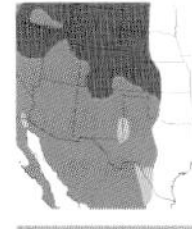

Uncommon in barren areas with rocks and scree, often where little else lives. Elusive, walking most places, it will suddenly perch on the top of a rock. Looks around before excitedly bobbing up and down with exaggerated movements, frequently giving its ringing call note. From a distance, fairly nondescript brown above, buff below, with indistinct supercilium and pale-tipped tail. But look closer.... Juv: plumage held briefly, duller with unstreaked breast. Occasionally wanders e.

Canyon Wren *Catherpes mexicanus* **CANW** L 5.75in

Another uncommon w. rock lover, often on canyon walls. Its beautiful, clear descending notes and frequently heard in its territory (Willow Warbler-like for Europeans). Creeps along rocks, somewhat BAWW-like in its movements, chiseling out food. Often elusive, it suddenly appears right in front of you. A richly colored bird with conspicuous white throat, duller cap, and very long downcurved bill. Juv: plumage held briefly, duller with uniform breast.

Carolina Wren *Thryothorus ludovicianus* **CARW** L 5.5in

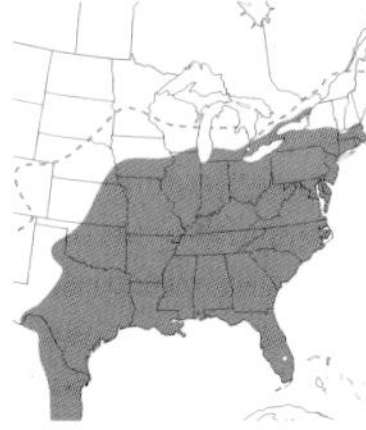

Common and would be very familiar if it allowed itself to be seen. Usually found in dense tangled vegetation and trees, particularly near homes, but also more extensive woodlands. It's amazing how something so small can make so much noise: you are sure to have heard it. A large repertoire of songs, calls, and scolds, *cheery cheery cheery* perhaps the best known. If you are not sure of a sound, it's probably this species. Skulks around, on or near the ground, tail nearly always cocked (straight when singing). Climbs trees nuthatch-like, often hanging upside down and entering holes. Inquisitive, it goes to odd places—garages, through the open window, under the car, in plant pots, and just about any nook it can get into: perhaps you have found one stuck somewhere. Sometimes bobs up and down. Hops rather than walks. Very aggressive, the yard boss. **ID**: Fairly large and chunky with a medium-length tail. Brown above and variably buff below with a large and bold supercilium. Bold triangular spots are suggestive of juv. Ageing difficult: juv averages duller with fluffy undertail coverts. Sometimes occurs out of range. Only BEWR is similar but is slighter with a longer tail and more subdued colors.

Bewick's Wren *Thryomanes bewickii* **BEWR** L 5.25in

Rare and declining in the E, fairly common in the S. E. birds found in scrubby vegetation, s. birds in drier areas with mesquite and other trees. Moves purposefully around branches, swishing its tail from side to side like a gnatcatcher. Often hangs upside down and twists from side to side on branches and trunks, only occasionally on the ground. **ID**: Slightly smaller than CARW with a slimmer build, and boldly marked long tail, often spread. Gray underneath, dull brown-gray above, lacking warmer tones of CARW and the white wing spots. Birds in the W average grayer. Very rare out of range.

Cactus Wren *Campylorhynchus brunneicapillus* **CACW** L 8.5in

Uncommon in desert areas, often near houses. Massive for a wren, it can be hard to place to family. Hops on the ground, never seems in a hurry and often oblivious to nearby people. Perches on cactus as if waiting to be seen. Large, domed nest a familiar site in a cactus crook. Often in loose groups. Vocal, a churring series of notes, a common sound in old Westerns. **ID**: Big and fat. Dark throat and striking head pattern obvious at distance. There are spots, streaks, and bars just about everywhere. Boldly patterned tail with extensive white tip; tail never cocked as in other wrens.

House Wren *Troglodytes aedon* **HOWR** L 4.75in

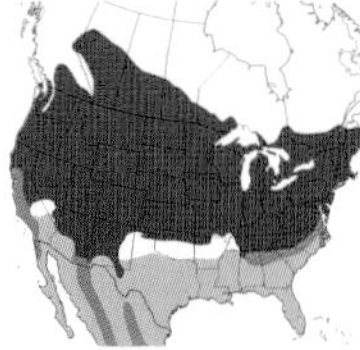

Familiar bird in parks and, particularly, gardens. Very vocal with a chattery song, invariably singing by the time you wake up. Quite content in your front yard, it often uses nest boxes, hanging flower baskets, or other man-made structures. Sits in the open to sing but is very small, so you still have to squint to see it. When quiet, is very hard to find. Likes to keep low in dense vegetation in fields or woods. Song: a descending series of rolling notes. Gives annoyed *chek* note when disturbed.

ID: Small and pallid with a narrow medium-length tail that is usually held straight and only half-cocked the rest of the time. Seems slim for a wren. Besides size and shape, look for overall pale brown or gray coloration with bland face pattern. Bold eyering, indistinct supercilium and pale throat—little else is of contrast. Subtle barring just about everywhere else. Much variation in color, birds in the W sometimes more rufous.

Winter Wren *Troglodytes hyemalis* **WIWR** L 4in

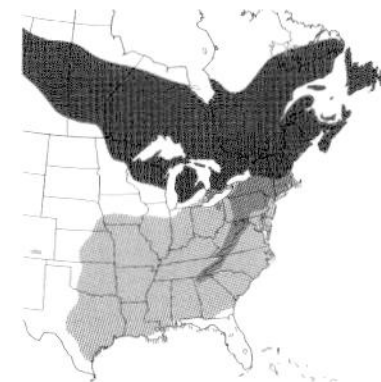

Recently split into 3 different species; 2 in N America: Winter Wren (*T. hyemalis*) and Pacific Wren (*T. pacificus*). Fairly common in dense coniferous forest in summer. In winter, in wetter and boggier areas, often with stumps and deep undergrowth. Frequently heard chipping and, in summer, gives an explosive warbled song that seems to go on too long, as if making up for the bird's tiny size. Can be very difficult to see, hugging the ground and staying in dense undergrowth—it blends in with background and is tiny to boot. Views tend to be brief, as it's always on the move. If you know the *chimp* or *chimp chimp* call note, you'll hear many more birds than you will see. **ID**: Tiny round ball of feathers with a cocked stub for a tail—nothing else like it! Variably colored: all shades of dark brown with a pale supercilium. Some more heavily barred than others (probably mostly juvs).

Sedge Wren *Cistothorus platensis* **SEWR** L 4.5in

Fairly common in the MW, rare in the E in overgrown grassy fields, particularly wetter areas with sedge and meadow grasses (often where LESP). Very hard to see. Unusually, some birds move e. late in summer to breed again. Wintering birds found in bushy coastal marshes, often with MAWR. A skulker, usually staying close to the ground and rarely high enough to be fully in the open. Usually active, patience is often needed to get a good view. Most often heard: 2–3 sharp *chip* notes followed by an accelerating chatter that is often repeated, particularly at twilight. Gives a sharp *tak* when not happy. **ID**: Its appearance is noticeably different from MAWR. Slightly smaller and slimmer, its overall warm appearance and orange flanks and bum stand out. Upperparts are strongly patterned with heavily barred wings and finely streaked back but lacks strongly capped look of MAWR.

juv.

ad.

Marsh Wren *Cistothorus palustris* **MAWR** L 5in

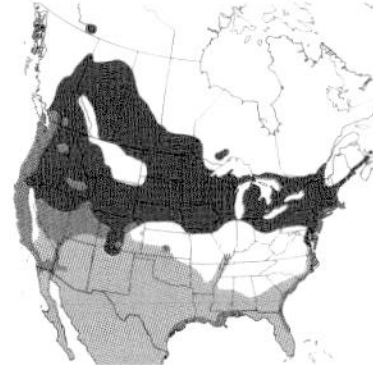

Locally common in cattail marshes. On the E Coast is common in estuarine marshes with phragmites or small bushes. Belts out its song, always giving 100%. Often sings in flight, a noisy gurgling jumble of notes. Usually stays well hidden as it actively moves from stem to stem, or stands, legs splayed between reeds, the hard *chack* note the only thing giving away its presence. Look carefully and you will often see its large ball shape nest suspended between reeds. Like the bird, the nest is rarely above head height. Always cautious, birds sneak in the back way to the nest. **ID**: Round and medium-sized with a spiky tail, which is usually cocked. Striking supercilium emphasized by bird's brown-centered dark cap. Upperparts always dark. Mantle streaked black and white but often tough to see. Wings and tail barred black and brown. Rufous rump and scapulars. Underparts variable from mostly white to strongly brown-flanked. Juv: similar to adult but with subdued markings, often lacks mantle streaks.

ad. ♂ br.

♀/nonbr.

Blue-gray Gnatcatcher *Polioptila caerulea* **BGGN** L 4.5in

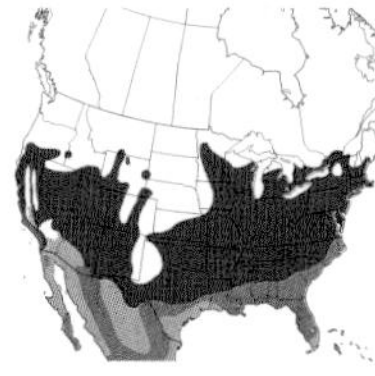

Common in most deciduous woodland. An early migrant, its small size and a long tail, that it swishes around, always draw attention. Very active, rarely on its own, as it constantly jumps and contorts on twig ends, tail seemingly everywhere. Hovers briefly to snatch insects. On migration often flies later in the morning than other species. Small, spaced groups high up, but very short rounded wings, long narrow tail, and strikingly pale look make birds very distinctive. Calls a lot, a distinctive, buzzy but nasal high-pitched *speeeez*. **ID**: Size and shape distinctive with enormously long bill for such a small bird. Paleish gray upperparts and white underparts with a bold eyering that pops, and bold white outer tail feathers; a distinctivelly pale overall appearance. Ad ♂ br: black eyebrow from bill to behind eye. Ad ♂ nonbr, ♀, and 1st-yr: lack the black eyebrow and are all similar. 1st-yr: often duller with some retained juv brown-fringed greater primary coverts.

Black-tailed Gnatcatcher *Polioptila melanura* **BTGN** L 4.5in

Uncommon gnatcatcher in arid scrubby areas of s.TX. Habitat is a great clue to identity, but look for black undertail with white tips. Breeding ♂ has striking black cap. In winter, this is reduced to short black line above eye (lacking in BGGN). Undertail is mostly black. Outer tail feather has narrow white fringe and large white tip: mostly white in BGGN. ♀ and 1st-yr: subtly browner upperparts than BGGN with narrower white fringes to tertials. Tail more graduated than in BGGN, giving rounded appearance. Bill narrower-based and shorter. Multiple calls: a hissing *psss*, *chips*, and others.

Northern Wheatear *Oenanthe oenanthe* **NOWH** L 5.75in

Breeds in remote ne. Canada, some flying nonstop to SE Europe to winter in Africa (1600 miles). Very rare, mostly fall in the NE. A tundra breeder that prefers open countryside on migration, where it hops, runs, and then quickly stops to look for food. Tilts forward as it stands on rocks, posts, and other perches. Long-legged and potbellied, stands tall when it stops moving. A buff bird, it sports a surprisingly bold white rump with inverted black 'T' at tip. Nonbr ♂: as ♀. Most recorded migrants are juvs with pale-fringed coverts.

ad. ♂

Golden-crowned Kinglet *Regulus satrapa* **GCKI** L 4in

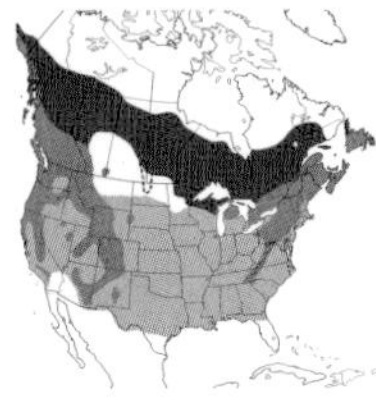

Fairly common breeder in dense coniferous forest. Difficult to see and often looks 'dirty' from residues on trees. Wintering birds and migrants also prefer conifers but will use other habitats, and often mix with other species. Sometimes feeds in weedy vegetation. Usually in small groups, fidgety, always moving and often flicking its wings nervously. Acrobatic, it hangs upside down and hovers to glean insects. Approachable but can get angry and even fluff up its crown feathers when agitated. High-pitched see notes often give it away. Song: a series of see notes finishing in trill **ID**: Very small and fat, though a little slimmer than RCKI. Striking head pattern, with yellow/orange crown stripe diagnostic. Bold wing pattern with broad greater covert bar accentuated by black base to secondaries—same as RCKI. Black legs, orange feet. ♂: orange center to crown, only seen in display or when agitated. All other plumages have yellow crown stripe.

Ruby-crowned Kinglet *Regulus calendula* **RCKI** L 4.25in

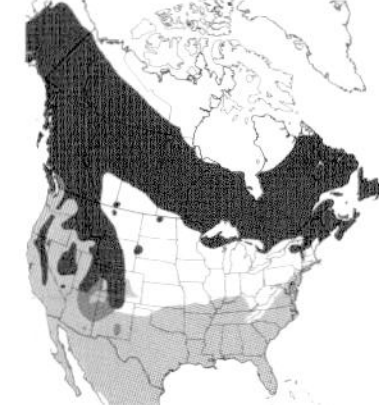

Common breeder in coniferous forest. Wintering birds and migrants found in a variety of trees, sometimes feeding in weedy vegetation, usually in mixed flocks. Hyperactive, often flicks wings and calls *chidit*. Hovers to pick and glean insects from underside of branches. Cheerful song from simple notes. **ID**: A small round ball of feathers with a short tail. Relatively nondescript except for bold eyering, which, on closer inspection, is only broad in front and behind the eye. Fine bill. Wings are boldly patterned as GCKI, flight-feather fringes being the brightest part of the bird. Although ♂ averages greener rather than grayer upperparts, ages and sexes broadly similar. Pale feet look out of place on black legs. ♂: has red crown that is only usually seen when bird is agitated or singing.

White-eyed Vireo *Vireo griseus* **WEVI** L 5in

Fairly common in deciduous woodland edges and hedgerows, particularly with tangles. Tends to stay lower in vegetation than other vireos, often at head height, but can be a devil to see. It moves a little faster than other larger vireos. Seems to lean forward more, perhaps to slip through its usually tangled world. Its pale eye always seems to have you in sight. Also growls at you, *spit and see if I care* or *come at me real quick*—a bird with real attitude. **ID**: Relatively bland with green upperparts, gray nape, and yellow flanks on pale underparts. 2 fairly narrow wing bars and white-fringed tertials. Although named after its white eye (dark in young fall birds), its bright yellow supraloral stripe is the most striking feature, creating spectacled look with narrower eyering. Adult: pale iris. 1st-fall: dark iris. Fresh fall plumage gives bird a more crisply marked appearance than in worn summer.

Gray Vireo *Vireo vicinior* **GRVI** L 5.5in

Uncommon and local in hilly arid areas with junipers. Active before midday heat, it moves longish tail like a gnatcatcher. Best found by typical vireo-like song. Best described as LGJ ('little gray job') for its blandness and lack of field marks. Gray, paler below. Weak wing bar and weak spectacles. Noticeably 'rounded' with potbelly and smoothly domed head. Habitat alone is a big clue!

Thick-billed Vireo *Vireo crassirostris* **TBVI** L 5in

Very rare in s. FL scrub. Similar to WEVI, but first impression should be of a more uniformly colored bird lacking contrast on underparts and neck sides. A larger bird with thicker bill, dark iris, broad yellow supraloral, and paler spot isolated behind eye.

Plumbeous Vireo *Vireo plumbeus* **PLVI** L 5.75in

Very rare in the W. Formerly conspecific with BHVI. From that species by gray color, lacking green tones except occasionally on flanks and rump. White fringes to flight feathers. Song and call slower, like YTVI. Beware, poor views of BHVI can sometimes suggest this species. Cassin's Vireo from the W is a very rare visitor. Very similar to formerly conspecific BHVI but slightly smaller and duller.

Hutton's Vireo *Vireo huttoni* **HUVI** L 5in

Very rare in w. TX (Edwards Plateau). Looks just like a RCKI but larger, with a thicker neck and a thick bill. Eyering broken at top and subtly different. Lacks black bar at base of flight feathers. Blue legs. If in doubt check for much slower movements and only occasional flick of wings. Different song and call.

Bell's Vireo *Vireo bellii* **BEVI** L 4.75in

Uncommon in scrubby areas in e. of range, mesquite in w. Very rare out of range. Active for a vireo, though hard to see in dense vegetation. Often flicks tail (tends to pump tail softly in e. part of range). E. populations tend to average brighter and shorter-tailed, leading to confusion with WEVI. First impression of BEVI is of a noticeably smaller, slimmer, and longer-tailed bird. Has narrower wing bars, paler bill, dark eyeline and pale 'eyelids' rather than 'spectacles.' Fall birds in fresh plumage/1st-w are greener with broader wing bars. Warbling song but still essentially vireo-like.

Black-capped Vireo *Vireo atricapilla* **BCVI** L 4.5in

Uncommon and local in rocky areas with oaks, notably the Edwards Plateau, TX. Most arrive early Apr. Small and active, though tough to see, staying hidden in bushes, usually at head height. Moves quickly for a vireo, but when you eventually lay eyes on it, you realize it was definitely worth the effort. White underparts with yellow flanks, green upperparts with 2 wing bars. The dark head pattern with bold white spectacles make this bird really striking. Head is black in ♂, gray in ♀/imm. Song: a warbling twitter, but characteristically vireo-like. Sings through much of the day.

Yellow-throated Vireo *Vireo flavifrons* **YTVI** L 5.5in

Fairly common in deciduous forest, particularly scattered oaks and near water. Stays high and moves slowly, even for a vireo. Usually alone, it will join mixed flocks on migration. Often heard first, slower, sweeter slurred phrases than other vireos, changing in pitch—*seeya later* or *see you see me*, maybe the only giveaway. Also gives a growling chatter. Once in sight, it is striking. **ID**: Large and long but with sturdy upper body with thick neck and smoothly domed head. Bright yellow breast, throat, and prominent spectacles in combination make ID straightforward. Green head and mantle contrast nicely with gray lower back and scapulars. Wing bars are always neat and bold. Ages and sexes similar.

Philadelphia Vireo *Vireo philadelphicus* **PHVI** L 5.25in

Uncommon vireo in open deciduous woodland, favoring willows and alders. On migration, usually found in mixed flocks, particularly with REVI. Small size, soft expression, and fairly small bill can make it look warbler-like (see TEWA). Song easy to pass off as REVI but slightly sweeter. **ID**: Small with average proportions and evenly domed head. In spring, tends to be duller with yellow strongest on central breast. In fall, upperparts greener and underparts more uniformly brighter yellow, though never brighter on breast sides than center. Dark primary coverts contrast with paler greater coverts. Dark eyeline extends to base of small bill. Similar to WAVI though usually, but not always, yellower. These 2 vireos are often confused, particularly when too much emphasis is put on color (see WAVI account for comparison).

spring

fall

Warbling Vireo *Vireo gilvus* **WAVI** L 5.5in

Fairly common in deciduous woods, particularly willows near water. Most breed farther s. than PHVI. Stays high, singing a lot—a fast and musical warble with scratchy tones but in many ways unlike a vireo. On migration, often joins mixed flocks. **ID**: Variable in coloration. Usually pale gray but some bright yellow. Open-faced and dark-eyed look with a hint of an eyeline just in front and behind eye. Supercilium is often broadest just behind eye. Pale base to bill enhances bland expression. Sometimes confused with PHVI, but WAVI has different facial pattern with pale lores, longer bill and tail (usually held looser), also lacks contrast between primary and greater coverts, yellow brightest on breast sides (center of breast in PHVI); usually paler appearance overall and very different song. All these features are good, but head pattern is usually the best place to start.

juv.

ad.

Blue-headed Vireo *Vireo solitarius* **BHVI** L 5.5in

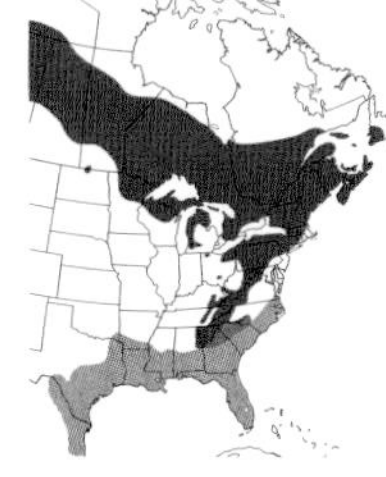

Common high up in deciduous and coniferous forest. Late fall and early spring migrant. Very slow, deliberate movements with upright posture. Song: somewhere between REVI and YTVI in speed with sweet notes: *seeyou seeme, waitup, comewithme, werin wereout.* Call: a nasty gremlin-like chatter. **ID**: Strikingly large. Seems dignified, moving slowly with erect posture and smart colors. Often tilts its head to check things out. Bold spectacles are striking on a gray-blue head that contrasts with olive back and bold wing bars. Underparts clean white with pale green flanks. Except for a brief time as juv, never appears scruffy! S. Appalachian population averages larger and darker. Ages and sexes similar.

1st-w.

ad.

Red-eyed Vireo *Vireo olivaceus* **REVI** L 6in

Very common. Found just about anywhere with deciduous trees. Often found in suburbs, it is the vireo to learn well. On migration, will join mixed flocks, and can be commonest bird. Although has typical slow vireo-like movements, it can move quickly between trees. When disturbed, often the first bird to come close, eyes looking right at you with head pushed forward. Calls frequently, an agitated slurred *weeer* (somewhat VEER-like) that is sure to bring other birds closer. Song: a series of short hurried undulating phrases continuing for a long time: *wereup weredown, werin wereout*. **ID**: You are always hit by the strong facial pattern. Flat-crowned, large-billed with a mean look. Black eyeline and lateral crown stripe bordering bold supercilium make it distinctive in most of range. Crown grayish on otherwise uniform olive upperparts that lack even a hint of a wing bar. From below, short-tailed with a yellow vent—this rules out most other vireos even with poor views. Underparts appear off-white, though yellow wash on flanks is sometimes obvious. Red eye in adults, dark in otherwise similar juv. Pointed head obvious in flight.

Black-whiskered Vireo *Vireo altiloquus* **BWVI** L 6.25in

Fairly common in limited range, particularly along the Gulf Coast, in mangroves or native hardwood hammocks. Occasionally out of range. Early migrant. Similar to REVI but larger, bigger-billed, and duller. Most importantly has distinct dark 'whisker.' At times whisker not easy to see but always there. REVI can briefly show a hint of one. Browner eye. Song: emphatic and fast, a repeated *chillip chilloo*.

Yellow-green Vireo *Vireo flavoviridis* **YGVI** L 6.25in

Very rare, mostly in TX, where has bred. Possible elsewhere. Similar to REVI and BWVI. First impression is of a brighter bird with yellow on flanks and breast, extending onto neck. Bill large and pale flesh. Head pattern subdued with little black. Song: reminiscent of a fast HOSP.

Tropical Parula *Parula pitiayumi* **TRPA** L 4.5in

Very rare, mostly summer, breeds s.TX. Like NOPA but subtly slimmer, longer-tailed, shorter primary projection, and thicker-billed. Face typically lacks bold white eye-arcs and has extensive dark mask. Underparts brighter and more extensive yellow. Some birds show intermediate characters. Songs similar.

Northern Parula *Parula americana* **NOPA** L 4.5in

Commonest in large, damp deciduous trees with lichens or Spanish moss, but also coniferous and other deciduous woodland. One of the commoner warblers on migration, and a great bird to know well. It fairly belts out its buzzy song, invariably from the topmost branches. Call: a short thin *tsee*. **ID**: Distinctive size and shape: small, fat and potbellied with a short, narrow tail. NAWA is most similar but lacks neckless, often pointed-headed, look of NOPA. Yellow breast and throat always stand out on otherwise white underparts; then check the tail with white bordered by black. Short, broad, often triangular wing bars are the stand out on the variably blue-and-green upperparts. Head pattern is also distinctive: pale eyelids on a blue head. Ad ♂: brightest with rufous-and-black breastband that is highly variable in color and size. Ad ♀/imm ♂: no or limited rufous breastband. Imm ♀: dullest and most uniform, always lacking breastband.

ad. ♂

1st-yr. ♀

ad. ♀/1st-yr. ♂

Blue-winged Warbler *Vermivora cyanoptera* **BWWA** L 4.75in

Fairly common in successional fields and second-growth deciduous woodland; the type of habitat common along power lines. Expanding n., pushing GWWA out or hybridizing with it. Stays lower than many warblers, often around head height and in the densest tangles and leaf clusters. Agile, often hangs upside down looking under leaves, chickadee-like, as if peeking closer for food than other birds. do An unusual song: a loud insect-like *fuzzzz buzzzzzzz*. **ID**: A somewhat sleek bird with average proportions. All plumages are bright yellow with a white vent and striking dark eyeline, pointed at rear. Black bill is so pointed it looks dangerous, typical of *Vermivora*. Unusually colored gray wings with 2 broad but poorly defined wing bars, white but often with yellow. Square-ended tail with white outer tail feathers. Ad ♂: extensive yellow forehead and broad white wing bars. Ad ♀/1st-yr ♂: limited yellow on forehead. 1st-yr ♀: green forehead with yellow supercilium. Wing bars typically yellow-tipped (this is not a sign of hybridization with GWWA, as often assumed).

Lawrence's backcross ♂

Lawrence's Warbler

Brewster's ♀

Brewster's ♂

ad. ♂

ad. ♀

Golden-winged Warbler *Vermivora chrysoptera* **GWWA** L 4.75in

Uncommon in the same habitat as BWWA but with a preference for wetter areas. Size, shape, behavior, and tail pattern also similar, perhaps in part reflecting why hybridization with BWWA is common where their ranges overlap. Declining rapidly in many areas due to encroachment from BWWA. Song: usually 4 high buzzes, the first a high note followed by 3 lower ones. BWWA and GWWA will sometimes sing each other's songs. **ID**: All plumages have striking color pattern. Chickadee-like black bib and dark ear coverts unique among e. wood-warblers. Yellow forehead and wing panel stand out on an otherwise pale gray bird. Ad ♂: solid black face pattern with bright yellow median and greater coverts. ♀: subdued version of ♂. 1st-yr: as same-sex adult. Brewster's Warbler (1st-generation hybrid): BWWA-like due to genetic dominance but with paler underparts and yellower wing bars. 2nd-generation backcross (2 types): Brewster's-like and rarer Lawrence's (recessive traits)—yellow with GWWA head and throat pattern. There are many intermediate in appearance and song type in hybrids.

ad. ♂ spring

1st-s. ♀

1st-w.

♀ spring

ad. ♂ spring

Tennessee Warbler *Oreothylpis peregrina* **TEWA** L 4.75in

A fairly common n. breeder in a wide variety of coniferous and deciduous forest. On migration, scarcer to the E, found in a variety of woodland habitats. Feeds in the treetops but is also found lower, even in weedy fields, or drinking nectar from flowering plants. Agile, like other *Oreothylpis* warblers, hanging upside down to glean food. Song: usually 3 series of repeated notes at different pitches—like a coin coming to a standstill on a hard top. Similar NAWA song is 2-parted, slower, and not so emphatic. **ID**: A fairly small warbler that appears slim with long wings and shortish tail. Smoothly domed flat crown. Bill is narrow, strikingly pointed. All plumages similar and relatively nondescript: upperparts green with indistinct wing bars. Head with bold supercilium and eyestripe. Underparts are uniform, varying from white to yellow with, importantly, contrastingly paler undertail coverts (similar OCWA has yellow undertail coverts that are never paler than underparts). Ad ♂: blue head contrasts with green back., duller in fall. Ad ♀: duller version of ♂. Adults show more yellow on underparts in fall. 1st-yr: fairly uniform yellow-green (white undertail stands out). Wing bar. Worn and pale by spring.

♂ spring

1st-yr. ♀

♂ nonbr.

Orange-crowned Warbler *Oreothlypis celata* **OCWA** L 5in

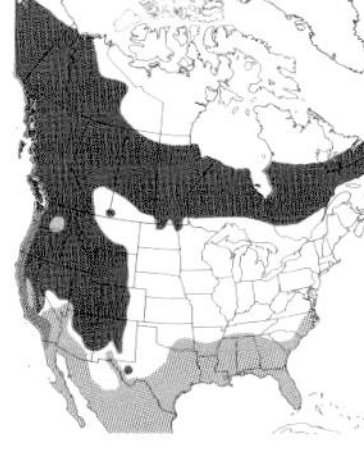

Widespread in the W, scarcer to the E. Breeds in small deciduous trees with shrubby undergrowth. A hardy bird with small numbers wintering in the cold NE. Found in overgrown field edges and tangles, its metallic *chip* note often the first sign of its presence. Often a loner, seemingly always on the move, invariably in dense tangles, vegetation blocking a clear view. Tends to stay lower than most other warblers. Rarely seen before Sep on fall migration. Song: a trill, lower notes at end. **ID**: Well proportioned with rounded crown and undercarriage. Best thought of as a fairly bland yellow-green bird with very subtle color variation, a dark eyeline, and pale supercilium. Faint diffuse breast streaking and yellow undertail coverts that are not found on similar TEWA. Upperparts are more uniformly colored. It is also subtly longer-tailed, shorter-winged with more rounded head. Slight plumage variation between ages and sexes. ♂: brighter than ♀, spring birds brighter than in fall/winter. Juv: similar to adults.

imm. ♀

ad. ♂

Nashville Warbler *Oreothylpis ruficapilla* **NAWA** L 4.75in

Fairly common breeder in a variety of habitats, both deciduous and mixed forest, particularly spruce bogs. These areas usually have dense understory. On migration, often seen in lower vegetation, second growth, and some in weedy fields. Active and fast moving, often at the tips of branches or on plants. Frequently pumps tail up and down with partial sideways movement. Sings a lot, a 2-part trill, second half lower. **ID**: A distinctive shape: nearly the smallest warbler, it appears the fattest with heavy round undercarriage accentuated by short narrow tail and big rounded head. Blue-gray head contrasts with olive upperparts, but it is the large, round eyering that really pops and is the striking feaure at any distance. From below is yellow but uniquely patterned: white on lower belly (around legs) bordered by yellow (a diagnostic feature used a lot when bird is directly above). Other warblers have bold eyerings, but none is so small and chunky. Ad ♂: brightest colors with yellow throat. Rufous cap difficult to see. Ad ♀/imm ♂: duller, often with pale throat. Imm ♀: dullest are olive with little contrast between head and back, underparts pale yellow but still white between legs.

ad. ♂ br.

1st-w. ♀

ad. ♀/1st-w. ♂

ad. late summer

Yellow Warbler *Dendroica petechia* **YWAR** L 5in

Very common in many areas and sometimes in backyards. Favors deciduous trees in open areas, woodland edges, and wet areas with willows. Usually head height or a bit higher, but also in treetops or low tangles. A solidly built warbler with average proportions and smooth contours: a good size and shape to use to compare all other warblers. First to leave, some heading s. by early Jul. Its song, *sweet sweet I'm so sweet*, is the sound of summer. Also gives a loud hard *chip* note. **ID**: Many shades of yellow: very slightly darker on the breast sides, greener above. Bold yellow edge to tail feathers is diagnostic and surprisingly striking. Yellow fringes to dark-centered wing feathers also stand out on a uniform bird. Diffuse eyering and black bill prominent on bland face. Ad ♂: bright yellow with red streaks in summer, mostly gone by fall. Ad ♀/1st-yr ♂: slightly paler than ad ♂ with thinner and less extensive breast streaks. 1st-yr ♀: dull, some with few yellow tones. Many ssp. The 'Golden' group, mostly in the W Indies, occurs in FL Keys. ♂ has bolder streaking, some ♀/imms are grayer, particularly on the head. 'Mangrove,' given full species status by some, breeds on S Padre Is., TX. Ad ♂ has chestnut head in summer.

Virginia's Warbler *Oreothlypis virginiae* **VIWA** L 4.75in

Very rare, mostly to the S and W. Very NAWA-like but smaller-headed and longer-tailed with exaggerated tail pumping. It is not yellow but pale gray with bold eyering and yellow patches on central breast and undertail coverts. Rufous crown patch and yellow reduced or absent on ♀/imm. Prefers scrubby vegetation.

Black-throated Gray Warbler *Dendroica nigrescens* **BTYW** L5in

Very rare visitor from the W, particularly fall. Like BTNW in behavior, structure, and call. Black-and-white color patterns more like BAWW. Bold face pattern with yellow spot in front of eye, usually dark throat and streaked flanks. Bold white wing bars and outer tail feathers on otherwise gray upperparts. All plumages have similar pattern, boldest in ad ♂ and weakest in imm ♀.

Kirtland's Warbler *Dendroica kirtlandii* **KIWA** L 5.75in

Rare breeder in young jack pine forest, mostly cen. MI. Increasing through conservation efforts (approx 3600 birds in 2008). Many recently discovered wintering on Eleuthera. Feeds mostly on sandy ground in scrubby habitat. Constantly pumps tail (also PAWA and PRAW). A large, chunky warbler with a long tail and big bill similar to PIWA, with bold eyelids to match. Upperparts boldly streaked black, white wing bars. Yellow underparts, white vent with heavily streaked flanks and spots across breast. Ad ♂: upperparts gray. Ad ♀/imm: brown mixed in upperparts. Song: a jumble of deep notes.

ad. nonbr. /1st-w. ♂
1st-w./ad. nonbr.
♂ br.
ad. ♀ br.

Chestnut-sided Warbler *Dendroica pensylvanica* **CSWA** L 5in

Common breeder in second-growth deciduous forest. Usually moves around mid-story and seems to have its tail perpetually cocked at 30 degrees. Appears stiff and not too agile—perhaps in part due to its posture. Emphatic and fast *pleased pleased pleased to meetcha* a common and welcoming song. **ID**: A fairly large, chunky warbler with a bulging chin that gives it a thick-headed look. Has a deep bulging belly as if a few months pregnant. Broad-based pale bill and broad wing bars. 2 distinct plumages. Ad ♂ br: yellow crown, black whisker stripe, chestnut 'sides,' black-and-gold upperparts create a very spiffy, unique look. Ad ♀ br: similar to ♂ with subdued pattern. Fall (ad nonbr and 1st-w.) birds all similar and also attractive. Uniquely lime-green upperparts that complement pale gray underparts. The bold white eyering is always the feature that stands out on such a clean background. Head-on, distinctively patterned with pale throat and lime cap that contrast with gray cheeks. Adults and 1st-w ♂ can show some chestnut on flanks.

ad. ♂ br.

1st-s. ♀

ad. ♀ nonbr./
1st-w. ♂

1st-s. ♂

1st-w. ♀

Magnolia Warbler *Dendroica magnolia* **MAWA** L 5in

Common breeder in second-growth deciduous and coniferous forest. Flits around lower vegetation, often flycatching and spreading tail, reminiscent of AMRE. Occasionally hovers. Stands horizontally with tail often cocked and spread. Song: somewhere between CSWA, AMRE, and HOWA, a weak but sweet *weeta weeta weechew.* **ID**: Slim with a long tail that often appears rounded. A beauty with striking color patterns. All birds can be instantly and easily identified by white undertail with broad black tip. Uppertail has a yellow rump and a white band when tail is spread. All have bright yellow underparts with white vent and that striking tail pattern! Ad ♂ br: bold black necklace, flank streaks, mantle, and mask through eye. Ad ♀ br: similar to ♂ with subdued pattern. Fall birds gray-headed with pale eyering. Streaking on underparts limited or absent. Most show pale gray necklace that seems out of place. Back mostly green with narrow white wing bars. Older birds and ♂ show more black streaking on underparts and mantle. When overhead, look for AMRE-like long rounded tail but with distinctive and easy-to-see undertail pattern.

Cape May Warbler *Dendroica tigrina* **CMWA** L 5in

Uncommon breeder in evergreen forest, particularly spruce. It is often at the top of the tree. On migration, uses many habitats, but favors cedars and spruce; often drinks nectar from flowering trees. Sometimes occurs in small groups. Varied and sometimes nondescript appearance may confuse. Song: short, a series of very high insect-like *seet* notes. On migration, often flies around in daytime giving buzzy *tzee* note. **ID**: Distintive shape. Fat, potbellied bird, neckless with a small head that looks pointed (often appears as an extension of the body). Bill is narrow, pointed and, importantly, slightly downcurved. Tail is short and narrow. All plumages have poorly defined eyestripe and supercilium with broken eyelids, a pattern shared by few other species. Underparts are very variable from off-white to yellow but always have streaks. One of only a few warblers with yellow-green on rump. Ad ♂ br: everything bold and bright; reddish ear coverts. Dull version in fall. ♀/imm ♂: duller with gray ear coverts, narrower wing bars. Imm ♀: sometimes confusing, drab gray and olive-toned with diffuse breast streaks and weak wing bars. Rump greener; but remember shape and head pattern!

Black-throated Green Warbler *Dendroica virens* **BTNW** L 5in

Common breeder in a wide variety of coniferous and mixed forest. Isolated se. birds in cypress swamps. Usually found in the mid- to upper story, often singing and feeding on insects from the end of branches. On migration, will feed lower, sometimes eating berries. Late fall birds often feed in weedy fields. Hover-gleans. Sometimes tame and approachable. Clicky call note is similar to W Coast, dark-throated warblers. Song: a buzzy *I'm so so layzee* with many variations but the same intonation. A relatively distinctive song and one of the commonest in many forests. **ID**: A stocky bird with a big head and friendly expression. This wing-barred warbler is always yellow-faced contrasting with olive crown and back. Within the yellow face is an indistinct eyestripe and dull ear-covert patch. Thick black streaks on breast sides. Yellow wash across vent, easy to miss unless you are looking for it. Ad ♂ br: extensive black on underparts. Ad ♂ fall/ad ♀/imm ♂: paler breast streaking and paler around throat. Imm ♀: underparts tinged yellow with few streaks and palest ear coverts. 1st-s: subdued pattern of adult with retained worn juv coverts.

Golden-cheeked Warbler *Dendroica chrysoparia* **GCWA** L 5in

Uncommon and localized around the Edwards Plateau, TX, in oak-/juniper-clad hills. Very early migrant arriving on breeding grounds in Mar. A close cousin of BTNW and easy to overlook as a BTNW. Song is deeper and slower. In all plumages is darker-backed. Has a cleaner yellow face with bolder eyestripe. ♀: duller with yellowish throat and dark olive back with black streaks. 1st-yr: as ♀ but lacks black streaking in upperparts. BTNW is scarce migrant in GCWA's range. Always focus on the face and back coloration. If in doubt, GCWA lacks yellow vent of BTNW.

Townsend's Warbler *Dendroica townsendi* **TOWA** L 5in

Rare visitor from the W, usually late fall. Prefers tall coniferous trees but found anywhere in cold weather Similar to BTNW in size, shape, and call, with smaller head and shorter tail. Most easily separated by dark ear coverts accentuated by yellow surround and dark crown. More extensive yellow on underparts and well-defined flank streaking. Often dark-centered upperpart feathers.

Hermit Warbler *Dendroica occidentalis* **HEWA** L 5in

Very rare visitor from the W. Usually gray-backed with unstreaked underparts and yellow surrounding ear coverts.,Ad ♂: yellow-faced, dark-throated. Imm ♀: dull with pale throat. Ad ♀ and 1st-yr ♂: intermediate. Hybridizes with TOWA; the offspring have yellow tones to breast and upperparts and streaks on breast sides. Example shown less obvious than many.

♂ Audubon's

ad. transitional

♀ Audubon's

ad. ♂ nonbr.

ad. ♂ br.

ad. winter

1st-w.

Yellow-rumped Warbler *Dendroica coronata* YRWA (MYWA/AUWA) L 5.5in

'Myrtle,' the n. and e. race of YRWA, is an abundant breeder in a variety of coniferous and mixed forest. W. race, 'Audubon's,' uncommon on w. edge of region, very rare in the E. Only winter warbler in many areas. MYWA arrives suddenly in Oct, after most other warblers have gone, often in thousands. Loves bayberry bushes but adaptable and tough, finding food on the ground or in treetops. Often flycatches for insects. Frequently in loose groups. Soft warbling song. Calls a lot, a hard *chep*. AUWA's song is slower and deeper. Call: distinct, softer and less emphatic *chup*. In flight, tail rounded at corners, slightly forked in center. **ID**: Large and solidly built with fairly long tail. Always has bright yellow rump. Ad br: bright yellow breast patches bordered by heavy black streaks on white underparts. Black cheeks, white throat. ♀ duller than ♂. Nonbr: a much browner bird with diffusely streaked buff underparts, white eyelids, indistinct supercilium. Ageing and sexing difficult. Nonbr AUWA is browner with yellow or yellow-buff throat (often with white border). Head uniform brown, emphasizing bold eyelids and throat. More white in tail. Breeding: bold yellow throat. Hybridizes where ranges overlap.

1st -s.
ad.
ad. ♂

Yellow-throated Warbler *Dendroica dominica* **YTWA** L 5.5in

Uncommon breeding bird in a variety of woodland habitats. In pine woods in the E, cypress swamps with Spanish moss in the S, and sycamores to the W. In winter, fairly common around suburbs with palm trees, often creeping around limbs like BAWW, and gets right inside palm clusters to root out insects with its long bill. In summer, stays higher in the canopy, moving more slowly and methodically than a lot of other warblers. Beautiful, sweet song with clear notes trailing off at end (CARW qualities). Early migrant in spring and fall. Occasionally tries to winter in the N. **ID**: A large flat-headed warbler with a very narrow tail. Massive bill really stands out: very long, though easy to overlook among the complexities of such a striking head pattern. All plumages similar, with bright yellow throat bordered by black that extends from ear coverts and onto streaked flanks. White eyelid, supercilium, and rear cheek. Uniform gray upperparts with wing bars. White outer tail feathers conspicuous in flight. !st-yr ♀: narrowest flank streaks and least black on head but differences slight. Race in the W, *albilora*, has white supraloral (yellow in e. birds) and shorter bill, but much variation and subspecific ID is tricky.

♂ nonbr.

ad. ♀ br.

1st-w. ♀

ad. nonbr.

ad. nonbr.

ad. ♂ br.

Blackburnian Warbler *Dendroica fusca* **BLBW** L 5in

Fairly common spruce-bog breeder, scarcer in mixed forest. In Appalachians, often breeds in deciduous forests. Migrants found anywhere but partial to evergreens. Breeding ♂ 'firethroats' always get a big "aah" from the crowd. Unfortunately they are often at the very tops of trees and rarely down low. Listen for the song, a series of accelerating high-pitched see notes, distinctive in the way it goes off the scale. Commoner in the W in spring and E in fall. Fairly early fall migrant. **ID**: A medium-sized warbler that often seems small. Has a long attenuated body that often makes the tail look short. Thick neck helps to give it a pointed front end. All plumages have dark triangular ear coverts, pointed at rear and bordered by yellow/orange. Throat is bright, flanks streaked, and has broad white wing bars. Particularly bold stripes to upperparts. Ad ♂ br: superb 'on fire' throat, white wing patches on mostly black back. Ad ♂ fall: similar to ♀/1st-yr with slightly brighter throat, 2 broad wing bars. Ad ♀ br: paler throat than ♂, diffuse flank streaking and narrower wing bars. Fall ♀ and 1st-yr: similar. Palest yellow mostly young ♀.

1st-w. ♀
1st-w. ♂
ad. ♀
ad. ♂ br.

Cerulean Warbler *Dendroica cerulea* **CERW** L 4.75in

Uncommon and local in the tallest deciduous trees usually near wet areas or rivers, some on ridge tops. Another warbler that loves the heights, it rarely comes low and usually only when offspring have just fledged. Song is series of buzzy notes in 2–3 parts, getting higher, somewhere between BTBW and NOPA in tone. **ID**: A medium-sized warbler with average proportions but a big belly. Ad ♂ : beautiful blue upperparts. Dark breast-band and bold flank streaking usually catch the eye. Ad ♀/1st-yr ♂: green upperparts with blue tinge. Diffuse flank streaking with yellow wash extending across breast. 1st-yr ♂: usually has bolder streaking on breast sides. 1st-yr ♀: dullest with yellow wash over most of underparts. 1st-yr BLBW very similar. Both are usually seen from below and share dark ear coverts, 2 bold wing bars, and pale underparts with flank streaks in most plumages. BLBW has more distinct ear coverts, brighter throat, darker wings/upperparts—a more boldly marked bird. An underrated feature is the under-tail pattern of CERW: white with dark tips. These appear as 2 spots in flight. BLBW has white outer tail feathers.

eastern br.

eastern nonbr.

western br.

western nonbr.

Palm Warbler *Dendroica palmarum* **PAWA (WPWA/YPWA)** L 5.5in

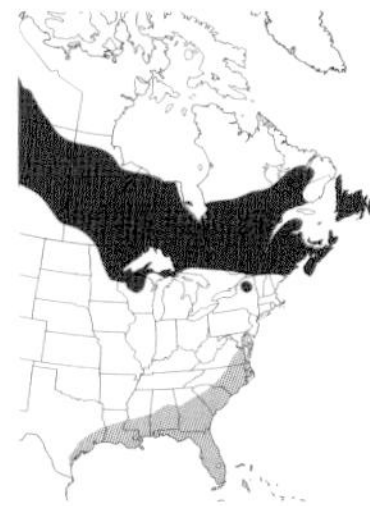

Common ground nester in spruce bogs, preferring open areas with small trees. On migration and wintering grounds, likes open weedy and grassy areas. This constant tail pumper feeds on the ground, frequently flying to fenceposts and bushes before dropping down again. Often forms small loose groups. Can seem friendly, even tame, and is found around humans away from breeding grounds. Migrates early in spring and late in fall, when usually in different habitat from other warblers. Calls *chek* a lot. Song: a buzzy trill. **ID**: A long, slim warbler with long legs and horizontal stance. Head is broad with a domed crown. Most striking feature is yellow undertail coverts. Green rump contrasts with browner back. All have a dark eyestripe, long downcurved supercilium, and diffusely streaked underparts. 2 groups: e. 'yellow' birds (YPWA), w. 'brown' birds (WPWA). E. birds separable by yellow underparts with rufous streaks and cap. By fall they are duller but with yellow tones to underparts and supercilium. W. brown birds duller and grayer. Yellow is mostly restricted to throat in summer. Occurs commonly in fall in the E and in winter in the SE. Ages and sexes similar.

ad. ♀/1st-w. ♂

ad. ♂

1st-w. ♀

Prairie Warbler *Dendroica discolor* **PRAW** L 4.75in

Fairly common in field edges, brushy areas, and mangroves. Often at head height, it feeds lower than many other warblers but not on the ground. One of 3 species (KIWA, PAWA) that always pumps tail, revealing its white outer tail feathers. Often the only warbler left singing in the midday sun—high-pitched buzzy zee notes steadily rising in pitch. Moves deliberately as though making sure it has enough time to pump tail. **ID**: A lightweight warbler with a slim feel due in large part to its long tail. Smooth lines, domed head, and pot-belly. Focus on distinctive face pattern: pale line above eye and dark arc below it. All have yellow underparts with streaked flanks, uniform upperparts with weak wing bars. Ad ♂ br: boldly marked with rufous on back. In fall, colors subdued and rufous often difficult to see. Ad ♀/ 1st-yr ♂: duller version of ♂. 1st-yr ♀: browner upperparts, paler yellow underparts with weak streaks.

1st-w. ♀

ad. ♀/1st-w. ♂

ad. ♂

Pine Warbler *Dendroica pinus* **PIWA** L 5.5in

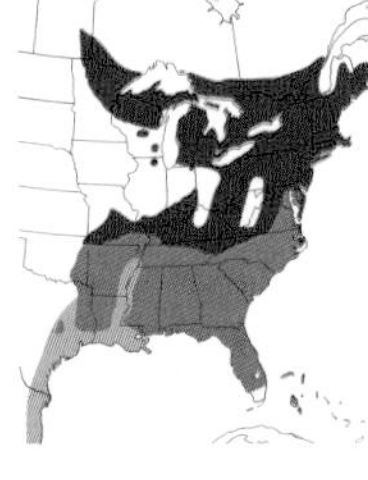

Found almost exclusively in pines, where it is often the commonest warbler. Feeds primarily on seeds. On migration can be found anywhere and will feed on the ground, often with PAWA. An early migrant in spring, late in fall. Wintering birds often in mixed flocks with warblers, bluebirds, and CHSP in parks, golf courses, and other fairly open areas. Song: a trill, with nicer tone than tinny CHSP. Frequently gives a hard *chip* note. **ID**: Variably colored and can be particularly nondescript and confusing. A large, sturdy warbler heaviest in the belly and with a long tail, often with a distinct fork (it recalls CHSP in flight). Large-billed, thick-necked, and round-headed. All plumages show broken eyering and pale lores on an otherwise bland face. Pale area behind ear coverts. Also bold wing bars, unstreaked back, and diffuse streaks on breast sides. Short primary projection and wing point. Ad ♂: yellow and green with faint streaking. Ad ♀/1st-yr ♂: duller yellow on breast and back mixed with brown. 1st-yr ♀: pale brown. ID from BBWA and BLPW straightforward: shape, head pattern, color, and always confirm with unstreaked mantle if in doubt. From drab CMWA by bold wing bars and shape.

Bay-breasted Warbler *Dendroica castanea* **BBWA** L 5.5in

Uncommon to scarce spruce-bog breeder in the tallest trees. Also prefers spruce on migration but found elsewhere. Often found with CMWA and BLPW. Call: a hard *zeet* or *zick* (similar to YEWA, BLPW, COWA—'the zickers'). Song: a few high-pitched *zee* notes, similar to CMWA and a short BAWW. **ID**: A large, powerful warbler, like BLPW with slightly rounder crown and broader-based bill. All plumages have large wing bars, streaked mantle, rufous wash on flanks, and buff wash to undertail coverts. Ad ♂ br: uniquely colored brown cap, throat and breast sides with black mask. Ad ♀ br: duller version of ♂. In fall, all birds have dark eyeline, pale supercilium, and pale area behind ear coverts. Rufous flanks are darkest in ad ♂ and palest in 1st-yr ♀. These pale birds are similar to pale BLPW but buffy rather than white on flanks/undertail coverts. They also lack strong yellow tones or diffuse streaks on breast, have greener upperparts with fewer streaks, dark feet, and broader wing bars. PIWA has different shape, and face and color pattern.

Blackpoll Warbler *Dendroica striata* **BLPW** L 5.5in

Very common breeder in coniferous forest. Migrants found anywhere. A late migrant spring and fall. At times the commonest warbler; on spring migration it can be heard in suburbs, or just about anywhere, singing a simple insect-like *tsee tsee tsee tsee tsee*. **ID**: Large, sleek, long-winged, an athlete designed to fly long distances over water. Plumages variable but all have long bold wing bars, long primary projection, green or gray upperparts with dark streaks, and white undertail coverts. Ad ♂ br: white cheek, black cap, whisker, and breast streaking—distinctive. Ad ♀ br: variable, with underparts from white to yellow with fine streaks. Upperparts gray to olive with dark streaks. 1st-s: similar to adult with some imm feathers. All fall birds similar with mostly olive upperparts, yellow wash on breast/throat with diffuse streaks. Ad ♂: has boldest black streaks on upperparts and fine streaks on flanks. 1st-yr ♀: the palest. Nearly all have orange or yellowish feet, often extending up legs.

1st-w.

ad. ♂ br.

ad. ♀/imm. ♂

ad. ♂ nonbr.

Black-and-white Warbler *Mniotilta varia* **BAWW** L 5.25in

Common and widespread breeder and migrant in deciduous and mixed forest. Scarce in winter. Distinctive behavior: creeps on or under branches, sometimes upside down, like a black-and-white mouse. On migration, joins feeding flocks that will often include several other warblers. **ID**: A medium-sized warbler with short tail, heavy belly, long, downcurved bill, and pointed head—a distinctive profile. Large feet for grasping branches. From below, black undertail coverts are a giveaway. Despite large variation in plumages, the bold black-and-white striped head pattern and coloring make it an easy warbler to id. Ad ♂ br: solid black cheeks, throat, and flank streaking. Throat and ear coverts paler toward fall. Ad ♀/1st-yr ♂: paler than ad ♂ with white ear coverts and throat. Flanks white or buff with narrower streaks. 1st-w ♀: paler and more buff. 1st-yr birds have brown greater primary coverts contrasting with black greater coverts.

Black-throated Blue Warbler *Dendroica caerulescens* **BTBW** L 5.25in

Fairly common in deciduous woods, usually in damp areas under the canopy. In late fall, often eats pokeberries. Tends to move slowly, not far, and often just appears in front of you at head height. Song: a slow buzzy *zi zee zee zee zee* ("*I am so lay-zee*"). Call: a distictive hard *tek*, junco-like. **ID**: A fairly large and robust, neckless warbler. Young birds are essentially the same as respective-sex adults. All have white square patch at base of primaries, reduced in younger birds, particularly the ♀. This is striking in flight, showing as a wing bar on already pale underwing. Ad ♂: striking black around face and flanks contrasts with blue upperparts and white underparts. 1st-yr ♂: green tips to upperparts, sometimes pale flecking in throat. Southern/Appalachian population can show dark mantle streaking, and mantle lacks green in youngsters. Ad ♀: dull green above, dirty buff below. Always look for distinctive face pattern: dark ear covers set off by pale border with pale eyelid. Pale base to primaries can be hidden or nearly absent, but diagnostic when present.

American Redstart *Setophaga ruticilla* **AMRE** L 5.25in

Very common breeder in deciduous forest. Often the commonest migrant, particularly in early fall. Always active, often fluttering 'butterfly-like' as it drops down to catch insects. Short broad bill, rictal bristles, and long tail that is fanned. Flits from branch to branch, leaning forward and rotating from side to side. Often partially opens wings and spreads tail to reveal patches of yellow or orange. Then it's off again. Song: variable, usually a number of sweet high-pitched notes trailing off at the end. **ID**: Striking shape and behavior. Bold yellow or orange patches at base of outer tail and flight feathers. All birds share the same color pattern except ad ♂. Ad ♂: black, orange, and white year-round. Ad ♀: underparts pale with breast sides yellow to orange. Gray head contrasts with browner back. Broad yellow bases to flight feathers. 1st-w ♂: overlaps with ♀ but often brighter orange breast sides and stronger contrast between head and back. In spring has black adult-like feathers on head and neck. 1st-yr ♀: palest breast sides, subtly duller above with narrow wing bar. The shape and color patterns make this an easy bird to id overhead and so provides a great starting point for learning warblers in flight.

ad. ♂ br.

ad. nonbr.

ad. ♀ br.

Prothonotary Warbler *Protonotaria citrea* **PROW** L 5.5in

Common in s. swampy woods, almost always near water, standing or moving. Early migrant spring and fall. Belts out its song: *tswit tswit tswit tswit* from mosquito-ridden areas, louder and more emphatic than other warblers. Usually found higher in smaller trees but often comes lower, occasionally dropping to the ground. It will hop slowly on logs, apparently careful to look in every hole. A cavity nester that will also use nest boxes. Moves more slowly than many warblers, almost vireo-like at times. **ID**: An oddly shaped warbler. Big and very fat with a potbelly and short tail. The bill is large and pointed—this stands out, just like its beady eye, against a background of uniform yellow. One color pattern: yellow with a green back and gray wings. Contrastingly white undertail tipped black, very prominent on such a yellow bird. Ad ♂ brightest, 1st-year ♀ dullest, ad ♀/1st-yr ♂ in between.

Worm-eating Warbler *Helmitheros vermivora* **WEWA** L 5.25in

Fairly common in a variety of deciduous forest. An early migrant like most s. breeders, it tends to stay in denser mid-level vegetation. Loves clumps of dead leaves where it snouts around for dinner. Very quiet and unobtrusive it suddenly appears from out of nowhere. On migration, often, but not always, joins mixed flocks. Song: a dry trill, more insect- than bird-like. Even CHSP's voice is fuller-bodied. **ID**: A large thick-set warbler with a broad tail and pointed head, enhanced in part by its large pointed bill. One plumage: khaki-olive above, tawny breast, duller on belly, and has dark-centered undertail coverts. However, focus on the striped head pattern. 1st-w: shows rusty fringes to tertials with very good views

Swainson's Warbler *Limnothlypis swainsonii* SWWA L 5.5in

Uncommon in swampy s. woodland or hardwoods with lots of leaf litter and a dense understory. Population in Appalachians found in rhododendron and laurel thickets. Regular overshoot to the N in spring, sometimes sticking around to sing for a while. Because of its inhospitable environment and skulky habits, feeding on the ground by turning over leaves, it is one of the more sought-after species. Sometimes listening for rustling leaves as it feeds is the best way to locate it. On migration, crawling in thickets in places such as the FL Keys would probably turn up more birds than expected. Sings from higher perches, moving only periodically, and can be tough to spot. Song similar to LOWA, a series of loud slurred, quick-ending notes (LOWA ends in several twittered notes). Loud strident *chip*. **ID**: A large potbellied warbler with a short tail and peaked crown. Long pointed bill. One plumage: drab brown above, buff below, slightly paler on throat and vent. Always focus on the brighter chestnut cap set off by pale supercilium and dark eyeline. Only vaguely similar warbler is WEWA, but latter has a strikingly different head pattern. Beware of CARW sneaking about!

Ovenbird *Seiurus aurocapilla* **OVEN** L 6in

Very common in deciduous woodland, though much harder to see than hear. Spends most of its time on the ground, particularly if there is lots of leaf litter. Mostly walks, or rather struts, on the floor and occasionally on limbs. Arguably no other bird has as much character as this punk rocker. Moving with exaggerated steps, the cocked tail is flicked upward while the head goes backward and forward as though the bird is tiptoeing through the forest. Usually stays in cover. Explosive song: *teacher teacher teacher* often surrounds you, but bird is nowhere to be seen. Also gives a loud deep *chup* note. With patience you can usually spot this thrush-like bird. The big eye is watching you. It will sometimes spike its crown as if to say "Who the heck are you?" **ID**: Large and really plump with long pink legs, short tail, and rounded head. One plumage: white underparts with bold black streaks, but it is the striking eyering that is hard to miss. Black lateral and chestnut crown stripe most obvious when head-on. 1st-w: rust-fringed tertials.

Northern Waterthrush *Parkesia noveboracensis* **NOWA** L 6in

Common breeder along streams and in wet wooded areas. On migration, also found in drier areas with leaf litter but still mostly on the ground. A ground dweller that walks (sometimes runs) purposefully, always bobbing its tail in rhythm. Stops to turn leaves over or pick at food. When disturbed, jumps on thick branches, bobs its body up and down while its tail is flicked up. When feeding, seems methodical, leaning forward carefully as if making sure it has seen everything. Song: a loud series of accelerating notes falling in pitch at the end. Call: a loud *chink*. **ID**: A large warbler with a small head, broad square-ended tail, and heavily streaked underparts. Streaks usually blend into each other forming rows on a generally buff background. However, some birds are white and far more visually striking. The bold supercilium is usually the same color as the breast, very long, and straight or downcurved behind the eye. Note that it stays the same thickness or narrows behind the eye. Upperparts uniform dark brown. 1st-w: rusty fringes to tertials. In flight, waterthrushes have noticeably dark underwings.

Louisiana Waterthrush *Parkesia motacilla* **LOWA** L 6in

Fairly common breeder along wooded streams. Also found in damp areas on migration, though look early in fall as most have left the NE by mid-Aug. Arrives in Mar. Behavior as NOWA. Song: a pleasing long series of slurred notes ending in a twitter. Call note flatter than NOWA. **ID**: Very similar and often confused with NOWA. LOWA is slightly larger and more potbellied bird with its center of gravity back by the legs. This, and a bolder white supercilium that broadens behind the eye, are often enough to id birds. Supporting features are: streaks on underparts noticeably larger, paler brown, more triangular and isolated (tend to form unbroken lines in NOWA) so stand out on white background. These streaks form fewer vertical rows (roughly 8 as opposed to 10 in NOWA). The flanks are usually buff (this field mark is often hard to see as is buff supraloral). Proportionally smaller and flatter head emphasizes larger bill which is longer and broader. Legs often brighter pink and sturdier. Upperparts subtly paler brown. Usually has an unstreaked throat, NOWA is occasionally unstreaked. Lots of things to look at, but some birds still cause head scratching!

Kentucky Warbler *Oporornis formosus* **KEWA** L 5.25in

Fairly common ground dweller in s. deciduous woodland with a thick understory, usually in wet areas. Typical of *Oporornis*, you see it well for a few seconds and then it does its magic act and disappears into thin air. Spends most time on the ground, hopping rather than walking, on long thick legs that are designed for the job. It tosses leaves looking for food. Body is held tilted forward with tail in the air. Typical of its genus, it sings from the same high spot in the tree for periods of time, always partially hidden, throwing its voice, and therefore making it tough to find. Tends to be a loner, its deep *chup* note or song the best way to find it. Song: similar to CARW but faster, a loud *cheery cheery cheery*. **ID**: Fairly large with a bulky body, short tail, and small head. Pink legs. The yellow-green plumage and striking bold yellow spectacles bordered by black are the only things you need to see to id this bird. Black ear coverts extend down neck sides. Olive above, completely yellow bellow. *Oporornis* species all have long yellow undertail coverts. All plumages similar. Ad ♂ averages more extensively black crown and ear coverts, 1st-yr ♀ the least; ad ♀/imm ♂ intermediate.

Connecticut Warbler *Oporornis agilis* **CONW** L 5.75in

Uncommon breeder in a variety of deciduous and coniferous forests. W. migration route in spring, e. in fall. Many fly long distances at sea. A much sought-after ground-loving species, that walks rather than hops. Stays hidden in tangles and field edges on migration. Habitat-specific, often in fields with ragweed—and lots of COYE; CONW is the bigger bird with strong flight and it doesn't drag its tail. Freezes briefly, then walks off, not to be seen again. On the breeding grounds blasts out its song from sparsely leaved trees, its round body the shape of the leaves. Throws its voice and invariably avoids detection until it moves. Explosive rolling song accelerates in pace. **ID**: Large and potbellied. Short tail (emphasizing long undertail coverts), thick neck, long wings for flight to S America. Long pink legs for much walking, and large bill. Besides shape, always start with bold complete white eyering that makes eye stand out. All have yellow underparts, olive upperparts with a hooded look. Adult hood is gray, some ♂s have black on lower border. 1st-w: browner hood with unbroken lower border and pale or buffy throat. Beware of variation in MOWA eyerings by double-checking bird's shape.

Mourning Warbler *Oporornis philadelphia* **MOWA** L 5.25in

Fairly common in secondary growth in mixed forest. Like other *Oporornis*, usually stays low, hopping rather than walking. Migrates late in spring and early in fall. Rolling song: *churry churry churry churry cheri cheri*, accelerating at the end, is a common sound in movie backgrounds. Call: a snappy *twik*. **ID**: Similar in color patterns to CONW. MOWA is slimmer and more stretched out with a longer tail and shorter wings. The primary projection is shorter (it does not make long-distance flights and does not undertake overhead daytime flight on migration). Importantly, the eyering is narrower and broken; these eye-arcs give it a smaller-eyed appearance. Underparts are generally darker yellow. Ad ♂: more extensive black on bib and dark lores. ♀/1st-yr: subtly paler supraloral spot and eyeline (CONW is plain-faced) with yellow throat. The hood is grayer in all ages, lacking brown earth tones of some CONW. COYE is paler-bellied with contrastingly yellow throat and undertail coverts, lacking uniformity of MOWA. Also COYE is shorter, smaller-bodied with a longer spiky tail.

MacGillivray's Warbler *Oporornis tolmiei* **MGWA** L 5.25in

Very rare visitor from the W in late fall after MOWA has gone. Similar to MOWA in behavior and appearance but slighter-bodied with longer tail. In all plumages, shows bolder eyelids, longer tail projection, past undertail coverts, and ♀/1st-yr are paler-throated. Call: a sharp *chick*.

Painted Redstart *Myioborus pictus* **PARE** L 5.75in

Very rare late fall and winter in the S. Plumages similar. Large and slim with a long tail. Typical redstart-like behavior, flitting around in its beautiful color patterns. Very loud and distinctive siskin-like *chu-wee* call.

Rufous-capped Warbler
Basileuterus rufifrons **RCWA** L 5.25in

Very rare to s. TX in drier brushy areas. Low or on the ground, it wafts its tail like a gnatcatcher. Distinctly marked face pattern.

Golden-crowned Warbler
Basileuterus culicivorus **GCRW** L 5in

Very rare to s. TX in woodland. Somewhat CAWA-like in shape and color patterns with tail often cocked. Always study the striped head pattern with broken eyering, and dark lateral crown stripes with yellow between.

Gray-crowned Yellowthroat
Geothlypis poliocephala **GCYE** L 5.5in

Very rare in s. TX. Similar to COYE but larger, longer-tailed, with big pale downcurved culmen. Bolder broken eyelids. Black lores (darker on ♂). Distinctive *peetaloo* call (never calls like COYE).

Common Yellowthroat *Geothlypis trichas* **COYE** L 5in

Very common breeder in marshes, also in bushy areas, overgrown fields, and tangles. On migration, loves dense weedy areas with lots of ragweed; can be common enough to flush in bunches. Sneaks around low, staying well hidden, rarely climbing much above head height. Though often hard to see, it is approachable. Seems reluctant to fly off, perhaps because it's a weak flyer on short rounded wings, and has a long tail that it drags behind, held slightly down—a familiar image on migration. Perched birds often hold tail raised. Hops rather than walks, movements jerky. Sings frequently, loud for such a small bird, a distinctive *witchety witchety witchety*, usually given from the top of a bush or reed. Breeding birds perform a high display flight, partially hovering and tail pumping as they sing. Nasal *chip* note. **ID**: Fat-bellied with a short neck but longish rounded tail that often looks spiky; distinctive. Importantly, all birds have yellow throat and undertail coverts contrasting with dull breast and belly, a unique combination. Ad ♂: striking black mask bordered above by gray. Ad ♀: olive-faced. 1st-yr ♂: initially like female, black flecks on face molted in through fall. 1st-yr ♀: drabber and paler yellow.

Hooded Warbler *Wilsonia citrina* **HOWA** L 5.25in

Fairly common breeder in a variety of habitats with dense understory, usually near water. Common in s. swamps. Although it will sing from high, it spends most of the time flitting through lower vegetation, and is quite happy jumping around fallen branches. Often flicks its tail open as it goes, as if to show off white outer tail feathers. Early migrant in spring and fall. Song: similar to MAWA and CSWA—*haweet haweet haweet haweetya*. Call a nice *tink*. **ID**: A heavy-bellied thick-set warbler that often looks slim because of its long but broad tail. Green upperparts and yellow upperparts. Ad ♂: bright yellow face masked in black can be simply stunning in nice light. 1st-year ♂: as ad ♂ with pale tips to black feathers sometimes visible. Ad ♀: variable amounts of black bordering yellow face. 1st-yr ♀: face bordered by olive (no black). Female WIWA has a different head pattern, shape, and behavior. Also lacks white in outer tail.

ad. ♂

♀

1st-w. ♀

Wilson's Warbler *Wilsonia pusilla* **WIWA** L 4.75in

Uncommon breeder in thickets, bushy areas, and other dense low vegetation, particularly near water. On migration, also stays in more open areas, rarely inside forest. A hyperactive warbler, always on the move, hopping on eye-level vegetation. Characteristically quickly flicks its long tail up high as it leans forward, even more exaggerated than in CAWA and much like BGGN. Song: a fast chattery trill, usually accelerating at end. **ID**: Really small, round, and neckless but may look bigger because of its disproportionately long tail. All plumages green above and bright yellow bellow. It has a soft expression, the large black eye prominent on such a featureless face. ID is straightforward when you can see the black cap. Ad ♂: solid black cap. Ad ♀/1st-yr ♂: reduced black cap, variable in size. 1st-yr ♀: little or no black in crown. Capless birds are similar to 1st-yr ♀ HOWA. WIWA is much smaller-bodied but longer-tailed (HOWA fans tail but does not flick it up), lacks white outer tail feathers, and its green ear coverts extend onto the neck.

♀

ad. ♂

1st-w. ♂

Canada Warbler *Wilsonia canadensis* **CAWA** L 5.25in

Fairly common in dense understory in mixed woods, often near water. Moves quietly, seems to always be peering forward, head stretched in a horizontal position before suddenly flicking its tail upright at great speed. Never lets you get too close, and the vegetation is usually too thick to follow. Song: a jumble of twittery notes with a 'friendly' quality. Call: a tangy *tlup*. **ID**: Long and slim with a lengthy tail. In all plumages has uniform gray upperparts, yellow underparts with contrastingly white undertail coverts. The dark necklace, bold eyering, and spectacles stand out. Ad $\male$ br: striking black necklace and head markings. Fall birds have reduced black, replaced by blue-gray. Ad $\female$: little or no black on head; paler streaks on breast. 1st-yr $\male$: intermediate between adult $\male$ and $\female$. 1st-yr $\female$: the palest, with only a few faint dull streaks across neck.

Yellow-breasted Chat *Icteria virens* **YBCH** L 7.5in

Uncommon but widespread in fields with hedges and brushy cover. Hardy, a few winter as far n. as Canada. Unlike any other warbler, a skulky bruiser. In spring, often sits out and slowly sputters its song of repeated notes. Frequently contorts body, sometimes leaning forward as if to scold you. In courtship flight, will fly into the air with exaggerated wingbeats, chest pumped out, before returning to base. Most of the time it skulks at head height in cover. Often sits quietly or moves slowly to avoid drawing attention; always a loner. Song: a wildly varied series of varying harsh notes, interspersed with chucks, squeaks, and clucks. **ID**: Massive—is it really a warbler? Thick bill and long tail. Brilliant yellow chest, white spectacles, and dark lores really catch the eye. All plumages similar with females averaging duller and paler-lored. 1st-yr: often has paler bill. W. populations average larger white malars and less olive upperparts.

1st-s. ♂

juv.

ad. ♂ (northern prairies)

ad. ♀ (East)

ad. ♂ (East)

Horned Lark *Eremophila alpestris* **HOLA** L 7.25in

Common and widespread. Many populations exist here, and throughout the world, with poorly understood subspecific variation. Breeds on prairies, short grass, dirt fields, airports, and tundra. In winter, forms large flocks, sometimes joining longspurs and SNBU. Feeds on roadsides and snow-laden arable lands, moving s. in bad weather. Stands hunched with legs bent, the walk more of a shuffle. Sings a lot, sometimes at night, a couple of notes followed by a sound like the jangling of keys. Call: a somewhat pipit-like *see-tew*. **ID**: Holds body horizontal, and is long-winged and slim with small head. All birds (except juvs) have dark breastbands and mustaches. Underparts are primarily white, often with yellow on the face. Upperparts are lighter, usually with a pinker rump and nape. Ad ♂: Devil-like horns. Boldly marked. Ad ♀: subdued pattern of ♂ with more uniform upperparts. Juv: scaly patterned, and can be confusingly different from older birds. E. and n. populations generally have more breast streaking than those in Interior.

ad. br.

ad. nonbr.

American Pipit *Anthus rubescens* **AMPI** L 6.5in

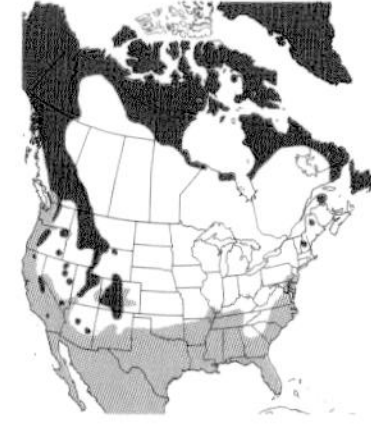

A common rocky tundra breeder. Formerly called Water Pipit, Buff-bellied Pipit in Europe. In winter, forms flocks, sometimes in the hundreds, on bare earth or short-grass farmland, beaches, and wet muddy areas. Walks and sometimes runs around feeding actively and will freeze when disturbed. Head jerks backward and forward as it moves. Pumps tail up and down when standing. Will 'up and away' considerable distances for no apparent reason, flocks staying fairly close together. Call: *psip* or *psip-it*. **ID**: Highly variable. A slim pipit with fairly long tail; pale-based narrow bill. All plumages have bold eyering, supercilium, and malar with uniform ear coverts but no eyestripe. White outer tail feathers. Upperparts pale gray-brown with diffuse streaks. Leg color variable, most black but some red or pink, particularly in winter. Ad br: underparts buff or orange-buff with streaking across breast and lightly down flanks. Upperparts paleish gray with diffuse streaks. Nonbr: underparts variably buff, some mostly off-white, with extensive bold streaking on breast and flanks. Upperparts brown with diffuse streaks.

Sprague's Pipit *Anthus spragueii* **SPPI** L 6.5in

Uncommon in grass prairies. Territorial ♂ uses bare patches on hilltops. Sings from high in sky, a cascading series of *shirl* notes, before dropping like a stone to the ground. In winter, found singly or in loose groups on damp grassy areas. Hard to find. Call: a *squeet*. **ID**: Fat-bodied and short-tailed, holds its chest forward with small head erect, looking around like UPSA. Large pale bill. Pale area around eye gives blank look. Broad wing bars. Breast streaking mostly restricted to upper breast, creating a visible band in flight. Boldly streaked back with extensive white on outer tail feathers. Doesn't bob tail.

Smith's Longspur *Calcarius pictus* **SMLO** L 6.25in

Uncommon breeder on damp open tundra. Winters in ankle-high grass. Rare vagrant farther E. On migration, look for large '1-ft-high' stubble fields with wet patches and lots of fox grass. Feeds quietly, tending to walk away rather than fly. Quiet rattled call in flight. Much more white in tail (like VESP) than LALO. Look for this when bird is flushed. **ID**: First impression is of a brown bird with bright orange-buff underparts. Look for pale spot in ear coverts and bold eyering to confirm ID. Other features include finely streaked breast, fairly long primary projection, and thin bill. Ad ♂ br: striking black-and-white head pattern, orange underparts. White lesser coverts. Ad ♂ nonbr/ad ♀/1st-yr: all similar.

McCown's Longspur *Rhynchophanes mccownii* **MCLO** L 6in

Uncommon in dry barren prairie. Winters in dry, short grass and in dirt fields, often with HOLA's. Usually found in summer by ♂'s superb parachuting display flight with wings held stiffly in 'V'. **ID**: Ad ♂: striking. Ad ♂ nonbr/ad ♀/1st-yr: all similar. First impression is of a pale bird with a bland head pattern. Differences from similar CCLO include 'T' tail pattern, long primary projection and wing point, and broad-based, usually pink bill. Median coverts are rufous in ♂, brown with buff fringes in ♀, CCLO has dark centers with pale fringes. Call: a dry rattle.

Chestnut-collared Longspur *Calcarius ornatus* **CCLO** L 6in

Fairly common in areas with long grass. Rare to the E. Stays hidden and is easiest to see perched on fences or at water holes. Call: a 2–3 note *kiddle*, also a rattle. Song: meadowlark-like but softer. **ID**: Ad ♂ br: striking. Ad ♂ nonbr/ad ♀/1st-yr: highly variable. Many have chestnut on nape similar to LALO, but have different breast pattern and lack rufous in wing. Duller birds from similar MCLO by dark rear ear covert patches, dark-centered median coverts, triangular dark center to tail, narrower base to bill, longer tail, shorter primary projection and wing point, and more heavily marked underparts.

♀ nonbr.

♂

ad. ♂ br.

ad. ♀ br.

Lapland Longspur *Calcarius lapponicus* **LALO** L 6.25in

Fairly common and widespread tundra breeder. Winters in arable fields, grass pastures, and upper beaches. The only regular E Coast longspur. Frequently mixes with larger flocks of SNBU and HOLA. Look for the smaller dark bird. At other times will be alone or in small flocks. Call: a dry rattle like other longspurs but has mellow *tew* notes mixed in. **ID**: A compact bird with fairly long tail. Long wings and the longest primary projection among longspurs. All plumages have rufous nape, greater coverts, and tertials. Underparts also different: dark markings on breast and flanks contrasting with white belly. Head pattern is bold, emphasized by dark-bordered ear coverts with dark patches at rear corners. White outer tail feathers. Ad ♂: bold face pattern lost in winter. Flank streaking and nape duller. Broad flank streaks. Ad ♀: duller version of ♂. 1st-yr: like ♀ but often with buffier underparts.

Snow Bunting *Plectrophenax nivalis* **SNBU** L 6.75in

Fairly common tundra breeder. Winters in arable fields, grass pastures, dune grass and on large beaches. Occurs in tight flocks from a few birds up to 100, sometimes with HOLA and LALO. Crouches to feed but then walks fast on short legs and can be deceptively tough to see on snow, sand, or earth. Often takes off in unison, a beautifully patterned mosaic of brown and white, and will move to another area only to return 10 minutes later. In flight, gives a soft chatter with *zrrt* notes (like NRWS). Also mellow *tew* like LALO. **ID**: Large-bodied with small head. Ad br ♂: solid black bill and upperparts. Ad ♀ br: like ♂ with dark markings on head and on back. Ad nonbr/1st-yr: similar when sitting, with warm brown breast patches, cap, ear coverts, and back (with black streaks). Bill mostly yellow-orange. In flight, age and sex sometimes become clearer. Ad ♂ has extensive white in upperwing, imm ♀ the least. By late winter, wear starts to produce darker, more constrasty, breeding plumage.

Bananaquit *Coereba flaveola* **BANA** L 4.5in

Very rare in s. FL. Small and chunky with a short tail, but always look for the downcurved bill. Adults are gaudy with striking supercilium, yellow rump, and yellow breast bordered by white. White patch at base of primaries. Juv.: very subdued version of adult. Nectar feeder. Call: a warbler-like *tsip*.

White-collared Seedeater *Sporophila torqueola* **WCSE** L 4.5in

Very rare resident along the Rio Grande river, TX, in long grasses or reeds. Occurs in small flocks. Tiny with stubby bill and paddle-shaped tail. Buffy underparts with pale neck sides and 2 wing bars. ♂: black-capped with dark upperparts. ♀: like ♀ INBU.

Yellow-faced Grassquit *Tiaris olivaceus* **YFGR** L 4.3in

Very rare in s. FL and s. TX. Feeds on or close to ground on grass seeds. Tiny but fat with a big head and fairly short tail. Olive above, grayer below with a black conical bill. ♂: unmistakable black-and-yellow face. ♀: nondescript. Some show just pale eyelids, others a trace of ♂'s face pattern.

Black-faced Grassquit *Tiaris bicolor* **BFGR** L 4.5in

Very rare in s. FL scrub. Size and shape as YFGR. ♂: black, paler on belly than YFGR with olive back. ♀: trickier! Like YFGR, but paler, grayer with uniform face pattern.

ad. ♀

ad. ♂

Eastern Towhee *Pipilo erythrophthalmus* **EATO** L 8.5in

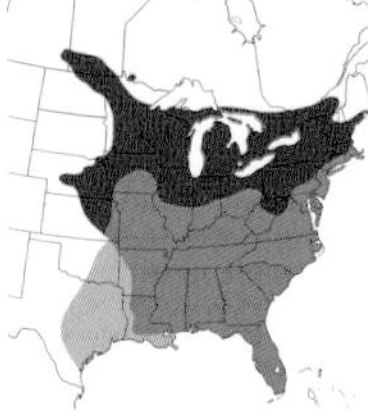

Uncommon in secondary-growth woodland and brushy scrub. Often on the ground in leaf litter, can be heard scratching around, and will jump forward with both feet and send leaves and dirt flying backward. Prefers to stay hidden and is more often heard than seen, calling *tow-hee*. Sings a higher *drink-your-tee*, the last part a higher-pitched trill. Formerly conspecific with SPTO. Occasionally hybridizes where ranges overlap. **ID**: Large, bigger than sparrows and other birds with similar behaviors. Chunky and thick-necked with a long rounded tail. All plumages have same color pattern with dark upperparts and throat, rufous flanks, and white belly. White corners to tail and bases to primaries are easiest to see in flight. Extensive white on undertail. Ad ♂: upperparts and throat black. Eye color red in the N. S. birds frequently have white eyes and less white in wing and tail. Ad ♀: black replaced by brown. 1st-yr: almost as adult.

1st-w. ♂
ad. ♀
ad. ♂

Spotted Towhee *Pipilo maculatus* **SPTO** L 8.5in

Uncommon in the W. Very rare winter visitor to the E. Formerly conspecific with EATO. Very similar and sometimes hybridizes. Most striking difference from EATO is heavily spotted and streaked upperparts, boldest on the wings. However, SPTO does not have a white patch at base of primaries. ♀: darker and grayer. As towhees are often heard before seen, listen for harsher and buzzier call, *zareeee*. Hybrids show intermediate characters.

Olive Sparrow *Arremonops rufivirgatus* **OLSP** L 6.25in

Common in the Lower Rio Grande Valley, TX, in dense wood thickets and scrubby grass areas. Shy, staying on or near the ground; creeps around with tail cocked. Listen for rustling leaf litter. Best found by listening for its buzzy *tzeee* call. Also gives a short *tsip*. Song: a series of accelerating notes trailing off at end. **ID**: Distinctive. Fairly large and well proportioned. Simple color pattern, olive above, grayer on the head and below, but always look for brown-striped head.

juv.
1st-w. ♀
ad. ♀/1st-w. ♂
ad. ♂

Dark-eyed Junco *Junco hyemalis* **DEJU (SCJU/ORJU/PSJU/GHJU)** L 6.25in

Distinct populations: Slate-colored, Oregon, Pink-sided, and Gray-headed. Slate-colored is a common breeder in mixed and evergreen forest, often at edges. In winter, forms large flocks, sometimes with other sparrows, warblers, or bluebirds, in areas with short grass and trees. Feeds mostly on the ground, moving slowly, usually in tight groups. Easily flushed, usually into trees—always shows striking white outer tail feathers in flight. Gives a quiet but hard *tik* call. Song: usually given from a tree, is a trill, like CHSP, but softer and more musical. **ID**: Round-bodied, neckless, small-headed with a gentle expression. Obvious long tail emphasized by bold white outer tail feathers. Ad ♂: striking. 1st-yr ♀: brownest and palest. Ad ♀, 1st-yr ♂, intermediate: often difficult to age and sex. Some have narrow white wing bars. PSJU and ORJU, scarce winter visitors in the W, rare in the E. GHJU very rare. PSJU from 1st-yr ♀ DEJU by defined lower border to hood, dark lores, pink flanks. ORJU dark hooded, also with well-defined lower border, brown back, salmon flanks. Races interbreed.

Green-tailed Towhee *Pipilo chlorurus* **GTTO** L 7.25in

Uncommon winter visitor in the SW, very rare in the E. Another skulker that stays on or near the ground in dense shrubbery. Often visits feeders, particularly out-of-range birds. **ID**: A distinctive small sparrow-like towhee with a long tail and small head, often with peaked crown. Mostly olive and gray like smaller OLSP. Always check the head pattern: rufous cap and white throat divided by black malar. Cap is slightly duller in winter with gray tips to feathers. 1st-yr: usually duller head pattern with smaller crown. Call: a cat-like *mewee*.

Canyon Towhee *Melazone fusca* **CANT** L 9in

Uncommon in brushy habitat and rocky canyons, but often near humans. Usually low in bushes or scratching around on the ground. Quite happy to walk in the open; occurs near houses and can be quite tame. **ID**: Fairly large with long tail; crown feathers often raised. Buff around throat, but otherwise colder gray tones. Markings can be subtle, but always look for chestnut cap and undertail coverts. Bold eyering, fine streaks bordering throat, and spot in center of breast that is sometimes easy to miss. Call: a *de-yup* and variety of squeaks.

Cassin's Sparrow *Peucaea cassinii* CASP L 6in

Fairly common in suitable habitat: tall grass with bushes or cactus. Very rare visitor farther e., mostly fall. Very difficult to see, except when singing from perch or in skylarking display flight, usually 4 whistles or trills at different pitches: *please come with me*. **ID**: Large and fat with a long rounded tail and small-headed appearance emphasized by narrow-based bill. First impression is of a bird with very complex upperparts; try describing each feather's pattern! Underparts with subtle buff breastband of fine streaks, gray belly, and pale throat with malar. Flanks with bolder streaks. Upperparts have well-defined pale fringes, small complex black internal markings, and relatively dark ear coverts.

Botteri's Sparrow *Peucaea botterii* BOSP L 6in

Scarce in s. TX, in similar habitat to CASP, often near the coast. Behavior also similar though doesn't skylark. **ID**: Similar to CASP in size and shape but with flatter-headed look, in large part due to larger bill. Overall impression is of a plain sparrow, rufous above with black streaks, gray and buff below. Differences from CASP include uniform underparts without streaking, upperparts without broad pale fringes to feathers, darker relatively unstreaked crown, and streaked uppertail coverts (anchors in CASP). Has indistinct eyering, throat lacks dark malar, and has more uniform underparts. Song: a series of notes accelerating like a bouncing ball with a couple of grace notes before and after.

Bachman's Sparrow *Peucaea aestivalis* **BACS** L 6in

Scarce in open pine forest with scrub palmetto and a thick grass understory. Usually found by listening for its beautiful, variably pitched, whistled trills. Call: a weak *tsip*, often given when disturbed. In winter it disappears into wetter and more open areas, where it is very hard to find. Similar to CASP and BOSP though ranges don't overlap. **ID**: A bulky sparrow with large body and long rounded tail. Boldly patterned. Diffuse streaks on buff breast contrast strongly with white belly. Dark eyeline and crown emphasize gray supercilium. Boldly patterned upperparts.

Rufous-crowned Sparrow *Aimophila ruficeps* **RCSP** L 6in

Scarce on rocky or grassy slopes with scattered bushes. Feeds on the ground, particularly scree slopes, occasionally sitting on perches, mostly to sing its fast twittered song. Often forms small groups in winter. **ID**: Typical *Aimophila* shape with long bill. Bold head pattern on an otherwise bland bird. Rufous cap and eyestripe on gray face. Broad black lateral throat stripe contrasts with white malar. Bold eyering. Underparts uniformly gray.

American Tree Sparrow *Spizella arborea* **ATSP** L 6.25in

Fairly common. Breeds in scrub. A tough sparrow with many wintering in snowy landscapes in weedy fields, hedges, and roadsides and at feeders. Found on or close to the ground in small flocks. In areas of deep snow, concentrations occur wherever they can find grass heads and other food. Often jumps off ground to grab seeds from overhanging grasses. In some locations it is the only regular sparrow. Call: a *teedle eet*. **ID**: Large-bodied with a long tail and rounded head. It has a cute friendly expression. Gray-headed with bold chestnut cap and eyeline. Boldly patterned upperparts with 2 wing bars. Underparts strongly buff in breast sides with a rufous patch near elbow, but it is the dark spot in the center of the breast that is distinctive. Bicolored bill. FISP can cause confusion where range overlaps, but latter shows less contrasting plumage, has all-pink bill, and lacks breast spot. CHSP is duller brown above with different head pattern and grayer breast lacking breast spot.

Field Sparrow *Spizella pusilla* **FISP** L 5.75in

A fairly common sparrow in overgrown fields, second-growth woodland edges, and any open area with bushes and other cover. Perches up on brambles and weedy stems but can then disappear for long periods. Will often sit quietly, low in bushes when can be hard to find. In winter, and on migration, forms flocks, sometimes with other species but most often not. Feeds on seed heads, and spends more time on the ground in winter. Familiar song: long clear notes accelerating like a bouncing ball, repeated at different pitches. Weak *tsee* call note. **ID**: A distinctively slim, long-tailed sparrow, though can look chunky at times. Plain-faced look emphasized by bold eyering and pale lores. Unstreaked underparts. Overall color usually appears buff, but some birds are grayer. Lacks strong contrast of ATSP and CHSP. If in doubt, always look for the all-pink bill.

juv.

1st-w. /nonbr.

ad. br.

Chipping Sparrow *Spizella passerina* **CHSP** L 5.5in

Fairly common breeder in parks, gardens, forest edges, and open areas near trees. Its mechanical dry trill is a good song to learn. Changes behavior in winter. Forms flocks, often with similar CCSP, DEJU, EABL, PIWA, and other species. Usually feeds on the ground but looks for cover in trees. In flight, tail is flared at the tip and distinctly forked. **ID**: Different plumages and widespread distribution make this a good bird to learn well. Small, fairly long-tailed with a strongly forked tail tip. Stocky, thick-necked but small-headed with a flat crown—but the long tail makes it appear slimmer. Strong contrast in upperpart markings. Bill fairly small and pointed. Importantly, all plumages have dark lores and gray rump. Underparts have grayer tones in summer, browner in winter. Ad br: bold chestnut cap and black eyeline. Ad nonbr: browner cheeks with reduced eyeline, streaked cap with rufous, nape grayer. 1st-w: similar to adult nonbr but has duller brown cap and buffier underparts. Juv (late summer): streaked breast. CCSP and BRSP are similar; focus on bolder color patterns and dark lores.

ad. br

1st-w. /nonbr.

Clay-colored Sparrow *Spizella pallida* **CCSP** L 5.5in

Fairly common in tall grass and brushy fields with bushes. In winter, joins flocks with other sparrows. Regular on the E Coast in fall. In summer, habitat different from similar CHSP. Song: several buzzes at one pitch. Call: a weak *tsee*. **ID**: Similar to CHSP and BRSP. From CHSP in breeding plumage by habitat, grayer upperparts, black crown lacking rufous and lateral cheek stripe. Nonbr/1st-w: buffier, paler, and less contrasting appearance overall including brown rump. Always confirm ID with pale loral area. Pale gray nape stands out but is not diagnostic.

Brewer's Sparrow *Spizella breweri* **BRSP** L 5.5in

Rare winter visitor and migrant. Often with very similar CCSP and CHSP. Slighter and longer-tailed with higher crown that often seems spiked. A gray, finely marked bird. Focus on bold eyering, pale lores, small bill, and finely streaked mantle, crown, and nape—plain-looking compared to CCSP and CHSP.

Vesper Sparrow *Pooecetes gramineus* **VESP** L 6.25in

Fairly common breeder, scarcer in the E, in grassland, long or short, but usually away from shrubs. In winter, prefers drier and barer areas than many other sparrows. Usually feeds on the ground, sometimes in small groups away from others. Has a shuffling gait, sometimes hopping, at other times running. Seems to walk with feet back and chest thrust forward. Commonly sings from barbed wire fences and posts, a long rich jumble of notes descending at the end. **ID**: Large and particularly fat with a large head and fairly short tail. A real chunker! Often seen in flight first, spreads its tail making sure you see the striking white outer tail feathers. Distinctive head pattern highlighted by bold eyering. Also large bill, and pale-centered dark ear coverts. Rufous lesser coverts not usually visible. Color and amount of streaking highly variable, so focus on the above.

1st-w.

ad.

Lark Sparrow *Chondestes grammacus* **LASP** L 6.5in

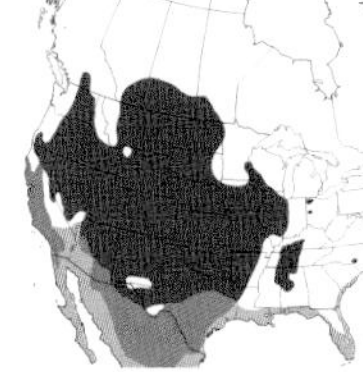

Fairly common in the W, regular fall migrant in the E, commonest in Aug and Sep. Usually found on the ground in open areas such as lawns, fields, parks, and often near trees and shrubs. Also sits on fences, utility wires, or treetops. In the midday sun small groups usually sit quietly inside trees, when much easier to overlook. In winter, forms small flocks, sometimes mixing with other sparrows. Has a horizontal gait as it shuffles along the ground. In flight, look for the striking white tail pattern. Call: a *tsip*. Song: usually 2 loud notes followed by a concoction of sweet notes and melodious trills. **ID**: A large, muscular but sleek sparrow. Long tail appears more rounded than it is due to pattern of white in outer tail feathers. Head often looks pointed, in part due to the conical bill. Bold head pattern is the key ID feature. Plain underparts have a dark breast spot. Fall migrants in 1st-w plumage sometimes have duller head pattern, darker breast sides, and retained juv streaking; others have nearly fully molted into adult-like plumage.

Black-throated Sparrow *Amphispiza bilineata* **BTSP** L 5.5in

Uncommon in arid rocky countryside with cactus or bushes. Very rare in the E. Usually on the ground feeding, often in pairs, but perches on bushes or rocks. Can be quite tame. Song: 2 clear notes followed by a trill. **ID**: Face pattern and large black bib make ID straightforward. Juv: plumage often held into fall and can confuse: look for streaked breast, long bold supercilium, and dark lateral cheek stripe. Briefly held juv plumage lacks black throat and is streaked on the breast.

ad. ♂ nonbr.
ad. ♂ br.
ad. ♀
1st-w.

Lark Bunting *Calamospiza melanocorys* **LARB** L 7in

Fairly common breeder on open prairie, particularly sagebrush. In winter, forms tight, sometimes large, flocks, often in agricultural areas. Very rare in the E. Commonly seen sitting on fences. ♂s sing in display flight. Song: mockingbird-like batches of trills and rattles. **ID**: Distinctively large and big-bellied. Head is blocky and has a large conical bill. Large white wing patches obvious in flight. Ad br ♂: solid black-and-white plumage in summer, but a patchwork of brown and black in spring and fall. Nonbr: ♀-like with black primaries and darker throat/face. ♀: brown, usually with white throat.

ad.

ad. Ipswich

Savannah Sparrow *Passerculus sandwichensis* **SAVS** L 5.5in

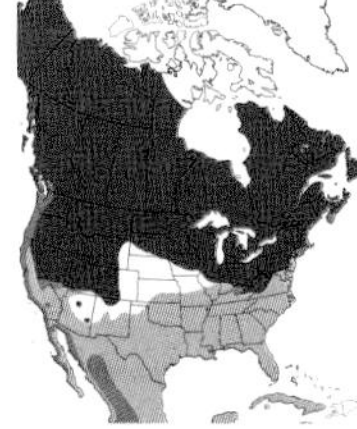

Very common in grassy areas year-round. Mostly on the ground, but perches on small bushes and posts. Often found in small loose groups. Commonly heard is a thin *tsip* call. Song: accelerating chips leading to a buzz: *seee yer*. **ID**: Much subspecific variation in color and streaking, so really learn its distinctive size and shape. Small, chunky, and neckless with a short, narrow, forked tail. Head often slightly crested. Always a crisply marked bird. Underparts are streaked on breast and flanks contrasting with white belly. Streaks vary from black to brown and are usually buff-fringed on the flanks. Malar region is strongly patterned. Upperparts are neat with a striped appearance, sometimes with pale lines down the back. They vary from gray to brown but always have broad rufous fringes to greater coverts. Supercilium is often yellow at the front, a great ID feature when birds show it. Many races. 'Ipswich Sparrow,' nests on Sable Is, NS, winters down the E Coast in sand dunes. Distinctively large and pale, with blurry brown streaks but easy to overlook as SOSP. The latter is larger than all races of SAVS with longer and broader tail rounded at tip; also broader and more extensive breast and mantle streaking.

Grasshopper Sparrow *Ammodramus savannarum* **GRSP** L 5in

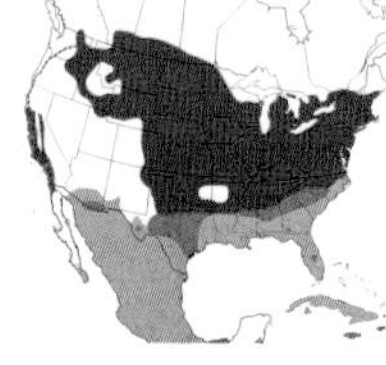

Fairly common in grassland, with or without bushes, overgrown fields, and palmetto scrub (FL). Rarely in wetter areas. Often sits up on grass stems or wire fences to sing a couple of notes followed by a mechanical insect-like trill. Sometimes a jumble of notes. Usually secretive, staying well hidden in grasses, where seeing it is nearly impossible. If flushed, it flies like SAVS but with weaker wingbeats, showing spiky tail and different color patterns. From other *Ammodramus* sparrows primarily by color patterns. On migration can show up on lawns and other short-grass areas. It shuffles on crouched legs rather than walks. Often approachable. Usually alone. **ID**: Typical *Ammodramus* sparrow, secretive, small and with large head, spiky tail, and intricately patterned upperparts. Large bill, and lack of forehead, give head a pointed look. Always look for unstreaked orange-buff breast and flanks. Bold eyering really stands out on a mostly orange face with gray supercilium. Black-spotted upperparts are beautifully marked with rufous and gray. LESP has large bill with streaking on sides of breast and flanks. Orange supercilium and gray ear coverts, reverse of GRSP

Baird's Sparrow *Ammodramus bairdii* **BAIS** L 5.5in

Scarce and local breeder in native prairie grassland with bushes and weeds. Secretive and very hard to see. Fairly big with a long bill and relatively long, square-ended tail. General appearance is a sparrow with an ochre-buff head, white belly, and isolated streaks across breast. No eyeline but double spot at rear of ear coverts. Very similar to HESP in all patterns, but is paler, lacks stronger green tones around head, upperparts are less strongly marked. SAVS common in same habitat, but shape, behavior, and breast-streaking pattern are very different. Song: deep clear notes with a descending trill.

Henslow's Sparrow *Ammodramus henslowii* **HESP** L 5in

Scarce in grass fields with lots of weeds, also reclaimed strip mines. Winters in wet weedy areas and understory of pine woods. Another skulker that will freeze or walk away rather than fly. Most easily seen singing its weak song, a *tss-zlik*, when often sits on stems but not fully in the open. Sings until Aug.

ID: Large head with broad-based bill, strong lime-green tones around face (when seen well). Boldly scaled upperparts with rufous wings. Underparts have buff breastband and extensive bold black streaks. White eyering. Double spot at rear of ear coverts. Can look surprisingly like SOSP at distance.

Le Conte's Sparrow *Ammodramus leconteii* **LCSP** L 5in

Uncommon in wet grassy meadows and marsh edges. Song: a single note followed by a forced-out *tszzzzzz*, flatter than SAVS. Hard to see, when flushed flies a short distance and then runs away or, if you are lucky, perches on a bush. Small and lightly built with a fairly long spiky tail; this and orange face pattern can be seen in flight. Sitting bird has a bright orange-buff face and crisply patterned appearance. Back and head appear striped. Easily confused with NESP, but latter lacks chestnut nape spots, pale crown stripe, strongly patterned back, and pale face with clean black eyeline.

Sharp-tailed Sparrow *Ammodramus caudacutus* **STSP** L 5.25in

Formerly conspecific with NESP. Uncommon in coastal salt and brackish marshes. Often found with NESP and commoner SESP. Larger than NESP with flatter head and tougher expression. The bigger, longer bill is best feature to start with. Head pattern is bold: gray ear coverts surrounded by deep orange. Breast is contrastingly pale. Bold streaks across breast and well-defined back pattern all stand out, compared with muted pattern of NESP. SESP is larger, longer-billed, and grayer. Some, mostly juvs, have orange tones, but these are not much stronger on face than body like STSP.

ad. *subvirgatus*

ad. *nelsoni/alterus*

Nelson's Sparrow *Ammodramus nelsoni* NESP L 5in

Fairly common in wet meadows and coastal marshes. Formerly conspecific with similar STSP. Secretive, putting its head up to sing its squeezed-out *p-shhhhh*. LCSP is often found in the same marshes, so look carefully. Call: a *tsee*, softer than STSP. In winter, found in coastal marshes with other 'marsh sparrows' in denser vegetation. Pishing will often bring them to the top of reeds. At low tide will creep around the muddy edges of creeks. At high tide you should look for them on the highest ground. **ID**: 3 populations: Interior (*nelsoni*), Hudson Bay (*alterus*), and coastal (*subvirgatus*). Interior population brightest with most contrasting plumage, and is only race to winter on the Gulf as well as Atlantic Coast. Coastal birds are the dullest. Hudson Bay population overlaps with both, Id from STSP by smaller size, more rounded head, and, most importantly, smaller bill that gives a gentler appearance overall. Head and breast are fairly uniform orange-buff with a well-defined breast-band and blurry streaks. Back has diffuse streaking.

Seaside Sparrow *Ammodramus maritimus* **SESP** L 6in

Fairly common in coastal marshes. Behaves like other marsh sparrows. Sits on bushes or reeds, usually in the cooler hours of summer. Often flies short distances across the marsh, sometimes being chased by another. Low flight is weak, with rounded wings and short spiky tail dragged along. Spends a lot of time feeding on the ground on exposed mud and gets pushed to high ground at high tide. Sings a lot, day and night, a RWBL-like *teep-zherrrrrr*. Call: a hard *tup*. **ID:** Large chunky body, big head, large bill, short spiky tail. Bigger than other marsh sparrows and much dingier. Yellow supraloral and white malar create a 'quarter moon' on the face. In winter, has lines of buff patches on the underparts. By summer, the feathers are a worn uniform gray and brown. Juv: buff breast with streaks, recalling STSP, but still has quarter-moon pattern and lacks orange surround to ear coverts. Gulf Coast birds average bolder streaks. Rare Cape Sable race (*mirabilis*)in s. FL has distinct streaking on white underparts. All show substantial variation in plumage.

Fox Sparrow *Passerella iliaca* **FOSP** L 7in

Breeds in dense bushy woodland edges and brushy taiga. Arrives in the S in Nov, when found in gardens, field edges, and hedgerows. A hardy bird, it will belt out its loud melodious song from deep cover on even the coldest winter day. Feeds on the ground in leaf litter or anywhere there is bare earth when it is snowy. Visits feeders and is happy to join in mixed flocks. Digs holes, kicking soil backward with both feet like a towhee. Typically stays well hidden. Song: a series of explosive warbled notes unlike other sparrows. Call: a distinctive loud *tchak*, most similar to BRTH. **ID**: Easily the largest and most robust sparrow with a thick neck and medium-length tail. Its size and rich colors are more thrush- or towhee-like. The overall rusty red complexion is nicely complemented by gray in the face and back. The breast streaks, on close inspection, are more arrow-like, and are concentrated in the center of the breast, often appearing as a single spot. Juv: averages slightly duller brown than adult, but there is much overlap.

juv.

ad.

Song Sparrow *Melospiza melodia* **SOSP** L 6.75in

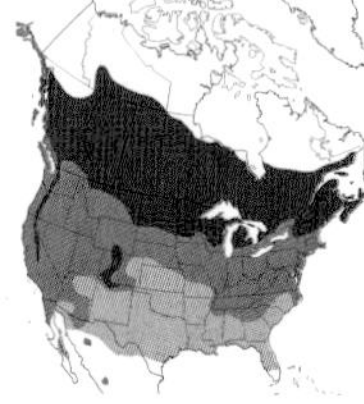

The common streaked sparrow in many areas. Occurs in a variety of habitats from gardens and forest edges to weedy fields. A familiar bird for most of us and often approachable. Sits on posts or bushes as it sings, a cheery range of up-and-down trills. Never flies far, seeming to move slowly and methodically through vegetation or on the ground, often giving its *chimp* note. On migration, will form groups with other sparrows. It is then that you can see the great variation within the many populations. **ID**: A large sparrow with a fairly long tail that is rounded at the tip. Not so potbellied as some, the chest seems to melt into the neck. All plumages have thick streaks, usually a warm brown matched by the upperpart color. The dark malar and central breast spot stand out. Head is rounded and quite small, an array of lines; the gray supercilium is the one that stands out. Often appears brown-capped. Upperparts have broader and more diffuse streaks than similar sparrows. Wings and tail sometimes more rufous. LISP is smaller, slimmer, grayer and with finer streaking.

Lincoln's Sparrow *Melospiza lincolnii* **LISP** L 5.75in

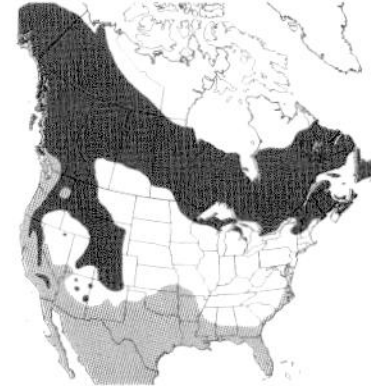

Fairly common in the W, scarcer to the E. Breeds in wet thickets, but found in drier habitat—usually with grass, bushes, and brush—the rest of the year. Secretive, tending to creep around in cover or fly off to hide. Feeds on the ground, moving slowly, usually while partially hidden. Generally alone, it invariably flies to bushes for cover, sitting quietly or carefully moving to thicker parts. Sometimes looks back and raises its crest as if to show it is worth following. Song: a series of full-bodied musical trills, the longest in the middle. Call: an emphatic *chip*. **ID**: Superficially like SOSP, but smaller and slighter. Rather cute with a peaked crown, small bill, and distinct eyering. A bird of fine streaks, on crown, nape and, most importantly, breast; actually just about everywhere if you look closely. Mostly a dull gray-and-brown bird, which makes subtle buff breast and flanks stand out. Like SOSP and SAVS, has a dark central breast spot. Briefly held juv plumage more similar to SOSP; focus on size and shape.

ad. br.

1st-w.

ad. nonbr.

Swamp Sparrow *Melospiza georgiana* **SWSP** L 5.75in

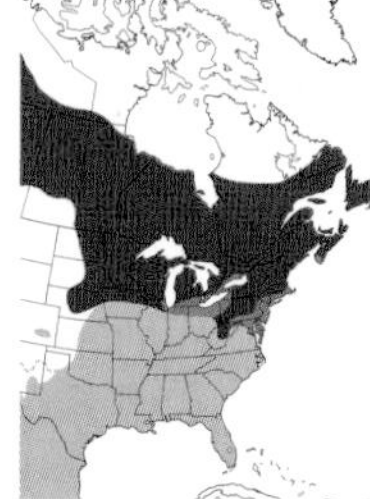

A fairly common sparrow that breeds in swamps and wet marshy woods. In winter and on migration, found in weedy fields, brush, and other dense vegetation, sometimes in large numbers. Usually secretive, walks on the ground. A 'pish' will often bring a bagful to the surface in what looked like an empty field. Often hangs from plant stems, contorting its body into any position to get seeds: much more agile than most sparrows. Tail is often cocked. Weak flight with rounded wings and spiked tail dragged behind. Song: a rattled trill. *Chip* call note like EAPH. **ID**: A really plump sparrow that looks neckless. Short-winged with a spiky tail. Bold rufous wings are the most prominent feature in all plumages. All plumages also have a capped appearance with rufous, and flared black eyeline with bold gray supercilium and hindneck bordering ear coverts. Underparts usually gray with brown flanks. Ad br: rufous cap and wings on a mostly gray bird. Nonbr: brown ear coverts and flanks; gray supercilium and streaked cap. 1st-w: as nonbr with little or no rufous in crown, duller supercilium and malar but often with more breast and flank streaking.

1st-w.

ad. br white-striped

ad. tan-striped

White-throated Sparrow *Zonotrichia albicollis* **WTSP** L 6.75in

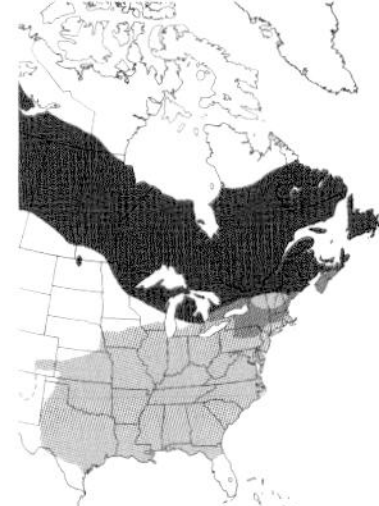

Breeds in mixed forest and bogs. Common winter visitor to parks, gardens, and woodland edges with undergrowth. Like other *Zonotrichia*, can sing in winter, slow clear whistles: *so wee wee wee wee wee*. Call: a sharp *chink*. Found in flocks, often at feeders. Feeds on the ground or low in the undergrowth, and quick to move to denser cover. **ID**: A heavy-bellied sparrow with a thick neck, domed crown, and fairly long tail. Upperparts appear mostly rufous-brown with large black smudgy streaks but has a uniform rump. 2 wing bars. Underparts are always browner on the flanks, palest on belly, and grayer across breast, often with smudgy streaks. The throat ranges from striking white to dingy buff and usually has a black border. Supercilium varies from dark tan to gleaming white, variably yellow above the lores. Color patterns are highly variable. Usually treated as 2 morphs: white-striped and tan-striped, however, better thought of as clinal with many intermediate. Further complicated by ♂ averaging brighter than ♀. Nonbr: duller, with white-striped birds becoming more like tan-striped. Brighter after spring molt. 1st-w: dull and similar to dull nonbr but usually with heavier streaking on underparts.

1st-s.

1st-w. *leucophrys*

1st-w. *gambelii*

ad. *leucophrys*

ad. *gambelii*

White-crowned Sparrow *Zonotrichia leucophrys* WCSP L 7in

Fairly common in the W, scarcer to the E. Found in hedgerows, brush piles, open areas with cover, feeders, and farmland. Usually feeds on the ground, near cover, for a quick getaway. Forms small groups, but will mix with other sparrows. Song: a clear whistle followed by several notes, fastest in the middle: *see, see what you can do with this*. Call: a metallic *chink*. **ID**: Slim with a long tail and small head. Crown feathers often raised, creating peak on rear crown. Adult: bold zebra-like head pattern set against gray neck and underparts. Beautiful and hard to miss. Back dark brown with contrasting flakes of buff or pale gray. Uniform rump and 2 wing bars. Wings usually appear rufous. Colors slightly duller in winter. 1st-w: a brown bird, appears relatively uniform with brown cap, has a paler crown stripe with good views. E. population (*leucophrys*) has pink bill and dark lores. Birds in w. of region (*gambelii*) have pale lores and orange bills. Many intergrades, and judging these characters in the field is often not easy.

Golden-crowned Sparrow *Zonotrichia atricapilla* **GCSP** L 7.25in

Very rare winter visitor, sometimes at feeders. Usually in dense vegetation on or close to the ground. Similar but larger and duller than 1st-w WCSP with dingier underparts. Appears grayer-headed in winter with a very small brown cap. Forehead is yellowish. Bill is duller. Ad nonbr: shows more black on crown. Breeding birds are boldly marked.

Harris's Sparrow *Zonotrichia querula* **HASP** L 7.5in

Uncommon in the Great Plains, very rare visitor to the E, often at feeders in winter. Found in hedgerows and brush in open areas, often with other *Zonotrichia* sparrows. Song: like WTSP, at one pitch. **ID**: Largest and heaviest sparrow, its bold head pattern stands out. Pink bill and white underparts give it a cleaner look than other sparrows. Adult: gray-headed with solid black crown and bib. Nonbr: brown-headed with patchy black on crown and bill. Shows much variation and often difficult to separate from 1st-w, which typically has whiter throat bordered by black malar stripe.

juv. ♂ transitional

ad. ♂ nonbr.

juv.

ad. ♀

ad. ♂ br.

House Sparrow *Passer domesticus* **HOSP** L 6.75in

Very common and familiar bird around humans as its scientific name suggests. A European import, unpopular with many, but tough and adaptable. The likeliest bird to encounter where you get your morning coffee, in the farmyard, the mall parking lot, nesting in the eaves of the nearby building (any cavity), or at your feeder. Often tame, usually in small groups feeding on the ground or sunning themselves on a hedge. Several chirping notes, sometimes repeated constantly. Flight powerful and straight with fast steady wingbeats. **ID**: Robust and thick-set with a broad tail. Dull unstreaked underparts, boldly streaked upperparts, and short wing bars. Sexes different. Ad ♂ br: nicely marked with gray crown, black bill, mostly brown upperparts (including wing coverts), pale cheeks, and black bib. Nonbr ♂: a much duller version of same pattern with paler bill. ♀: pale dingy brown, indistinct supercilium contrasts with darker cap and eyestripe. Pale bill. Juv: as ♀ with buffier underparts and fringes to coverts, soft-textured feathers, and rounded tips to flight feathers. 1st-w: molts into adult-like plumage through fall.

Brambling *Fringilla montifringilla* **BRAM** L 6.25in

Old-World vagrant. Very rare at n. feeders during winter. Large attractive finch with a striking white rump. Plumages vary but all have strong doses of orange on breast, wing bars, and scapulars. Gray-headed in winter with yellow bill, both black in summer. Feeds in trees and on ground, often with sparrows.

Gray-crowned Rosy-Finch *Leucosticte tephrocotis* **GCRF** L 6.25in

Very rare winter visitor from w. mountains, often at feeders in winter. Superficially dark brown with a gray cap, dark forehead, and yellow bill. Birds in breeding plumage have black bills. With closer views note pink mixed in underparts and wing.

Cassin's Finch *Carpodacus cassinii* **CAFI** L 6.75in

Very rare fall and winter visitor to the W. From very similar PUFI by larger size with more elongated look, slimmer-necked and flatter-headed with longer bill. Primary projection also longer. ♂ has brightest red on crown. Streaked undertail coverts and white eyering, both lacking in PUFI. Squeaky *chil-eee-up* call.

Eurasian Tree Sparrow *Passer montanus* **ETSP** L 6in

Uncommon and local, introduced around St Louis, MO. Has spread to cen. IL and se. IA. Found around trees in urban and agricultural ares. Nests in tree holes. A smaller, neater version of breeding ♂ HOSP. Sexes alike: all-brown cap, dark cheek patch, and full white collar. Juv: dull version of adult.

♀

1st-w. ♂

ad. ♂

Evening Grosbeak *Coccothraustes vespertinus* **EVGR** L 8in

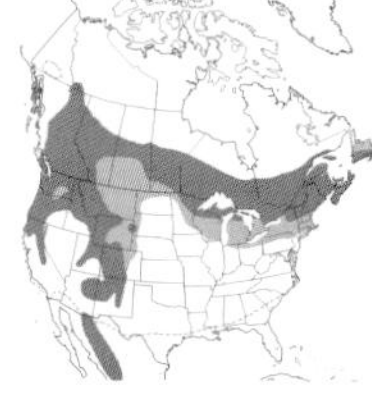

Hardy n. bird that breeds in mixed forest. Winters in snowbound lightly wooded areas, having favorite trees and feeders. Irruptive, though numbers have declined drastically recently, and former large movements s. appear to have stopped. Easiest to find in winter, forms flocks that can often be spotted perched like Christmas lights on a tree. Flies in tight flocks, bold white wing patches always striking. Often tame and approachable. Noisy, the call a high-pitched *clear*, sometimes mixed with other chattery notes. **ID**: Large head and bill with a short tail, a hulking bird with dark through the eye that gives it a mean look. However, at times tilts head, making it seem comical and toy-like. Ad ♂: pale bill and striking thick yellow supercilium on dark head contrast with yellow body. Secondaries all white. 1st-w ♂: as adult with white tips to greater coverts and retained juv primary coverts. ♀: milky gray with yellow collar. White in wing extends to base of primaries but duller than in ♂.

russet ♀/1st-w. ♂

ad. ♂

♀

Pine Grosbeak *Pinicola enucleator* **PIGR** L 9in

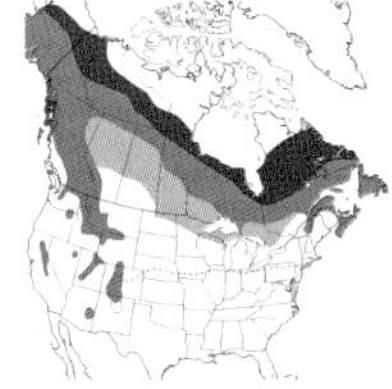

Uncommon hardy n. bird that breeds in mixed forest, particularly with spruce and larch. Irruptive, will move in winter to deciduous trees at lower latitudes, feeding on buds, berries, crabapples, and other fruit. Sometimes at feeders. Tameness, soft expression, and lovely 'fall' colors make this bird popular. Usually in small tight flocks, eating quietly in trees, but can be embarrassingly easy to overlook due to its slow movements. Direct undulating flight. Very quiet, mellow *pew* call given mostly in flight. Song: a musical warble. **ID**: Large and bulky with a longish forked tail that seems disproportionately narrow. Head is small, rounded, and uniform, except for a small dark area through eye. Bill small and stubby with curved edges. Bold wing bars and tertial fringes. Gray on back and underparts a beautiful complement to other colors. Ad ♂: usually extensive pinkish red. Ad ♀: more extensive gray with mustard-yellow. 1st-yr: as adult ♀, aged by retained juv wing feathers. Some ♀s and many 1st-yr ♂s show russet or similar color tones that extend onto breast.

1st-w. ♂

juv.

ad. ♂

♀

Purple Finch *Carpodacus purpureus* **PUFI** L 6in

A fairly common n. breeder in coniferous and mixed forest, often at higher elevations. Varying numbers head s. in winter to open woodland. Often found feeding on seeds and berries near treetops or hedges, but also on the ground in winter. Usually outnumbered by similar and more aggressive HOFI in lower areas. Regularly visits the same feeders. Very vocal, sings its long rich warbled song from treetops. Flight undulating, during which will give distinctive gentle *pik* call. **ID**: Distinctively pot-bellied and large-headed with short tail that is markedly forked; a striking profile in flight. Bill has straighter edges than HOFI's, giving it more of a pug-faced look. All plumages show strong facial contrast with dark ear coverts, cap, and malar. Ad ♂: many have strong crimson and pink tones, others are more red. Colors are usually distinctive from HOFI. Ad ♀: boldly marked brown and white with strong face pattern and breast streaking. 1st-yr: as adult ♀, though some ♂s have purple tones in upperparts. From HOFI by shape, head pattern, straighter edges to bill, color tones in ♂, and bolder streaking in other plumages.

ad. ♂

♀

House Finch *Carpodacus mexicanus* **HOFI** L 6in

The common and familiar finch in most areas. A sw. species. E. birds derive from released NY pet trade stock; population then spread like wildfire. Found just about anywhere, suburbs, feeders, wooded and scrubby areas, usually around humans. One of the most common feeder birds. Conjunctivitis, most easily seen as swelling around the eyes, drastically reduced numbers in the 1990s, but populations now recovering. Often in small groups feeding on seed heads on the ground or in tangled vegetation. Easy to see flying overhead. Diurnal migrant in flocks late fall. Call: a *qwee-er*. Song: 3 loud warbled phrases, fastest at the start. **ID**: Much slimmer and longer-tailed (square-ended) than other *Carpodacus* finches, and lacks strong contrast of other species. Edges to bill noticeably curved. Plumages show a lot of variation, but all have extensive blurry streaking on flanks and diffusely marked upperparts. Ad ♂: red (not pink) usually restricted to head, breast, and rump. Red occasionally shows as yellow or orange where diet is atypical. Ad ♀: relatively dull and nondescript; underparts diffusely streaked on white or buff. Weak face pattern. 1st-yr: similar. The commonest 'unknown' feeder bird!

Red Crossbill *Loxia curvirostra* **RECR** L 6.25in

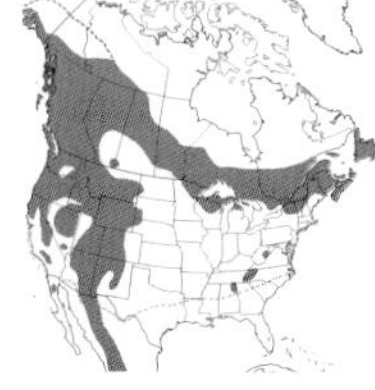

Uncommon denizen of evergreen forest with pine cones; has crossed bill for prying out seeds. Nomadic, moving to other areas when food is scarce. Able to breed throughout the year if food supplies are good. Often feeds quietly in small flocks, hanging from any angle parrot-like. Most easily found flying overhead giving HOFI-like *chip* notes. In winter, will sometimes come down to salted roads. In N America, there are 10 'types' (6 in the E), all possibly full species, averaging different bill sizes and shapes. Call types are separable using sonograms. Types will often mix, and field ID is still evolving. Parallel situations in Europe had interbreeding populations. **ID**: Crossed bill is diagnostic except for confusable WWCR. Chunky with oversized head and short forked tail. All have dark wings and tail. Ear coverts subtly dark, often appearing as thick eyeline. Some show weak wing bars. Ad ♂: uniform dark red on underparts, head, and back. Undertail coverts white with dark centers. Ad ♀: mostly olive-green but variable and can show yellow, orange, or red. 1st-yr ♂: patchy green and red. Juv: heavily streaked underparts.

juv.

ad. ♀

ad. ♂

White-winged Crossbill *Loxia leucoptera* **WWCR** L 6.5in

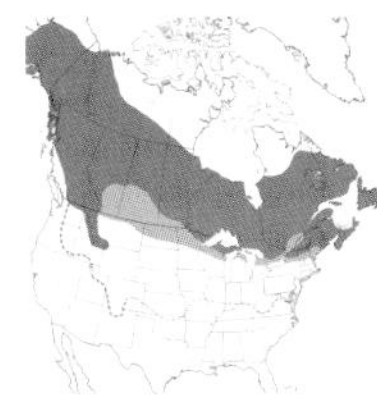

Fairly common in evergreen forest. Behavior much as RECR, though WWCR's bill is adapted to eating smaller cones of trees such as larch and spruce. Rarely mixes. There have been recent large movements into the NE, but birds disappear within a few months. Sometimes found in flocks of up to 100. Both species of crossbill vocal in flight and travel in tight flocks. Compact shape, with forked tail and broad-based wings, makes them distinctive; look for wing bars to separate species. Sings from the treetops, a combination of mechanical trills and rattles at different pitches. Some years will take over forests, singing everywhere, and then they are gone for many years. When not singing, found lower down and will sometimes feed on the ground. *Chip* notes recall sonar of a submarine. Also redpoll-like chatter notes. **ID**: From RECR by slimmer build with smaller head and longer tail. Smaller and narrower-based crossed bill. Strikingly larger double wing bars with white-tipped tertials on darker wings. Darker lores. Ad ♂: pinkish red. Ad ♀: diffusely streaked underparts, usually greenish gray with some yellower or orange-tinged. Juv: heavily streaked; quickly molts to adult-like plumage.

ad. ♂

♀/1st-w.

Common Redpoll *Carduelis flammea* **CORE** L 5.75in

A common breeder in tundra scrub. Nomadic and irruptive 'winter' finch, ventures s. in variable numbers. Winters in open fields, woodland, and around birdfeeders. Feeds on seeds in trees and weeds, often hanging upside down like a chickadee. Found in flocks, occasionally in 100s. Takes off in a tight, noisy, 'bouncing' pack giving rapid *chid chid chid* calls. Often flies around until it finds a suitable tree to alight in, and will gradually drop to a birdfeeder or to the ground, only to take off again for no apparent reason. Although easily spooked, can be tame. **ID**: Appears as a small ball of feathers with a narrow but strongly forked tail. Small head, stubby bill with black at the base, and red cap give it a cute look. Streaked upperparts with a white rump. Underparts are pale, with blurry streaks, mostly on the flanks. Undertail coverts streaked, fewest on adult ♂. Ad ♂: pink on breast, extensive on some but highly variable. ♀/imm: lacks pink and are more heavily streaked. Juv lacks red on forehead and black in face. *Flammea* widespread. Greater Redpoll (*rostrata*), from Baffin Is. and Greenland, can sometimes be identified, with care, by darker, bolder streaking, larger size, and bigger bill.

Hoary Redpoll *Carduelis hornemanni* **HORE** L 5.5in

Remarkably hardy n. bird. Occurs farther s. in smaller numbers than very similar CORE (1:150 in US). Almost always found in CORE flocks. Id with care! Usually picked out by paler gray upperparts. Look for fewer, better defined streaks on underparts, less contrasting and paler face, few or no streaks on undertail coverts, and all- or mostly white rump. Sometimes has shorter 'nipped-in' bill, and more feathering around legs. Age and sexing as CORE. ♂: only lightly pink-breasted and more lightly streaked. *Exilipes* widespread. Scarcer *hornemanni*, from Baffin Is and Greenland, is noticeably larger and bigger-headed, in comparison, also averages paler. Some birds are very similar to CORE and best left unidentified.

Lesser Goldfinch *Carduelis psaltria* **LEGO** L 4.5in

Fairly common but local in scrubby open habitat. Visits feeders. Practices vocal mimicry of some other species. Similar to AMGO, but differences include smaller size, drab-colored wing bars, white at base of primaries, dark underwing coverts, and usually yellow undertail coverts. Ad ♂: black-backed with some birds to the w. appearing black-capped with green mixed in mantle. ♀: lacks contrast of AMGO with dull green upperparts and yellow underparts, never with bright white undertail coverts. Song: a slow jumble of high wheezy notes.

ad. ♂ nonbr.

ad. ♀ nonbr.

ad. ♀ br. (worn)

ad. ♀ br.

ad. ♂ br.

American Goldfinch *Carduelis tristis* **AMGO** L 5in

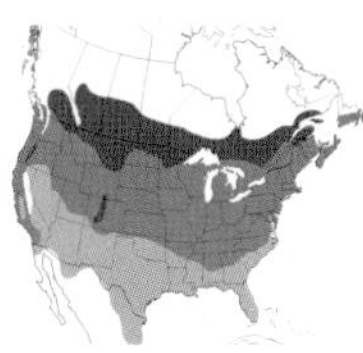

Common, familiar, and beloved bird, found at feeders, on thistles, sunflowers, and hedgerows, at roadsides, on farmland, and in weedy fields. Feeds quietly, often hidden in tall grasses or on the ground, before suddenly flying off, usually a short distance away, always twittering as it goes with its characteristic, exaggeratedly bouncy flight. Mostly found in pairs or small groups. In late fall many n. populations move s., traveling during the day in compact flocks. Small broad wings and forked tail, combined with undulating flight, make it distinctive at long range. Song: a long fast ramble of jumbled sweet twitters. Call: a *choo leee*. Flight call: a *te de de*. **ID**: A small neat-looking bird. Plump with a short forked tail and triangular bill. Bold wing bars. Ad ♂ br: familiar black and yellow. Wing feathers fade quickly and wing bars become worn. Molts in fall into a very different-looking plumage with duller bill. Nonbr ♂: broad wing bars, brown back, pale yellow on head and underparts. Ad ♀ br: pale yellow below and variably brown or green above with an orange bill. Nonbr ♀: similar to nonbr ♂ but lacks yellow head.

1st-yr.

ad. ♂

Pine Siskin *Carduelis pinus* **PISI** L 5in

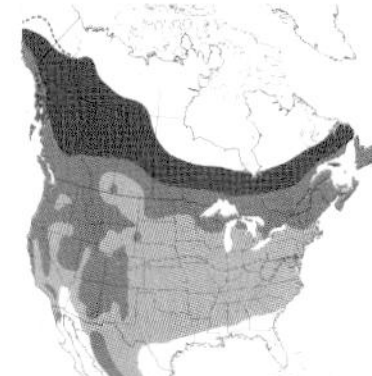

Fairly common breeder in mixed forest. In winter, many move s. to more open wooded areas, feeding in trees, brush, and on the ground. Common visitor to thistle feeders in some areas, often with AMGO and other species. Like crossbills, often flies around in tight bouncy flocks. Noisy, giving an array of whinny and squeaky notes and trills at different pitches. Tips at 45 degrees as it peers down. **ID**: A streaky bird with a complex wing pattern and a bill as sharp as a rose thorn. Small-headed and long-necked for a finch. Forked tail. In flight, yellow bases to flight feathers form a strong panel that really pops in flight. All show same basic pattern. Ad ♂: whitest underparts and boldest streaks, with the most yellow in wing bar. A few birds have less extensive streaking and stronger yellow tones, most noticeable on underparts. Ad ♀ similar but duller. Many are not reliably sexed. 1st-yr: duller with diffuse streaks on buff background. and buff wing bars.

ad. ♀

juv. ♂

ad. ♂

Northern Cardinal *Cardinalis cardinalis* **NOCA** L 8.75in

The most popular bird in N America. Common and widespread in woodland edges, hedges, and suburban gardens. A frequent visitor to feeders. Sits in the open, in ones or twos, rarely moving quickly. Very vocal, sings much of the year, often with raised crest. Will hop around on the ground looking for seeds or fruit, which it picks up with its incredibly powerful bill. Often chase each other in flight, which is direct with very rounded wings and obvious long tail. Call: a metallic *chip*. Song: a series of 3–6, slightly descending, clear-whistled *whits*. **ID**: Large-bodied with a long rounded tail and large red conical bill. Ad ♂: all-red with gray fringes to upperparts and flanks. Black face and bib. Ad ♀: gray bird with varying amounts of buff on underparts. Restricted black on face. Red in crest, tail, and wings always stands out. Juv: similar to a dull ♀, but note black bill. Juv becomes adult-like through summer and most easily aged in fall by any retained black in bill.

Pyrrhuloxia *Cardinalis sinuatus* **PYRR** L 8.75in

Found in desert and scrubby areas with mesquite, in s. TX. Often hops on the ground or sings from bushtops, a higher-pitched version of NOCA's song. Range overlaps with similar ♀ NOCA, but PYRR has a more rounded stubby bill, longer crest, and is slimmer. If in doubt PYRR is the one with the pale bill. All are gray with a pink underwing that's obvious in flight. Ad ♂: red face and line down breast. ♀: uniform gray-buff underparts. Juv: like ♀ but with duller bill.

Black-headed Grosbeak *Pheucticus melanocephalus* **BHGR** L 8.25in

W. species found in mixed woodland. Rare but annual in the E late fall/winter, often at feeders. W. counterpart of RBGR,. Call and song similar, but latter slower and deeper. Call: an *ik*, slightly less squeaky than RBGR. **ID**: A clean-looking bird, boldly patterned above; but it is the orange-buff underparts that stand out and are key to ID. Size and shape as similar RBGR. BHGR has more extensive buff underparts, fine streaks restricted to breast sides, grayer bill, usually with darker upper mandible. 1st-yr ♂: RBGR has similarly marked orange-buff breast, always has crimson underwing. All BHGR have yellow underwings. Hybridizes with RBGR, offspring show features of both species.

♂

1st-w. ♂

♀

1st-w. ♀

1st-s. ♂

1st-s. ♀

ad. ♂ br.

ad. ♀

Rose-breasted Grosbeak *Pheucticus ludovicianus* **RBGR** L 8.25in

Common bird of mixed forest. In summer, lives high in trees and is often easier to hear than see. Has lovely sweet whistled song. Most easily found at other times by distinctive call: sounds like the squeak of a rusty bed-spring. Moves slowly, often sitting for long periods in the same place. On migration, found lower down and is frequent visitor to birdfeeders. Will look around imperiously as if sizing up the situation. In flight, has distinctive shape and bold color patterns, and gives a mellow *pew* call. **ID**: Solidly built with a large head and powerful bill. Ad ♂ br: striking black and white with 'rose' apron. In late summer, many black feathers replaced by duller brown-edged feathers with streaks on rump and flanks. Crimson underwing. ♀: brown and white with a boldly striped head and spotted wing bars. Yellow-buff underwing. Underparts whiter in spring, buffier at other times, streaked across breast. Imm ♀: similar. 1st-w ♂: similar but more orange-buff with red underwing coverts.

Blue Grosbeak *Passerina caerulea* BLGR L 6.75in

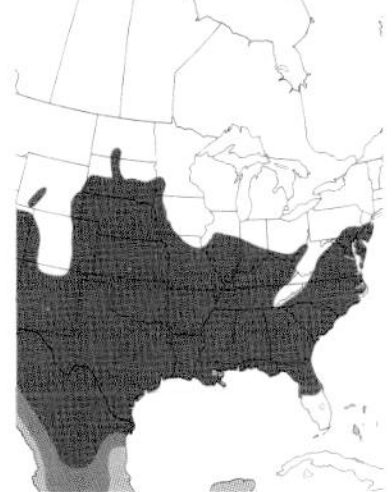

Fairly common in overgrown fields with hedges or scattered trees. Also found in riparian corridors in the MW. Rare visitor n. of normal range. Feeds in weeds, or on the ground. Can be identified at distance by its behavior: slowly swishes its tail from side to side with an uneven rythm as if to emphasize it. Also fans tail, the rounded tip even more noticeable. Sings a lot, usually from the tallest perch, tree, or plant: a long rolling warble that fluctuates in pitch. ♀ is less conspicuous. When quiet, is hard to find. Gives a hard waterthrush-like *chink*. **ID**: A large bunting with a grosbeak's bill. Slim but powerful with a large head and long broad tail with rounded tip. Distinctive broad orange-buff wing bars. Often confused with commoner INBU, but latter not so large, smaller-billed, has different tail shape, and lacks broad rufous wing bars. Ad ♂: violet-blue—INBU lacks any purple tones. Ad ♀: pale gray-brown, sometimes with blue on coverts, tail, and/or head. 1st-w: similar to ad ♀ but browner; no breast streaks. 1st-s ♂: molts in large patches of blue.

Lazuli Bunting *Passerina amoena* **LAZB** L 5.5in

Uncommon in the W. Very rare in the E. Uncommon w. counterpart of INBU with which it sometimes hybridizes. Same size, shape, and behavior. Similar call, but LAZB's song slightly higher-pitched and faster. Ad ♂ br: striking blue head, orange breast, bold wing bars. Much duller in fall with brown fringes to lighter blue upperparts. Ad ♀/1st-w: from similar INBU by lack of streaking on orange-buff across breast; fringes to coverts and tertials broader, less buff and bolder; throat often grayer and lacks hint of malar. Early fall juv breast streaking is sharper and thinner than the blurry streaks on INBU.

Varied Bunting *Passerina versicolor* **VABU** L 5.5in

Uncommon in s. TX in arid areas with dense, tall vegetation, often where rocky. Size and shape as INBU, though sometimes appears slimmer, standing more erect. The culmen is more curved than in INBU. Ad ♂: incredible colors, can look black at distance. Ad ♀: similar to INBU but blander with unstreaked pale underparts and more uniform upperparts with narrow indistinct wing bars. Buffier in winter. Bill pale. 1st-s ♂: as ♀ with some darker feathers. Call: a similar, *spik*. Song: shorter, deeper phrases.

juv.

ad. ♂ nonbr.

ad. ♂ br.

ad. ♀

Indigo Bunting *Passerina cyanea* **INBU** L 5.5in

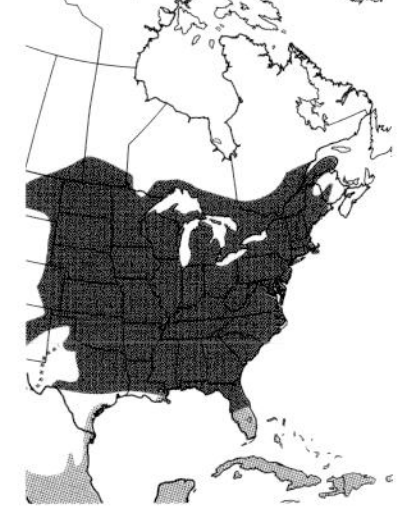

Very common and familiar bunting of open countryside with fields and trees. Usually found alone; sits on trees or wires, belting out its song with head held back, often during heat of the day. Feeds quietly on seed heads on grass verges and in fields, occasionally on the ground. On migration sometimes occurs in large numbers in fallouts. Song: a series of doubled-up high-pitched notes trailing off at end. Call: a *spik*, like other similar buntings. You will hear it a lot if you know it, as it is often overhead, most looking nondescript and sparrow-like. **ID**: A small, nicely proportioned bunting. Conical bill appears pointed but not large like BLGR's. Ad ♂ br: superb indigo blue (dark at distance), darker around lores, wings, and tail. In early spring and fall, has varying amounts of nonbr brown feathers mixed in. Ad ♂ nonbr/1st-s ♂: a mixture of brown and blue feathers. Ad ♀: brown above with two narrow wing bars; paler underneath with indistinct streaking often visible. Some ad ♀s have a little blue on head, wings, and tail. 1st-w: like all-brown adult ♀ but with more pronounced streaking on underparts.

1st-yr. ♀

ad. ♂

ad. ♀/1st-yr. ♂

Painted Bunting *Passerina ciris* **PABU** L 5.5in

Uncommon in woodland edges, overgrown hedges, and dense tangles. Very rare in spring and fall in the N, often at feeders. In summer, stunning ♂ sings from high perches and utility wires. More typically, is found skulking low in brush piles and other denser vegetation. Clearly wary, always seems to know how close you are to it. Movements are slow and deliberate for a bunting; often sits quietly for periods of time. Warbling song intermediate between INBU and WAVI. **ID**: Size and shape similar to INBU but slightly bulkier and larger-billed with curved culmen. Colors alone should always be conclusive. Ad ♂: dramatic patchwork of varied shades of red, blue, and green. Back can be luminescent in some lights. Ad ♀/1st-w: distinctively colored, variable shades of green, palest in 1st-w ♀; variation makes ageing and sexing difficult. Typically, underparts paler and often yellower than upperparts. 1st-s ♂: some, but not all, get a few darker adult-like feathers; check to see if it sings!

♀/1st-yr. ♂

ad. ♂ br.

juv.

Dickcissel *Spiza americana* **DICK** L 6.25in

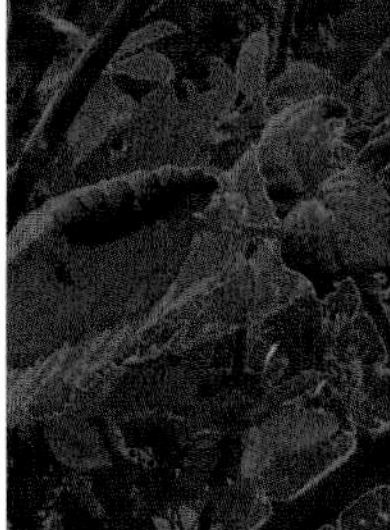

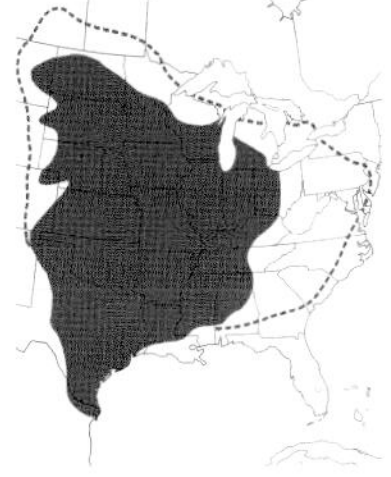

Common in the MW, a somewhat irruptive breeder farther e., and scarce but regular fall migrant in the E. Found in open grassland, particularly in weedy vegetation and hedges. Named after its song, a loud stuttered *dick-dick-dickcissel*. Moves in small, tight flocks, giving short, grunting raspberry sound: *prrrtt*. On migration, sometimes found in large flocks, though loners often hang out with HOSP or BOBO flocks. Can be frustratingly hard to see, and always flies to the next field when discovered. You will hear several flying over for every one you see perched. **ID**: A lean HOSP-like bird with conical bill and sometimes peaked crown with yellow patch in the center of the breast. Looks slim in flight. All plumages have streaked back, yellowish supercilium and malar on gray head, and yellow central breast on mostly pale underparts. Ad ♂ br: black bib borders throat, yellow breast, and rufous shoulder patch. Black on breast is obscurred by yellow feather tips in fall. Ad ♀/1st-w ♂: duller version of ♂, lacks black on breast and coverts are mostly brown rather than rufous. 1st-w ♀: very drab with little yellow and limited rufous on shoulders; fine breast streaking.

Blue Bunting *Cyanocompsa parellina* **BLBU** L 5.5in

Very rare in s.TX. Chunkier than INBU with broad-based dark bill and curved culmen. Ad ♂: dark blue with paler highlights on crown, cheek, elbow, and rump; often looks black. ♀: Richer buff-brown than INBU, uniform breast lacks streaks. A real skulker, mostly in thick low vegetation..

Crimson-collared Grosbeak *Rhodothraupis celaeno* **CCGR** 8.75in

Very rare in s.TX woodland with fruiting trees. Tough to see. A slim, medium-sized bird with an inquisitive look and unique color pattern. Ad ♂: black-headed, red underparts and neck with dark upperparts; unmistakable. ♀/1st-w: green version of ♂.

Hepatic Tanager *Piranga flava* **HETA** L 8in

Very rare in the SW. Most similar to SUTA but has dark bill, contrastingly gray ear coverts, back, and flanks. ♂s are orange-red, ♀s mustard-yellow on throat, crown, and tail, but overall a fairly drab bird. Call a HETH-like *chuck*. Beware of SUTA with a darkish bill.

Western Spindalis *Spindalis zena* **WESP** L 6.75in

Very rare to s. FL. Formerly called Stripe-headed Tanager. A small tanager, though not immediately obvious, with long tail, small head, and a stubby bill. Ad ♂: striking; back color but variable depending on home West Indian island. ♀: dull with pale malar. Upperparts dark olive with pale edges to wing feathers and white patch at base of primaries. Distinctive shape.

Summer Tanager *Piranga rubra* **SUTA** L 7.75in

Fairly common s. tanager in open, deciduous—particularly oak—woodland. Despite bright coloring can be very difficult to see, presence often given away by its 2- or 3-note *nip-n-tuck* call. Sings from within the trees, an explosive series of warbled, AMRO-like notes. Sits quietly, moves slowly, if at all, making it easy to overlook. A loner or found in pairs. **ID**: A large and bulkier bird than SCTA with a broad head, and often peaked crown. Tail is fairly long and broad. Bill is large, with a strongly curved culmen and is distinctively pale (horn), SCTA's is smaller and darker. Birds in the far W have larger bills, are paler with contrastingly paler fringes to wing coverts. Ad ♂: cardinal-red including fringes to wing feathers. ♀: uniform green upperparts, mustard-yellow underparts. Some have red, mostly in tail, wing coverts, and head. SCTA has much greener tones with contrastingly darker wings. 1st-w: as ad ♀. 1st-s ♂: variable mixture of red (head and body) and green (tail and wings). SUTA never shows SCTA's strong contrast between wing and body.

Scarlet Tanager *Piranga olivacea* **SCTA** L 7in

Common woodland bird found in all types of deciduous forest. Breeding ♂ is an unforgettable red, yet can be surprisingly difficult to find. Behaves much as SUTA. In characteristic strongly undulating flight, tanagers look potbellied, compact, and broad-winged. Song: a series of 5 harsh burry notes changing slightly in pitch. Call: a *chip-bree*. Flight call: a soft *phew*. **ID**: A compact tanager, with rounded head and undercarriage. Lacks bulk of SUTA and is smaller-billed. Ad ♂ br: deep red, even orangish in bright light, with black wings and tail. Ad ♂ nonbr: molts on breeding grounds into ♀-like green plumage but with black wings. Ad ♀: yellow-green below, uniform olive-green above with darker wings (not black). 1st-w ♂: like ad ♀ but with a few replaced black ad ♂ coverts. 1st-s ♂: usually red with some green and old coverts. 1st-w ♀: like ad ♀ but averages paler, and often shows narrow wing bars.

Western Tanager *Piranga ludoviciana* **WETA** L 7.25in

Scarce in the far W. Rare in the E, sometimes at feeders in late fall/winter. Found in pine or open mixed woodland, often feeding on berries in winter. **ID**: A small compact tanager with a very round head. Ad ♂: striking black and yellow, head is variably orange. Duller in fall/winter. Ad ♀/1st-w: easily confused with SCTA but with broader wing bars and gray back that contrasts with olive rump and head. Wing bars sometimes worn. Also differs from SCTA by paler bill, often paler underparts, yellow on underwing, and different call. Some orioles show similar color patterns!

Phainopepla *Phainopepla nitens* **PHAI** L 7.75in

Rare in the S, vagrant to the NE. A slim, long-tailed, crested bird of arid areas and tall bushes. Often sits on treetops, and flies around a lot as if to show off its large white wing patches. Overall, a dull-colored bird. Ad ♂: glossy black with a red eye. Ad ♀: gray with a red eye; wing feathers have white fringes. Distinctive call, a mellow rising *hoi*.

Scott's Oriole *Icterus parisorum* **SCOR** L 9in

Found on the Edwards Plateau, TX, and in a variety of trees in arid areas. Very rare to Canada and the E Coast. Large, with a long graduated tail. Large bill long, straight, and pointed. Ad ♂: striking black and yellow with bold wing bars and yellow median coverts. Ad ♀: upperparts dull gray-green with diffuse streaking. Underparts green, some birds with darker throat, very variable. Wing bars narrow. 1st-yr ♀: pale-throated. 1st-s ♂: black throat. Call: a hard *chack*. Song: has meadowlark-like qualities.

Audubon's Oriole *Icterus graduacauda* **AUOR** L 9.5in

Scarce in s.TX woodland. Often secretive, slow moving, and hard to see. Size and colors most similar to SCOR. Very long graduated tail and long straight bill. Sexes alike. Adult: unstreaked light green back, bright yellow underparts, and black hood. Broad white tertial fringes. 1st-w: unstreaked green upperparts and dull yellow underparts with blotchy dark spots as hood is molted in. Song: a series of flat whistles.

Altamira Oriole *Icterus gularis* **ALOR** L 10in

Uncommon along stretches of taller woodland in the Lower Rio Grande Valley, TX. Visits feeders. Largest oriole. All plumages orange. Black bib and lores. Sexes alike. Ad: mostly black upperparts, orange median coverts, white wing bars. Juv: all orange, molts to more adult-like plumage in fall but pale orange with narrower wing bars.

Spot-breasted Oriole *Icterus pectoralis* **SBOR** L 9.5in

An established escape. Scarce in suburban s. FL, particularly around Miami, in exotic plants and palms and at feeders. A large and conspicuous black-and-orange oriole with extensive white patch in tertials. The s. FL version of very similar ALOR: id by range and black spots on breast that border bib. 1st-yr: paler orange than adult.

Hooded Oriole *Icterus cucullatus* **HOOR** L 8in

Uncommon in s. TX in open wooded areas, palm trees, or gardens. Sometimes at feeders. Very rare as far n. as Canada. Early migrant spring and fall. **ID**: An orange oriole. Fairly small with long rounded tail; flat-headed with a large bill. Ad ♂: best told by black 'hood.' Black back in summer. browner in winter. Wings with large white median-covert bar. ♀: similar to ♀ OROR but larger, bigger-billed, longer-tailed, and generally paler and dingier above. 1st-s ♂: larger, more messy black bib than OROR. Call: a *wheet*. Song: a choppy array of rattles and whistles.

Orchard Oriole *Icterus spurius* **OROR** L 7.25in

A common bird in suburbs, villages, and areas with scattered trees bordering overgrown fields. In the W, also found in riparian woodland. Occurs out of range. Look for it in hedges and smaller trees, or even weedy fields. Sometimes sneaks into gardens, feeding low among flowers and fruiting trees. One of the first migrants in fall (Jul), usually in small flocks. Chattering call gives away presence. Pleasing song: a warbled array of sweet musical notes that it belts out a lot. **ID**: Distinctively small and slender. Narrow-tailed with a small head that appears pointed. Head sometimes seems rounded, and bird has a gentle expression. Small-billed compared to other orioles. Ad ♂: unmistakable chestnut and black. Ad ♀: yellow and green tones always lack orange of BAOR. Back gray-green, duller than head. Wing bars always narrow and well-defined. 1st-w: as adult ♀. 1st-s ♂: as ♀ with a black throat. Very vocal so commonly seen in summer.

1st-yr. ♀
1st-s. ♂
ad ♀ nonbr./1st-w. ♂
ad ♀ br.
ad. ♂

Baltimore Oriole *Icterus galbula* **BAOR** L 8.75in

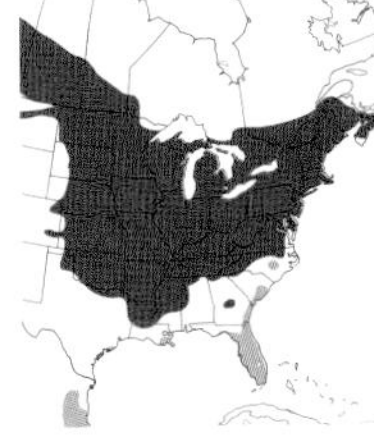

The familiar oriole in most areas, in part due to the male's beautiful colors and melodious song. Found in deciduous forest and areas with more scattered trees, invariably perched in the upper canopy. Woven hanging nest suspended under limbs, often a giveaway. Sometimes at feeders. On migration, can be found anywhere, usually travelling in loose chattering groups, often sitting alertly on treetops before dropping to more sheltered places. Fall birds frequently found eating berries, and hanging at all angles. Flies from tree to tree giving *wink* call. Flight is direct and powerful though unhurried. Song: a mixture of slow, slurred, fluty notes interspersed with faster chattering. **ID**: A sturdy oriole with a shorter tail than most of its cousins. Great variation in color. Ad ♂: striking black and orange. Appears hooded with bold wing bars. Ad ♀br/1st-s ♂: highly variable, from dull version of ♂ with white median-covert bar, to having paler upperparts and mostly orange below. In fall, orange-yellow below. 1st-yr ♀: similar but paler, often gray on belly. Late fall birds paler, like BUOR, but lack yellow supercilium, and distinct dark eyeline and so look plainer-faced. Darker and more streaked mantle.

Bullock's Oriole *Icterus bullockii* **BUOR** L 8.75in

Formerly conspecific with BAOR. W. woodland species. Very rare as far e. as the coast. **ID**: Size, shape, and behavior as BAOR; hybridizes commonly in contact zone. Ad ♂: large white wing patch and black eyeline through orange face. Ad ♀/1st-yr: very similar to pale 1st-yr ♀ BAOR. Overall impression is of a paler yellow-and-gray bird. Most important difference is that supercilium and hindneck are yellow. The malar is usually brightest yellow area (breast on BAOR). The back is grayer, dark centers to median coverts are usually pointed. 1st-s ♂: as ♀ with diagnostic black chin. Song: more chattery than BAOR's; calls similar. Beware of late fall BAOR. The latter often atypically pale yellow and gray—perhaps due to diet.

Red-whiskered Bulbul *Pycnonotus jocusus* **RWBU** L 7in

Scarce introduced species around Miami, FL. Found in suburbs in trees or on telegraph wires, often in small groups. A brown-and-white bird with red undertail coverts and whiskers. The crest is the most obvious feature on this species.

1st-w.
juv.
ad. br.
ad. nonbr.

European Starling *Sturnus vulgaris* **EUST** L 8.5in

Very common and familiar. Despite its bad reputation, EUST has a subtle beauty. Nests in man-made cavities but also competes for natural holes, often kicking out woodpeckers and other species. Forms large waxwing-like flocks in winter, flying around in tight balls, often with a confused raptor nearby. Stomps around, sometimes inquisitively, at other times marches in a hurry, grabbing food with a big open mouth. Song: a mishmash of whines, whistles, and imitations all wrapped into one. **ID**: Stumpy with a short square tail. Distinctive profile in flight with broad-based triangular wings. Flight is a series of rapid flaps, followed by short glides. Bill fairly long, straight and pointed. Adult: brightest in fall/winter, fresh feathers with large white spots on iridescent purple and green, though often looks black. Dark lores give mean look. By spring, spots are worn and iridescence becomes more obvious; bill turns yellow. Juv: brown, palest on throat, dark lores, and striking pale fringes to wing feathers. Molts through summer to very different adult-like feathers, giving a pied appearance.

Eastern Meadowlark *Sturnella magna* **EAME** L 9.5in

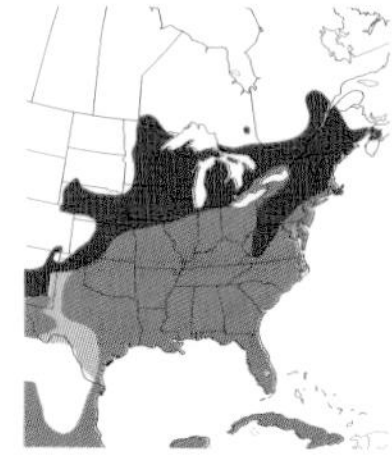

Uncommon and widespread in open grassland. Declining in many regions. Spends most of the time walking on the ground, long strutting strides with body tipped forward watching closely for food—a good impression of EUST. Sometimes perches nervously on bushes and fenceposts, neck outstretched, and short tail cocked and flicked. When relaxed, the tail comes down, and the bird often sings, your only chance of seeing it when the grass is long. Shy and usually unapproachable. It will hunker down, then quickly fly off. Tends to be common where the habitat is suitable and forms tight flocks in winter. In flight, the shape, along with rounded wings and rapid wingbeats, followed by short glides, is distinctive. Song: beautiful, short melodic whistles descending in pitch. Call: a sharp *tzeet*, and a rattle. **ID**: Large body, short square tail, and flat head with pointed bill. Much variation in plumage. Ad br: bright yellow underparts, black necklace with buff flanks and dark streaks. Boldly patterned head and complex upperpart markings with rufous tones. White malar. Ad nonbr/juv: most much duller and strongly buff with reduced necklace. Striking white outer tail feathers.

juv.

♀

♂

Western Meadowlark *Sturnella neglecta* **WEME** L 9.5in

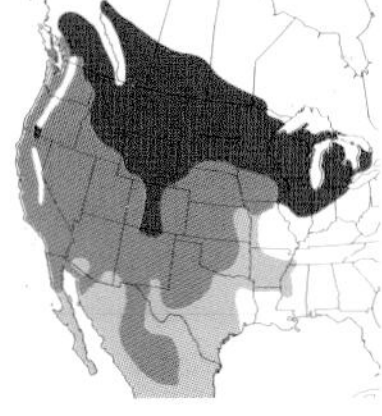

Common w. counterpart of EAME. WEME forms tight flocks in winter in arable areas and at roadsides: a behavior rarely shown by EAME. Both species are very similar, and many individuals are not safely identifiable. Song is by far the best way to identify this bird, a sound that belongs in the Amazon jungle. Incredibly variable, a melodious, descending, gurgled ramble of notes: *weet wet weetidleoo*. Songs are learned, so overlap between species is possible. Calls: a rattle and diagnostic *chuck*.

ID: Notoriously difficult, complicated by individual, subspecific, and seasonal variation. WEME tends to be a paler and less contrasting bird than EAME, has whiter flanks with spots rather than long streaks, less rufous tones on upperparts, and blander face pattern. The malar is usually mostly yellow, strongest in males, white in EAME, a feature that is quite easy to see with good views. White in outer tail feathers less extensive than in EAME.

ad. ♂ transitional

♀/nonbr./juv.

ad. ♂ br.

Bobolink *Dolichonyx oryzivorus* **BOBO** L 7in

A fairly common grassland species; find one and you will probably find many. On migration, often occurs in reedbeds. Usually found in flocks, can be heard calling *ink ink* as it flies overhead. Much harder to find on terra firma. On the breeding grounds, males will perch up on grass stems, bushes, or fences. Spend lots of time flying around in display flight with upperpart feathers punked up. The rapid jingling song of varied notes is a common feature of summer grasslands in the N. Despite being so territorial, they are always gregarious; yet ♀s stay hidden and are hard to find. After breeding, they form large flocks and head s. In flight, long and sleek with pointed wings; a powerful flyer. DICK is similar and also flies in flocks in the MW. **ID**: A somewhat sparrow-like bird that sits upright with pointed tail and bill. Ad ♂ br: glossy black with bold white rump and scapulars. Golden buff nape. By Jul, they have molted on the breeding grounds into ♀/imm-type plumage, some with a few breeding feathers remaining. ♀/nonbr/1st-w: yellow-buff with a striped head, pale nape, and boldly striped upperparts with 'tram lines.' Underparts are uniform except for well-defined streaks on the flanks.

1st-yr. ♂

ad. ♂ nonbr.

ad. ♂ br.

♀

Red-winged Blackbird *Agelaius phoeniceus* **RWBL** L 8.75in

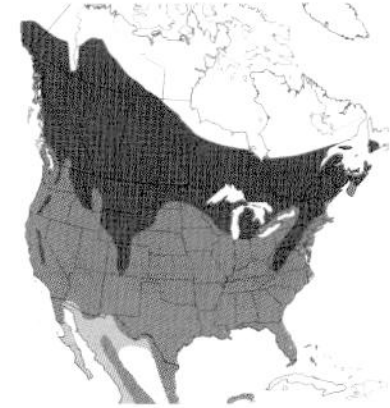

Abundant just about everywhere. Breeds in marshes and wet, scrubby areas. In winter, forms large, sometimes massive, flocks, often with other blackbirds and EUST, in agricultural areas, gardens, and at feeders. They spend most of the time feeding on the ground, doddering along on short legs. Common summer sight singing on bulrushes. Leans forward precariously with tail spread and wings half open, as if to show off its mind-blowing headlights—the red-and-yellow wing patch. Song: a whistled *get-me-teeee*. **ID**: Stocky with a head that looks too small and pointed for the body. Pointed bill an extension of head shape. Medium-long tail. Ad ♂ br: jet-black with stunning red-and-yellow coverts. Ad ♂ nonbr: muted version of breeding ♂ with pale fringes. Imm ♂: highly variable, most as nonbr ♂ but with rustier fringes and paler underwing coverts. Ad ♀: strikingly different from ♂. Dark brown upperparts and underparts completely streaked brown; buff face. Imm ♀: similar to ad ♀ but whiter around face; paler underwing coverts. ♀-type RWBLs are source of much confusion for inexperienced birders. Learn the bill shape, and you will be halfway there!

juv.

1st-w. ♂

♀

ad. ♂

Yellow-headed Blackbird *Xanthocephalus xanthocephalus* **YHBL** L 9.5in

Uncommon and local in summer in MW marshes, most reedbeds hold several pairs. Feeds in nearby farm fields and cattle feed pens, particularly near activity, be it tractor work or feeding cattle. ♀ tends to sit low while ♂ sings from top of reed, displaying his flamboyant plumage. That's if you call it singing; a varied array of clucks, whistles, squeaks, as if given with a sore throat—perhaps penance for delivering so many bad notes. ♂ frequently swaps perches, flashing its white wing patches. Scarce elsewhere throughout year, particularly Aug/Sep, when migrating birds are often found with other blackbirds. **ID**: A large sturdy blackbird; ♀ is smaller. The only blackbird with a yellow throat. Ad ♂: striking yellow head, black body with white primary coverts. Ad ♀: duller than ♂, yellow mixes with brown on head. Lacks white on primary coverts. Imm ♂: aged by white-edged primary coverts. Similar to ad ♀, but larger, with darker lores and more extensive yellow. Imm ♀: duller overall. Juv: plumage held briefly.

♂ br. (spring)

ad. ♂ br.

♀ nonbr./juv.

♂ nonbr.

Rusty Blackbird *Euphagus carolinus* **RUBL** L 9in

An uncommon n. blackbird of wooded, swampy areas. In winter and on migration, usually forms single-species flocks that can be seen sitting in the tops of trees, suddenly dropping en masse into wet woods to quietly snout around under leaves for food. In flight, distinctively slim and well-balanced proportions with narrow wings and rounded tail, nipped in at the base. Often flies in evenly spaced single-species flocks—unlike other blackbirds. Distinctive *chup* call. Song: the thinnest and scratchiest of the blackbirds. **ID**: Structure and behavior usually draw attention. Yellow eye (appears white) in all plumages. Shape like BRBL with slightly larger neck and head, subtly narrower-based bill. Different behavior and habitat is usually key. Ad ♂ br: glossy blue-green with purple head. Ad ♂ nonbr: patchy appearance with a hodgepodge of brown and black feathers and with rusty fringes to wings (including tertials, never on BRBL). Ad ♀ br: uniform dull brown with dark mask and pale iris. ♀/imm: very difficult to age and sex. Warm brown overall, brightest on fringes to wing feathers; gray tail and rump. Dark mask emphasizes pale iris, supercilium, and throat, giving slightly sinister look.

1st-w. ♂

♀

ad. ♂

Brewer's Blackbird *Euphagus cyanocephalus* **BRBL** L 9in

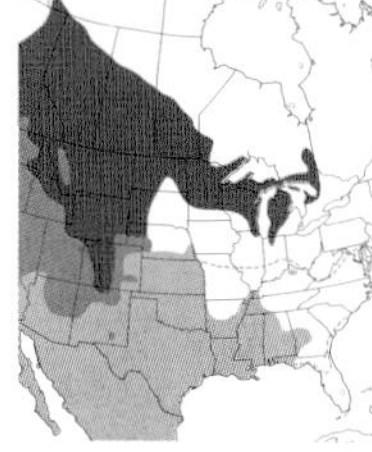

W. species that decreases toward the E. Found in open areas, parks, and gardens, particularly around livestock. A ground-dwelling blackbird that walks with purpose, chest pushed out and feet kicked high as if in the marching band. Sometimes tame and under your feet, its nondescript plumage makes it easy to overlook. Mixes with other blackbirds but equally happy alone. Likes cattle, a great place to look for out-of-range birds. Song: short, buzzy *t-zeee*. Call: a hard *tek*. **ID**: Well-proportioned, much like RUBL but with subtly smaller head and broader-based bill. Ad ♂: often appears black with bright yellow iris. A closer look shows beautiful glossy greens and blues and purple head. Ad ♀: nondescript and uniform with cold gray/brown plumage, in some respects dipper-like. Most birds are slightly darker on wings. Brown iris appears black. No other blackbird is so uniform. 1st-w ♂: variable but usually intermediate between adult ♀ and adult ♂. Often shows contrasting dark around eye.

ad. ♂ FL
ad. ♂ Interior
ad. ♂ FL
juv.
ad. ♂ eastern
♀

Common Grackle *Quiscalus quiscula* **COGR** L 12.5in

Very common and widespread in open areas: agricultural fields, cattle lots, marshes, but also your garden feeders and suburban areas. Nests in trees and will stalk around in wet woods like RUBL. In fall and winter, forms flocks, creating lines or 'clouds' of birds. Often dominates feeders or patrols grassy areas such as lawns. Size, color, facial expression, and pale iris result in 'bad boy' look. Closer views reveal a bird of stunning iridescent colors (♂). In flight, large size, slender build, and long, flared tail help it stand out in blackbird flocks. Song: a variety of forced wheezy notes. Call: a hard flat *chek*. **ID**: Large and slim with long, graduated tail. Sturdy all-dark bill. Colors vary across populations. ♂ (e. population): variably blue/purple. Interior birds have brown body with green/blue head and neck. In FL, green-and-bronze body with purple-and-blue head and neck. ♀: slightly smaller and shorter-tailed than ♂. Generally brown with bluish head, lacking strong iridescence of ♂. Juv: dull brown, often darker-tailed. Iris darker and bill often paler than in adult.

1st-fall. ♂

1st-w. ♂

♀

ad. ♂

Boat-tailed Grackle *Quiscalus major* **BTGR** ♂ L 16.5in, ♀ 14.5in

The grackle of coastal marshes. Forms large single-species flocks in winter overtaking playing fields, and readily perches on rooftops and trees, often in towns. Spends most of its time walking purposefully on the ground in short-grass areas or in marshes. Proudly holds its head high, sings and spread its huge tail. Has a pathetic song for such a bruiser of a bird: a ringing *bent bent*. Also variable piping whistles and *chuks*, but not so varied or strident as GTGR. Very vocal and aggressive. Living in a saltwater environment, loves rain pools and other fresh water for drinking and bathing. **ID**: Much larger than COGR with longer, more bulbous-ended tail. Bill longer and more pointed. Head shape variable from rounded to quite pointed. Atlantic Coast population has pale iris, FL and Gulf Coast birds have dark iris (see GTGR). Ad ♂: head variably purplish/blue, darker than green/blue body. Pale iris. ♀: smaller. Head and underparts tawny brown. Paler throat and supercilium contrast with darker ear coverts and lores. Upperparts dark brown. Juv: as ♀, but paler and duller. Imm ♂: juv. is ♀-like, grows in darker feathers through fall becoming all dark. Pale iris. Shorter-tailed than ad. ♂ with no iridescence.

Great-tailed Grackle *Quiscalus mexicanus* **GTGR** L ♂ 18in, ♀ 15in

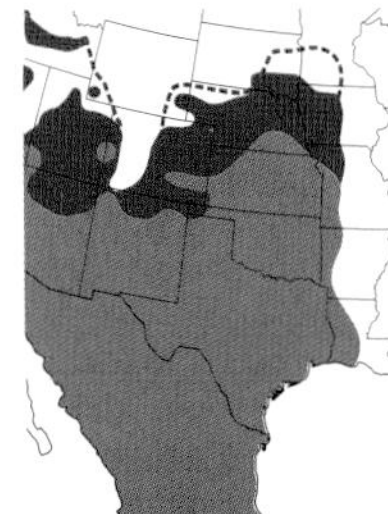

Common in agricultural and other open areas and towns. Also found in marshes alongside BTGR. Usually in flocks, strutting along the ground looking for food. Often throws its head back to sing or call. Demands attention, spreading its huge tail in a large 'U' and holding its wings half open. Points bill skyward as if to say "I am it"—given its size, iridescent colors, and exuberant behavior, perhaps it is! Roosts in big noisy gatherings, often in village and town squares, and causes mayhem all night long. A deafening range of loud chuckles, wheezes, and trills. **ID**: Very similar to BTGR but slightly larger and more angular with longer, diamond-shaped tail. Head is flatter and bill larger, giving a more pointed look. Where ranges overlap, GRGR has pale iris and BTGR dark iris. Ad ♀: colors as BTGR but averages darker on back. ♀: similar to BTGR but darker brown, with more contrasting face pattern. Juv: similar to ♀ with diffuse pale streaking on underparts. Imm ♂: shorter-tailed and duller than ad ♂. Imm: can keep dark iris through first year of life.

Bronzed Cowbird *Molothrus aeneus* **BROC** L 8.75in

Scarce but locally common in open areas, particularly near cattle and feed lots. The largest cowbird, a weird bird that seems to have a variety of postures. Large-chested with a pointed head, it will suddenly hunch its back, puffing its neck feathers out to form a ruff. ♂: hovers in unique breeding display. **ID**: The most thick-set blackbird with broad, muscular neck. Heavy, pointed bill extends onto forehead. Diagnostic bold red iris striking on all but youngest birds. Ad ♂: very dark with contrasting eye and glossy blue wings and tail. Ad ♀: a browner and duller version of ♂ with very little contrast in plumage. Juv: similar to ♀. ♂: grows adult-like feathers through first winter.

Shiny Cowbird *Molothrus bonariensis* **SHCO** L 7.5in

Scarce visitor to s. FL, very rare farther n. Found in dry areas and gardens, at feeders, and often with other blackbirds. Slimmer than BHCO with a more pointed bill. Ad ♂: uniformly glossy blue/purple, greener on wings. ♀: similar to BHCO but browner and more uniform with less contrasting throat and darker ear coverts, and with weak supercilium. Concentrate on shape.

Brown-headed Cowbird *Molothrus ater* **BHCO** L 7.5in

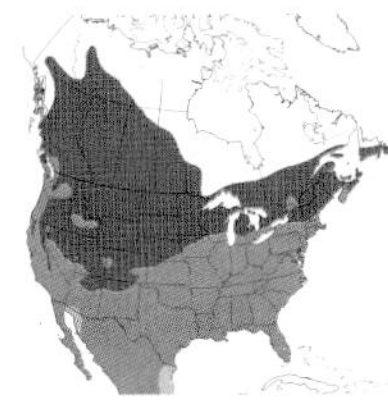

Very common in open areas near woodland edges. Feeds on the ground in short-grass areas. Tail often distinctively cocked high in the air as it feeds, and sometimes points bill skyward as if sniffing the air. Joins other blackbird flocks in winter. Frequently around animals, often walks near hooves, looking for insects or catching a ride. Not liked for its nest-parasite behavior. Newly hatched chicks are demanding and take full attention of host parent. Song: gurgles followed by thin whistles. Long, drawn-out, high *tseeeee*—a common flight call. **ID**: Noticeably small, chunky, and big-headed. Seems short-tailed though not so. Pointed wings in flight. Ad ♂: body often appears black, though is actually green/blue. Head and neck brown, creating a distinctive hood. ♀: dull pale brown with gray tinges. Slightly darker on upperparts with barely visible diffuse streaking on underparts. Head appears paler than body with indistinct supercilium and weak malar stripe on throat. Juv: brown with pale-fringed upperparts giving scaly effect. Underparts have well-defined streaks. Starts molting adult-type feathers through fall. Imm ♂: a patchwork of dark grayish and brown feathers.

ACKNOWLEDGMENTS

In life and with any major project, it's always a strong foundation that counts. For that, I thank my wife, Debra, and daughters, Sophie and Samantha. Their endless patience during the months of travel and hours of writing this book are appreciated.

This book would have been difficult to complete without the work of my right-hand man, Ciprian Patulea. His meticulous skills with Photoshop, all-around technical support, and positive attitude were irreplaceable.

My parents, Brian and Margaret Crossley, as always supported and encouraged me from the start. My father's artistic influence has been the cornerstone of my appreciation for color and patterns.

Robert Kirk of Princeton University believed in this book wholeheartedly. His help and camaraderie have been above and beyond. He was given excellent support from the Princeton staff including Jessica Pellien, Stephen Edwards, Dimitri Karetnikov, Maria Lindenfeldar, Jason Alejandro, and Ellen Foos.

Paul Lehman for maps and endless patience on 'where to find' questions—an encyclopedic wealth of knowledge on bird distribution.

Killian Mullarney for encouragement in the early days and advice in the later.

Reviewers that provided great help included Katy Duffy, Paul Lehman, Brian Gibbons, Killian Mullarney, Jeff Gordon, Tony Leukering, Jerry Ligori, Jessie Barry, Michael O'Brien, Mary Gustafson, Alvaro Jaramillo, Tom Johnson, and Steve Howell.

It is impossible to name all the people who have helped with this project—on so many levels. A book like this involves business, birding, photographic, design, and many other skills. These have been acquired over a lifetime from so many people. A lot of these are close friends, others with a big influence were just people I met in passing who were generous enough to help with advice or information. The sum of it all is this book! I apologize for the names omitted from the following list. Some of the helpers include:

Jonathan Alderfer, Jon Allen, Scott Angus, Jackie Barnes, Scott Barnes, Jessie Barry, Louis Bevier, Wes Biggs, Adrian Binns, Paul Bithorn, Tom Blackman, Jeff Bouton, Bill Boyle, Joseph Brin, Ned Brinkley, Jen Brumfield, Alan Brady, Michael Brothers, Tink and Corey Bryan, Richard Chandle, Bill Clark, Paul Cook, Dan Cooper, Jill Conrad, Vinnie Carrissimi, Cameron Cox, Mike Crewe, John Crossley, Kit Day, France Dewauge, Chris Dooley, Jim Dowdell, Peter Dunn, Megan Edwards, Vincent Elia, Scott Elowitz, Bill Elrick, Sean Farrell, John Feenstra, Shawneen Finnegan, Bob Fogg, Don Frieda, Mike Freiberg, Marci and Terry Fuller, Hugh Gallagher, Sam Gallick, Mark Garland, Kimball Garrett, Brian Gibbons, Bill Glaser, Jeff and Liz Gordon, Mary Gustafson, Pete Gustas, Bernie Haas, Chris and Lee Hadjuk, Eamon Harrington, Patti Hodgets, Paul Holt, Steve Howell, Julian Hough, Alvaro Jaramillo, Laura Kammermeier, Roy Karo, Lars Jonsson, Kenn Kaufman, Mike Lanzone, David LaPuma, Tony Leukering, Jerry Liguori, Pat Lonergan, Derek and Jeanette Lovitch, Karl and Judy Lukens, Bruce MacTavish, Larry Manfretti, Tom McGrarian, Clive Minton, David Mizrahi, Charlie Moores, Art Morris, Stephen Moss, Alan Murphy, Brennan and Bryn Mulrooney, Michael O'Brien, Robert O'Toole, Brian Patteson, Jim Pawliki, Brian Perkins, David Povey, James and Debbie Provenzano, Peter Pyle, Don Reapy, Martin Reed, Will Russell, Carmen Saginario, Shigeta San, Greg and Diane Scarano, Bill Schmoker, Keith Seager, Steve Shunk, David Sibley, Gerry Smith, David Sonneborn, Lloyd Spitalnik, Brian Sullivan, Roy Sutton, Dave and Kathy Tetlow, Arnoud van den Berg, Chris Vogel, Ron Walker, John Watkins, Scott Whittle, Chris Wood, Jonathan Wood, Rick Wright, Louise Zemaitis.

PHOTOGRAPHY CREDITS

I am grateful to the following good friends for completing the 'picture' in some of the plates. The following people have one or more images in plates for the following species.

Adrian & Jane Binns:	GRSK, LPCH, RBTR, MABO, RFBO, PAJA, SWHA, WWPA, BRNO, BLNO, BTHH, KIWA, SWWA
Bill Clark:	APFA, HBKI, MIKI
Bob Fogg:	LTJA, POJA, BEPE, VIRA
Brian Gibbons:	ELOW, LPCH
Brian Sullivan:	ANMU, SAGU, SPSK, PAJA, FEHA, THGU
Bruce Mactavish:	MABO, RFBO, TBMU, BLGU RUGR, WWCR
Hugh Gallagher:	BLKI, LIST, WWTE, GBMA
Jerry Ligouri:	GOEA, PRFA, GYFA
Jonathon Wasse:	ROGU
Killian Mullarney:	ROGU, GRSK, RBTR, SPRE, EUSP, FEPE, BRBO, ARTE, ATPU, ROTE, WFSP
Laura Kammermeier:	KIWA
Michael Brothers:	ANMU
Michael O'Brien:	LPCH
Paul Cook:	PFGO, RTLO
Richard Chandler:	RUFF, CUSA, SPRE, FTFL, SNBU, ETSP
Scott Whittle:	BTHH
Steve Howell:	YBLO, RFBO, ZTHA, BEPE, ANMU, CBHA, SNKI
Stuart Elsom:	CEWA, DOVE, LBMU, PFGO, SBGU,
Tom Johnson:	BLGO, PFGO, ROPT
Michael Engelmeyer:	Cover portrait

INDEX: SHORTHAND (ALPHA CODES)

INDEX: SCIENTIFIC NAMES

INDEX: COMMON NAMES

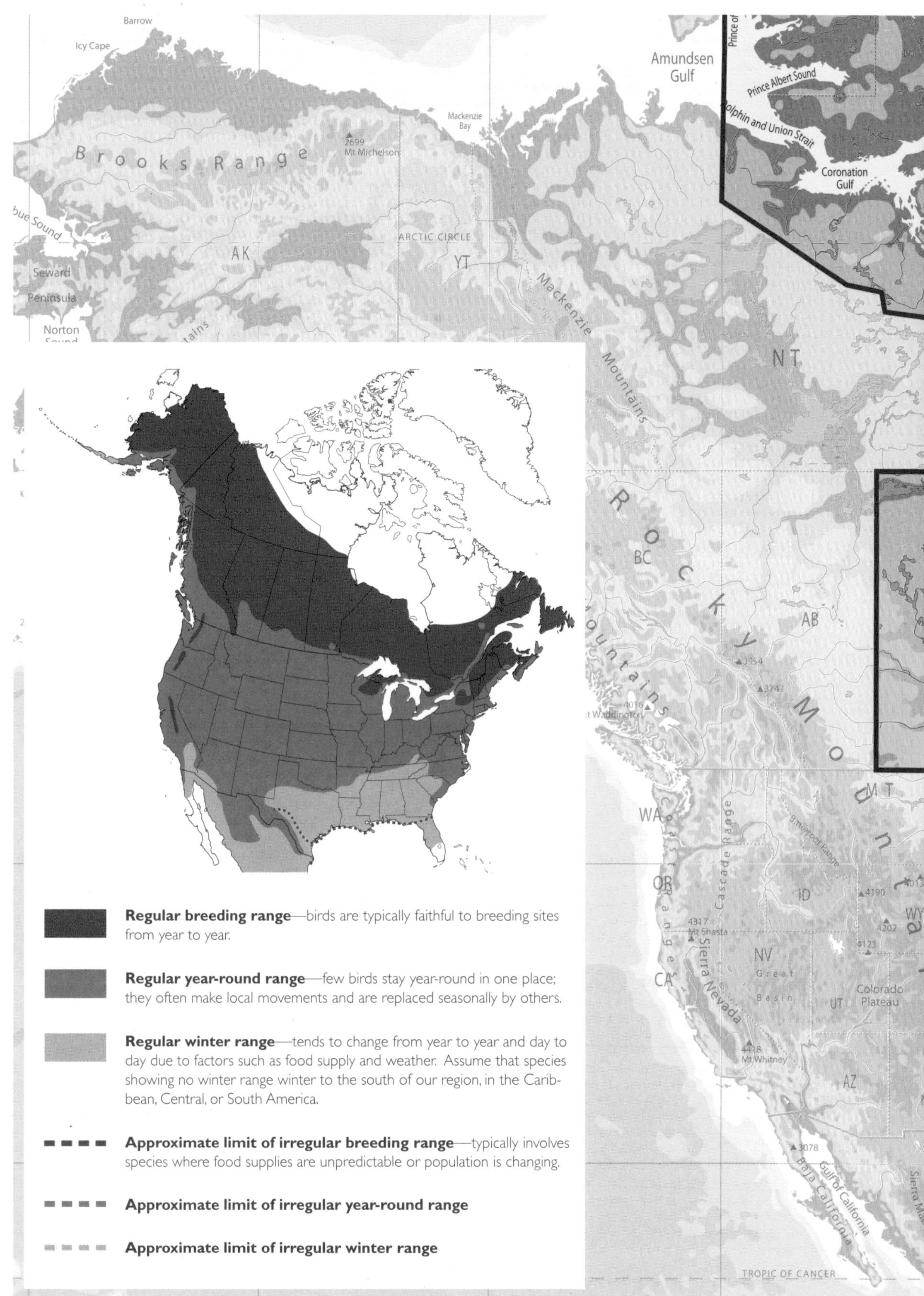

Barrow
Icy Cape
Amundsen Gulf
Prince Albert Sound
Dolphin and Union Strait
Coronation Gulf
Mackenzie Bay
Brooks Range
2699 Mt Michelson
Seward Peninsula
Norton Sound
ARCTIC CIRCLE
AK
YT
Mackenzie Mountains
NT
Rocky Mountains
BC
AB
WA
OR
CA
NV
ID
UT
AZ
MT
WY
Coast Ranges
Cascade Range
Bitterroot Range
Sierra Nevada
Great Basin
Colorado Plateau
4347 Mt Shasta
4418 Mt Whitney
Mt Waddington
Gulf of California
Baja California
TROPIC OF CANCER
Regular breeding range—birds are typically faithful to breeding sites from year to year.
Regular year-round range—few birds stay year-round in one place; they often make local movements and are replaced seasonally by others.
Regular winter range—tends to change from year to year and day to day due to factors such as food supply and weather. Assume that species showing no winter range winter to the south of our region, in the Caribbean, Central, or South America.
Approximate limit of irregular breeding range—typically involves species where food supplies are unpredictable or population is changing.
Approximate limit of irregular year-round range
Approximate limit of irregular winter range